Cyclic Nucleotides and the Regulation of Cell Growth

Cyclic Nucleotides and the Regulation of Cell Growth

Edited by
Morad Abou-Sabé
Rutgers University

Dowden, Hutchinson & Ross, Inc.
STROUDSBURG, PENNSYLVANIA

Library of Congress Catalog Card Number: 76-28686
ISBN: 0-87933-264-6

78 77 76 1 2 3 4 5
Manufactured in the United States of America.

LIBRARY OF CONGRESS CATALOGING IN PUBLICATION DATA
Main entry under title:
Cyclic nucleotides and the regulation of cell growth.
Includes index.
1. Cellular control mechanisms. 2. Cyclic nucleotides. 3. Cell cycle.
4. Cell proliferation. I. Abou-Sabé, Morad A.
QH604.C92 574.8'761 76-28686
ISBN 0-87933-264-6

Exclusive distributor: **Halsted Press**
A Division of John Wiley & Sons, Inc.
ISBN: 0-470-98953-X

Preface

The understanding of the mechanism(s) which are involved in the control of cell growth has for a long time been a black box, and still remains obscure at the present time. However, recent advances in cyclic nucleotide research have introduced a considerable body of information implicating cyclic AMP and cyclic GMP in the regulation of cell growth. The actual mechanism(s) through which these cyclic nucleotides take part in the control of cell growth remains largely unknown at the present time.

Due to the widening scope of these investigations, both at the metabolic and mechanistic levels, a specialized conference, representing a variety of views and dealing with different cell systems, both eukaryotic and prokaryotic, was held for a discussion of the recent advances in this area of cyclic nucleotide research.

The proceedings of this conference, published herein, present the current view of cyclic nucleotide involvement in the regulation of cell growth, as well as an up-to-date review of the field in a number of principal areas of cell growth.

For an appropriate sequence of information, this book is divided into five sections, three of which deal with the role of cyclic nucleotides in the regulation of bacterial growth, cell cycle, and cell proliferation; and two dealing with cyclic nucleotide metabolism and the regulation of cell growth, and mechanisms of growth regulation.

It is hoped that this book will provide for the student and the research scientist an updated view of the field with a detailed account of original discoveries in the area of cyclic nucleotides and the regulation of cell growth.

I would like to take this opportunity to acknowledge Rutgers University Research Council, Charles and Johanna Busch Memorial Fund to the Bureau of Biological Research, Roche Institute of Molecular Biology, Johnson & Johnson, E. R. Squibb, Merck and Company, Ortho Pharmaceutical, Ethicon Inc., and Pharmacia Laboratory for their support which made this conference possible.

MORAD ABOU-SABÉ

Contents

Preface v.
Contributing Authors ix

PART I: CYCLIC NUCLEOTIDES AND THE REGULATION OF BACTERIAL GROWTH

cAMP and the Regulation of Bacterial Growth 3
M. Abou-Sabé
The Mechanism of Regulation of *Escherichia coli* Adenylate Cyclase 27
A. Peterkofsky
Cyclic Nucleotides and the *Caulobacter* Cell Cycle 37
N. Kurn and L. Shapiro

PART II: ROLE OF CYCLIC NUCLEOTIDES IN THE REGULATION OF THE CELL CYCLE

Role of Cyclic AMP in the Regulation of the HeLa Cell Cycle 57
D. L. Friedman, R. A. Johnson, C. E. Zeilig, J. B. Kurz, K. V. Kumar, and P. Gray
Cyclic Nucleotides and the Cell Cycle of the Hematopoietic Stem Cell 81
J. W. Byron

PART III: CYCLIC NUCLEOTIDES AND THE REGULATION OF CELL PROLIFERATION

The Positive Regulation of Cell Proliferation by a Calcium-Cyclic AMP Control Couplet 97
J. F. Whitfield, A. L. Boynton, J. P. Mac Manus, R. H. Rixon, P. R. Walker, and U. Armato
Cell Surfaces, Cyclic Nucleotides, and Unrestrained Cell Division: A Natural Role Is Proposed for the Cell Surface Receptor of *Vibrio cholerae* and *Escherichia coli* Enterotoxins 131
H. C. Kiefer and H. S. Kantor
Cyclic AMP Metabolism in BHK21 Cells Infected or Transformed with Adenovirus Type 12 173
K. Raska, Jr.

PART IV: CYCLIC NUCLEOTIDE METABOLISM AND THE REGULATION OF CELL GROWTH

Cyclic GMP Metabolism in Transplantable Tumors 191
F. Murad
Regulation of Cyclic GMP Metabolism in Splenic Cells 207
M. K. Haddox, L. T. Furcht, M. E. Moser, C. C. Bull, J. H. Stephenson, and N. D. Goldberg
Cyclic AMP-Mediated Nuclear Translocation of Cytoplasmic cAMP-Dependent Protein Kinases: Identity of the Nuclear and Cytoplasmic Enzymes 225
R. A. Jungmann, M. L. Christensen, J. S. Schweppe, M. I. Mednieks, and A. M. Spielvogel

Cyclic AMP Systems in Selected Cells of Erythroïd and Lymphoid Origin 253
H. Sheppard, C. R. Burghardt, and W. H. Tsien

PART V: CYCLIC NUCLEOTIDES AND MECHANISMS OF GROWTH REGULATION

Growth Regulation of Cyclic AMP in Mouse Lymphoma Cells 267
P. Coffino, J. W. Gray, I. Lemaire, and P. A. Insel
Membrane Enzyme Studies Using Drug Resistant Cell Lines 279
J. R. Sheppard, L. LeLievre, and A. Paraf

Index 293

Contributing Authors

M. Abou-Sabé	Department of Microbiology, Rutgers University, New Brunswick, New Jersey 08903
U. Armato	Tissue Culture Laboratory, Department of Human Anatomy, Faculty of Medicine, University of Padua, I-35100 Padua, Italy
A. L. Boynton	Animal and Cell Physiology Group, Division of Biological Sciences, National Research Council of Canada, Ottawa, Canada K1A OR6
C. C. Bull	Departments of Pharmacology and Laboratory Medicine and Pathology, University of Minnesota, Minneapolis, Minnesota
C. R. Burghardt	Department of Cell Biology, Roche Research Center, Hoffmann-La Roche Inc., Nutley, New Jersey
J. W. Byron	Department of Pharmacology and Experimental Therapeutics, University of Maryland School of Medicine, Baltimore, Maryland
M. L. Christensen	Department of Biochemistry, Northwestern University Medical School, Chicago, Illinois 60611
P. Coffino	Department of Microbiology, 412S; and Department of Medicine, Division of Clinical Pharmacology, University of California, San Francisco, San Francisco, California 94143

D. L. Friedman — Department of Molecular Biology, Vanderbilt University, Nashville, Tennessee 37232

L. T. Furcht — Departments of Pharmacology and Laboratory Medicine and Pathology, University of Minnesota, Minneapolis, Minnesota

N. D. Goldberg — Departments of Pharmacology and Laboratory Medicine and Pathology, University of Minnesota, Minneapolis, Minnesota

J. W. Gray — Biomedical Division, Lawrence Livermore Laboratory, University of California, Livermore, Livermore, California 94550

P. Gray — Department of Molecular Biology, Vanderbilt University, Nashville, Tennessee 37232

M. K. Haddox — Departments of Pharmacology and Laboratory Medicine and Pathology, University of Minnesota, Minneapolis, Minnesota

P. A. Insel — Department of Medicine, Division of Clinical Pharmacology, University of California, San Francisco, San Francisco, California 94143

R. A. Johnson — Department of Physiology, Vanderbilt University, Nashville, Tennessee 37232

R. A. Jungmann — Department of Biochemistry, Northwestern University Medical School, Chicago, Illinois 60611

H. S. Kantor — Division of Infectious Disease,
Department of Medicine,
University of Health Sciences,
The Chicago Medical School,
North Chicago, Illinois

H. C. Kiefer — Department of Biochemistry,
Northwestern University Medical and
Dental Schools,
Chicago, Illinois

K. V. Kumar — Department of Molecular Biology,
Vanderbilt University,
Nashville, Tennessee
37232

N. Kurn — Department of Molecular Biology,
Division of Biological Sciences,
Albert Einstein College of Medicine,
Bronx, New York
10461

J. B. Kurz — Department of Pharmacology,
University of Colorado Medical Center,
Denver, Colorado
80220

L. LeLievre — Institut de Biologie Moléculaire,
Centre National de la Recherche Scientifique,
Université Paris VII,
2, Place Jussieu,
75005 Paris, France

I. Lemaire — Department of Medicine,
Division of Clinical Pharmacology,
University of California, San Francisco,
San Francisco, California
94143

J. P. Mac Manus — Animal and Cell Physiology Group,
Division of Biological Sciences,
National Research Council of Canada,
Ottawa, Canada
K1A OR6

M. I. Mednieks — Department of Biochemistry,
Northwestern University Medical School,
Chicago, Illinois
60611

M. E. Moser — Departments of Pharmacology and Laboratory Medicine and Pathology, University of Minnesota, Minneapolis, Minnesota

F. Murad — Division of Clinical Pharmacology, Departments of Internal Medicine and Pharmacology, University of Virginia, Charlottesville, Virginia 22903

A. Paraf — Institut National de la Recherche Agronomique Station de Virologie et d'Immunologie, 78850 Thiverval-Grignon, France

A. Peterkofsky — Laboratory of Biochemical Genetics, National Heart and Lung Institute, National Institutes of Health, Bethesda, Maryland 20014

K. Raska, Jr. — Department of Microbiology, College of Medicine and Dentistry of New Jersey, Rutgers Medical School, Piscataway, New Jersey 08854

R. H. Rixon — Animal and Cell Physiology Group, Division of Biological Sciences, National Research Council of Canada, Ottawa, Canada K1A OR6

J. S. Schweppe — Department of Biochemistry, Northwestern University Medical School, Chicago, Illinois 60611

L. Shapiro — Department of Molecular Biology, Division of Biological Sciences, Albert Einstein College of Medicine, Bronx, New York 10461

H. Sheppard — Department of Cell Biology, Roche Research Center, Hoffmann-La Roche Inc., Nutley, New Jersey

J. R. Sheppard — Dight Institute of Human Genetics,
University of Minnesota,
Minneapolis, Minnesota
55455

A. M. Spielvolgel — Department of Biochemistry,
Northwestern University Medical School,
Chicago, Illinois
60611

J. H. Stephenson — Departments of Pharmacology and
Laboratory Medicine and Pathology,
University of Minnesota,
Minneapolis, Minnesota

W. H. Tsien — Department of Cell Biology,
Roche Research Center,
Hoffmann-La Roche Inc.,
Nutley, New Jersey

P. R. Walker — Animal and Cell Physiology Group,
Division of Biological Sciences,
National Research Council of Canada,
Ottawa, Canada
K1A OR6

J. F. Whitfield — Animal and Cell Physiology Group,
Division of Biological Sciences,
National Research Council of Canada,
Ottawa, Canada
K1A OR6

C. E. Zeilig — Department of Pharmacology,
University of Minnesota Medical School,
Minneapolis, Minnesota
55455

Cyclic Nucleotides and the Regulation of Cell Growth

Part I

CYCLIC NUCLEOTIDES AND THE REGULATION OF BACTERIAL GROWTH

ABOU-SABÉ
cAMP and the Regulation of Bacterial Growth

PETERKOFSKY
The Mechanism of Regulation of Escherichia coli *Adenylate Cyclase*

KURN and SHAPIRO
Cyclic Nucleotides and the Caulobacter *Cell Cycle*

cAMP and the Regulation of Bacterial Growth

Morad Abou-Sabé

Bacterial growth may be defined in terms of either cellular mass or number of cells. The two remain proportional under conditions of steady growth, but this ratio can vary with growth conditions. As defined, regulation of bacterial growth thus becomes a reflection of the effect of growth conditions, presence or absence of growth factors in the growth medium and/or the culture response to molecules which do not, in themselves, serve as growth factors. Regulation of bacterial growth by the latter category thus implies a direct or indirect effect on one or more aspects of the cellular growth machinery, i.e., its metabolic proficiency (anabolic and amphibolic). Of such factors are metabolic inhibitors, metabolic stimulants, and factors that affect the cell in both directions inhibiting some reactions while stimulating others, such as cyclic 3',5'-adenosine-monophosphate (cAMP). Studies of the role of cAMP in many biological functions in a variety of prokaryotic systems have shown the involvement of cAMP in the transcription of inducible enzymes, the development of competence in bacteria, the cell cycle and development of *Caulobacter* (see Kurn and Shapiro, this volume), the formation of flagella and the sporulation process as well as many other specific functions (for recent reviews, see 40,41). The role played by cAMP in these diverse functions does not follow a uniform pattern, e.g., while cAMP stimulates transcription it inhibits uracil uptake in *Escherichia coli* (16,28).

If, at the risk of generalization, one considers

growth a final expression of all reactions and interactions that occur in the bacterial cell, cAMP effect on cell growth would become an expected result of its diverse effects on the cell. The difficulty in this conception is the fact that no primary biochemical reaction directly influencing or responsible for growth determination has been identified in any living system. Thus, growth remains a final expression of many biological reactions. Consideration of DNA synthesis and replication as the ultimate precursor of cell growth (reproduction) still does not satisfy the criteria of a primary biochemical reaction directly responsible for growth since DNA synthesis and replication are dependent upon many other prerequisite reactions in the cell. In order to define a primary step for cell growth, therefore, one would have to consider the cell as a whole, in its environment, define its interactions with the growth medium (under normal growth conditions), and characterize the first result of such interactions upon which the final cell growth characteristics are determined. Once studies direct themselves at the internal interactions, they have already passed the primary step. Interactions at the membrane level thus become of crucial importance in defining what (if anything) regulates cell growth.

Of the many biological effects of cAMP in bacteria, two have received most detailed analysis. The role of cAMP in catabolite repression has received extensive attention, resulting in the identification of the requirement of cAMP for the transcription of mRNA in inducible operons (16,46). On a different level, the role of glucose in the regulation of the cAMP level in bacteria, has been studied at great length (see below). These two specific levels of investigation were directed at two sides of the same coin. On the one hand, are the variations in the cAMP level directly related to the phenomenon of catabolite repression? On the other hand, how is the cAMP level itself regulated? These two questions are thus directly related since whatever the mechanism of cAMP regulation, its ultimate level itself will regulate transcription (as well as all other functions sensitive to cAMP) influencing, in the final analysis, cell growth. As pointed out above, in relation to cell growth, the regulation of the cAMP level thus becomes an essential step which ultimately affects cell growth. With this concept in mind, we have directed our attention in the past few years toward the question of the regulation of cAMP in bacteria.

REGULATION OF cAMP LEVEL IN BACTERIA

Metabolism of cAMP in bacteria involves two basic reactions, one in which cAMP is synthesized from ATP by the enzyme adenylate cyclase and a second reaction in which cAMP is hydrolyzed to 5'-AMP by the enzyme cAMP phosphodiesterase (35,44,45). Localization of these enzymes in *E. coli* showed the adenylate cyclase to be membrane bound and the cAMP phosphodiesterase to be soluble, i.e., cytoplasmic (27,35,44,56).

Regulation of cAMP level in bacteria was first shown by Makman and Sutherland (33), they observed a glucose effect on the intracellular level of cAMP, where cells grown on glucose as a sole carbon and energy source had a lower level of cAMP than cultures grown in glycerol media. The level of cAMP also rose upon the depletion of glucose from the culture medium. This observation was explained to mean that in the presence of glucose cAMP was excreted from the cells at a higher rate, resulting in a lower level of intracellular cAMP. Glucose sensitive cells thus retain a low level of cAMP, and inducible enzyme synthesis, e.g., β-galactosidase, appears inhibited by glucose. Addition of exogenous cAMP to cultures grown in glucose was found to reverse the glucose effect on β-galactosidase synthesis (36). The question of glucose reduction of the level of cAMP in bacteria thus became a focal point in the understanding of the glucose affect (catabolite repression). Attempts by Ide (27) to show whether glucose or glucose metabolites had an affect on the adenylate cyclase in *E. coli* were unsuccessful. In 1970, we isolated a number of mutants of *E. coli* B/r (*rbs*$^+$) which were insensitive to glucose catabolite repression (1). Studies on one of these mutants (*E. coli* B/r ABDR01), showed that it was not only insensitive to catabolite repression, but also stimulated by glucose in its capacity to synthesize the inducible enzymes, β-galactosidase (β-gal) and tryptophanase (TPase) (2). Determination of the rate of cAMP synthesis in this mutant and its catabolite sensitive parent strain (*E. coli* B/r EE1001), showed that while cAMP synthesis was inhibited in EE1001 by glucose, it was synthesized throughout the culture's growth in glucose media at a rate one and one half times its rate of synthesis in glycerol grown cultures (Figures 1 and 2) (2). The availability of this mutant, ABDR01, facilitated our analysis of how the cAMP level is regulated in *E. coli*. Studies by Peterkofsky and Gazdar (37) on the effect of glucose on cAMP metabolism in *E. coli* showed that while cAMP formation depends on the absence of glucose, its excretion from the cells must be dependent upon some factor other than glucose. Determination of whether the adenylate cyclase and/or the cAMP phosphodiesterase activities or synthesis were subject to glucose presence (at different concentrations) in the culture media showed no significant alteration of either enzyme activity and the authors concluded that an explanation of the glucose effect on cAMP accumulation could not be based on changes in the activities of the synthetic or degradative enzymes of cAMP (37). Similar studies in Rickenberg's laboratory (9), however, supported Makman and Sutherland's idea of cAMP excretion in the presence of glucose.

REGULATION OF cAMP SYNTHESIS IN ESCHERICHIA COLI

It, therefore, appeared to us that a most direct way to elucidate whether the mechanism of glucose regulation

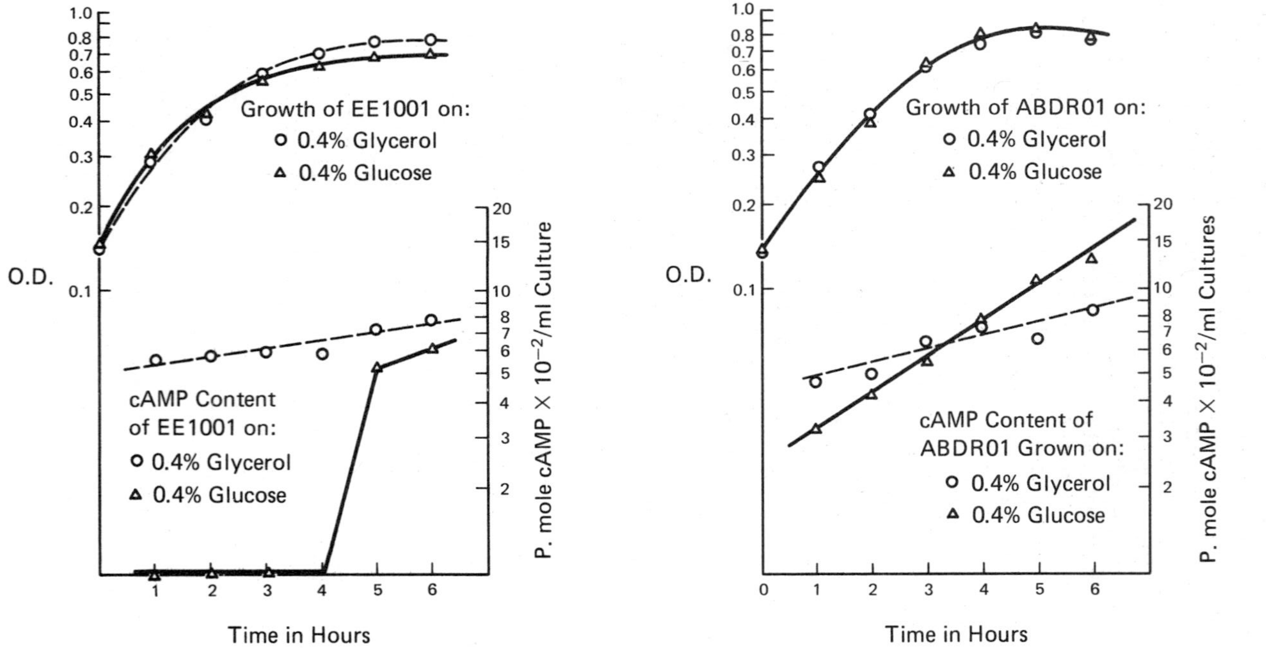

FIGURES 1 and 2. *cAMP synthesis in Escherichia coli B/r EE1001 and ABDR01 grown on minimal glycerol and glucose media (0.4%). Two ml samples were collected at 1 h intervals for cAMP assay, and 3 ml collected for turbidity measurements. cAMP assay samples were incubated in a boiling water bath for 3 min, and centrifuged at 1,000 rpm to remove cell debris. cAMP determination was performed by the Gilman binding assay (23). Figure 1 shows cAMP synthesis and culture growth in glycerol and glucose for EE1001, and Figure 2 shows cAMP synthesis and growth of ABDR01 in the same media. From Abou-Sabé (2).*

of the cAMP level occurs via excretion or synthesis, would be to test the effect of glucose on the membrane bound adenylate cyclase activity directly using membrane preparations of ABDR01 and EE1001. From the results of these studies (Table 1) it was immediately apparent that glucose stimulation of cAMP synthesis in the bacterial culture (ABDR01) was paralleled by a stimulation of the adenylate cyclase activity *in vitro* in the presence of glucose (3). The action of glucose on the adenylate cyclase activity was also found to depend on the integrity of the bacterial membrane preparations since solubilization of the enzyme abolished its sensitivity to glucose action.

The possible participation of the cAMP phosphodiesterase in the glucose regulation of the cAMP level was ruled out since the bacterial membrane preparations were devoid of phosphodiesterase activity (Abou-Sabé, unpublished results) and for the fact that glucose regulation of the cAMP level was reported in cAMP phosphodiesterase deficient mutants of *E. coli* (9,35). These findings thus demonstrated that glucose regulation of the cAMP level in *E. coli* occurs at the level of cAMP synthesis and not due to the excretion phenomenon. Thus, the interaction of glucose with the membrane bound adenylate cyclase, the enzyme activity may be inhibited or stimulated, i.e., regulated, determining the level of cAMP in the cell (3). In support of this concept was the finding of *in vivo* glucose regulation of the adenylate cyclase activity in *E. coli* (38), and by the finding that cAMP excreted in the culture medium is dependent upon the intracellular concentration of cAMP obtaining a constant ratio of intracellular to extracellular cAMP unaffected by the culture's growth conditions (18).

GLUCOSE PHOSPHOTRANSFERASE ENZYMES AND THE REGULATION OF ADENYLATE CYCLASE ACTIVITY

Now that glucose regulation of cAMP synthesis has been established, and that the membrane bound adenylate cyclase must be maintained in intact membranes (3), the question was asked whether the glucose transport enzymes mediate the glucose effect on the adenylate cyclase activity.

In *E. coli* glucose is transported via a group translocation enzyme system (phosphotransferase, Pts) which is composed of three main elements: a cytoplasmic enzyme, Pts E_I, which catalyzes the phosphorylation of a small molecular weight heat stable protein (HPr); a sugar specific membrane bound enzyme (Pts E_{II}) responsible for the phosphorylation of each specific Pts sugar such as glucose, mannose, fructose, etc., and a phosphate carrier protein (HPr) (32). For the fact that Pts E_{II} is sugar specific and is composed of two ubunits, one responsible for the specific recognition of each sugar (E_{II_A}) and the other required for the general phosphorylation of the sugars (E_{II_B}), the effect of the specific solubilization of these subunits should determine whether the Pts E_{II} (or

TABLE 1. *Effect of glucose on the adenylate cyclase activity*

Experiment #	Bacterial Strain	Enzyme Source	Additions	Adenylate Cyclase Activity (nanomoles cAMP/30 min/mg protein)	Total Activity per ml Pellet*	Adenylate cyclase Activity as percent of Control
(a)	ABDR01	10,000 rpm pellet	None	0.158	1.153	100.0
			Glucose	0.210	1.533	133.0
(b)	EE1001	16,000 rpm pellet	None	0.184	1.82	100.0
			Glycerol	0.202	2.0	109.8
			Glucose	0.149	1.48	81.0
	ABDR01	"	None	0.254	2.36	100.0
			Glycerol	0.313	2.91	123.0
			Glucose	0.452	4.20	178.0
(c)	ABDR01	40,000 rpm pellet	None	3.161	32.90	100.0
			Glucose	8.607	89.50	272.3

**In the preparation of the pellet fractions, a standard procedure was used in which, cells collected from 1500 ml culture at 80 Kletts were resuspended in 9 ml buffer, sonicated and centrifuged as indicated, and the pellet resuspended in 2 ml buffer (0.05 M Tris-HCl, pH 8.5, containing 1 mM dithiothreitol (DTT)). The various membrane preparations were used as the enzyme source and the adenylate cyclase activity determined as described by Krishna et al. (31). Reproduced from Abou-Sabé and Nardi (3).*

either of its subunits) is involved in the glucose regulation of the adenylate cyclase activity.

Solubilization of the membrane bound Pts E_{II} into its subunit components, followed by the determination of the glucose effect on the membrane bound adenylate cyclase activity, showed that both subunits are required for the glucose effect (4) (Table 2). While glucose had no effect on the adenylate cyclase activity of either of the solubilized fractions, reconstitution of the membrane was accompanied by a regain of the regulative effect of glucose on the adenylate cyclase activity (Table 3), suggesting the following mechanism:

$$E_{II_A} + \text{Glucose} \rightarrow E_{II_A} - G$$

$$E_{II_A}\,G + E_{II_B} \rightarrow E_{II_{AB}} - G$$

$$E_{II_{AB}} - G + AC^* \rightarrow \text{inhibition or stimulation of AC activity}$$

*AC = adenylate cyclase.

The mediation of the glucose Pts E_{II} of glucose regulation of adenylate cyclase activity was subsequently confirmed by the work of Harwood and colleagues (25) using mutants deficient in the glucose Pts E_{II} (*gpt*⁻).

TABLE 2. *Effect of the urea/butanol solubilization on the adenylate cyclase activity.*

Adenylate Cyclase Source	Protein Content (mg/ml)	Adenylate Cyclase Activity* Control Activity	Glucose (0.2 *M*)	% of Control Activity
40,000 rev/min membrane preparation	10.0	1.787	4.685	262
Solubilized pellet fraction	9.9	0.812	0.820	101
Solubilized supernatant fraction	1.8	0.211	0.200	95

Adenylate cyclase activity was determined as the nmole cyclic AMP produced/30 min/mg membrane protein. Reproduced from Abou-Sabê and Mento (4).

Although it was demonstrated above by the solubilization experiments and by the use of the *gpt*⁻ mutants that the glucose phosphotransferase E_{II} is essential for glucose regulation of the adenylate cyclase activity, further evidence showed that the direction of the regulation of the adenylate cyclase activity, i.e., inhibition vs. stimulation is determined by the nature of the glucose transport molecules and their conformation in the cell membrane. Analysis of the detailed properties of *E. coli* B/r (ABDR01), showed that glucose transport in this mutant was

TABLE 3. *Effect of the membrane reconstitution on the adenylate cyclase activity and its response to glucose regulation in vitro*

Reconstitution Mixture	Tube No. 1	2	3	4	5	6
Supernatant fraction (mg protein)	0.018	0.036	0.054	0.072	0.090	0.108
Pellet fraction (mg protein	0.250	0.250	0.250	0.250	0.250	0.250
Total protein content	0.268	0.286	0.304	0.322	0.340	0.358
Pellet:supernatant protein ratio	13.7	6.9	4.6	3.4	2.8	2.3
Expected adenylate cyclase activity* in:						
Supernatant fraction	0.0038	0.0076	0.0114	0.0152	0.0190	0.0228
Pellet fraction	0.2030	0.2030	0.2030	0.2030	0.2030	0.2030
Total reconstitution mixture	0.2068	0.2106	0.2144	0.2182	0.2220	0.2258
Observed adenylate cyclase activity in reconstitution mixture	0.0646	0.1118	0.279	0.1704	0.1034	0.1436
% Adenylate cyclase reconstituted activity	31.0	53.0	130.0	78.1	46.6	63.6
Observed adenylate cyclase activity in the presence of glucose	0.070	0.1187	0.380	0.3664	0.0915	0.1378
% of control activity	108.4	106.1	136.2	215.0	88.0	96.0

*Adenylate cyclase activity determined as the nmol cyclic AMP produced/30 min assay/assay/reaction tube. Reproduced from Abou-Sabé and Mento (4).

catalyzed by two transport systems (5). One transport system (low affinity) was found to be functional primarily at the early logarithmic phases of growth and catalyzes low efficiency glucose transport as determined by its high apparent K_m value (3.4×10^{-5} *M*). A second transport system (high affinity) which was found to be functional primarily at the late logarithmic and stationary phases of growth, catalyzes high efficiency glucose transport as determined by its low apparent K_m value (1.2×10^{-5} *M*). Under these conditions of glucose transport, the adenylate cyclase sensitivity to glucose regulation was variable. As shown (Table 4), glucose, while stimulating the adenylate cyclase activity in the early phases of growth (80 Kletts), inhibited the enzyme activity under conditions of late logarithmic or stationary phase growth (200 Kletts).

The suggested mechanism of glucose regulation of the adenylate cyclase, referred to above, can thus be expended as follows:

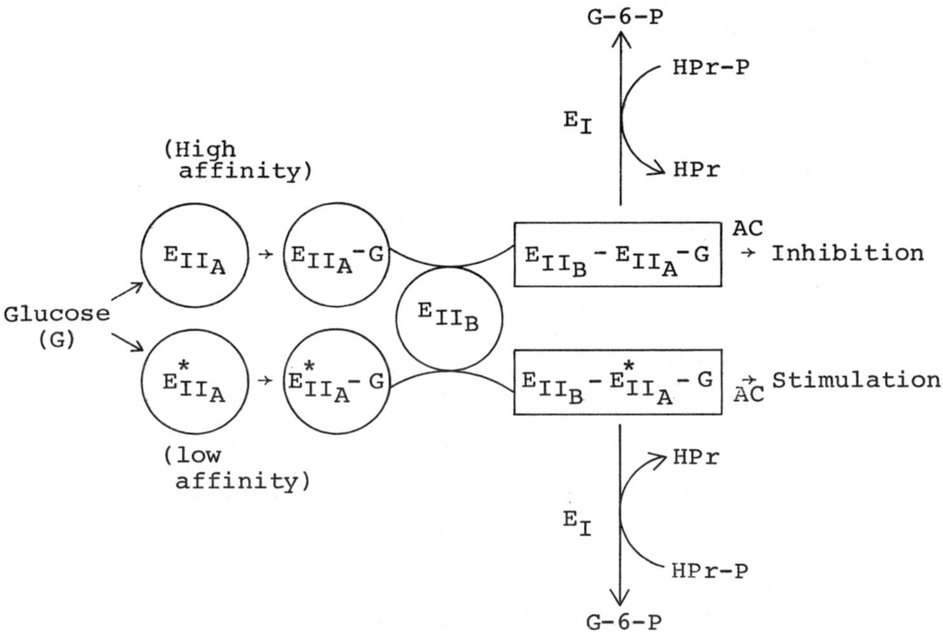

AC = Adenylate cyclase,

E_{II_A}, E_{II_B} = glucose phosphotransferase E_{II} subunits,

$E_{II_A}{}^*$ = modified conformation of the E_{II_A} subunit.

In summary, glucose interacts with the bacterial cell membrane and through its interaction with the glucose Pts E_{II} an effector molecule is formed which mediates two

TABLE 4. *Effect of culture growth phase on the adenylate cyclase regulation by glucose*

Culture Growth Phase	Adenylate Cyclase Activity* Control A**	Control B***	+ Glucose§ A	+ Glucose§ B	% Control Activity† A	% Control Activity† B
80 Kletts (early log)	1.78	0.80	4.68	1.29	2.62	1.62
200 Kletts (late log)	4.12	2.99	1.51	0.92	0.36	0.31

**Activity determined as nmoles cAMP produced/10 min/mg protein.*

***Assay performed on 40,000 rpm membrane preparation of sonicated cells.*

****Assay performed on 12,000 rpm membrane preparation of toluenized cells.*

§Glucose concentration used in each reaction mixture was 20 mM.

†Note the glucose stimulation of the adenylate cyclase activity of both preparations when made from an early log culture, and its inhibition of the adenylate cyclase activity when preparations were from late log or early stationary cultures.

functions: one, it leads to the phosphorylation and subsequent transport of glucose to the cell cytoplasm; and two, it either directly or indirectly regulates the adenylate cyclase activity, thereby determining the level of cAMP in the cell.

With this analysis of the glucose regulation of cAMP synthesis in bacteria, the question arises whether regulation of the adenylate cyclase is limited to glucose.

CORRELATION BETWEEN cAMP LEVELS AND BACTERIAL GROWTH

Determination of the levels of cAMP in a variety of *E. coli* strains grown on different carbon sources has been performed in many laboratories. One common result has been the finding that, as in mammalian cell cultures, an inverse correlation exists between the growth rate attained by the bacterial culture and the level of cAMP found in the cultures (5,9). As shown, for example (Table 5), cultures grown on rich carbon sources such as glucose are found to synthesize cAMP at a low rate, whereas those grown on poor carbon sources such as succinate assume a much higher rate of cAMP synthesis (5). Similar results, at the cAMP level, show higher intracellular cAMP concentration under conditions of poor growth and lower concentrations under conditions of fast growth (9). In contrast to these results,

TABLE 5. *Effect of carbon source variation on inducible enzymes and cAMP synthesis*

Carbon and Energy Source (0.4%)	Generation Time (h)		Differential Rate of Tryptophanase Synthesis*		Differential Rate of β-Galactosidase Synthesis*	Differential Rate of Cyclic AMP Synthesis***	
	EE1001	ABDR01	EE1001	ABDR01		EE1001	ABDR01
Glucose	0.90	1.35	8	26	18.2	0.05	0.82
Glycerol	1.10	1.00	19	18	14.9	0.49	0.52
Arabinose	1.45	1.40	37	27	19.6	18.33	8.20
Ribose	10.30	1.80	71	49	20.6	55.00	23.00
Succinate	14.60	12.00	146	130	28.9	320.00	371.00

**Units of tryptophanase/mg dry wt.*

***Units of β-galactosidase/mg dry wt.*

****nmol/mg dry wt.*

Overnight glycerol grown cultures were transferred to fresh 0.4% glycerol medium (5-10% inoculum) and incubated at 37°C on a gyrotory incubator shaker until culture turbidity reached 80 Klett units (blue filter). The culture was centrifuged at room temperature and washed with minimal salts medium (MS) and resuspended in twice the original volume in MS, divided into 50 ml portions in 250 ml flasks and immediately used for the experiment. To each flask the inducer and the carbon source were added. At time intervals over a 6 h growth period samples were removed for turbidity measurements, enzyme assays and for cyclic AMP determinations. Reproduced from Abou-Sabé et al. (5).

studies by Peterkofsky *et al.* (39) showed that the effect of carbon sources on the adenylate cyclase activity is subject to the growth of the culture on the respective carbon source, otherwise no effect on the adenylate cyclase activity was observed. Exception to this was the case with glucose, where regardless of the growth substrate, the adenylate cyclase activity was inhibited by glucose.

Consideration of the correlation between the level of cAMP and the culture's growth on a single carbon and energy source is often hampered by the phenomenon of catabolite repression, since the metabolism of many of these carbon sources involves an inherent requirement of cAMP for the transcription of their metabolic enzymes (inducible enzyme systems). Similarly, consideration of the requirement of cAMP for growth through the isolation of mutants deficient in their adenylate cyclase activity (*cya*$^-$) or lacking the cAMP receptor protein (*crp*$^-$), does not definitively answer the question either, since during the isolation of the *cya*$^-$ or *crp*$^-$ mutants, alternate molecules may be evolved to support the survival of these mutants. For example, a *cya*$^-$ deletion mutant was isolated and shown to grow almost as well as the parent strain on nutrient rich media. It, however, lacked the capacity to synthesize inducible enzymes (8). The isolation of this mutant was used as an argument against the requirement of cAMP for growth. In a recent report, however, *ara*$^+$ revertants isolated in a *cya*$^-$ *crp*$^-$ background were found to synthesize the inducible *ara* operon enzymes (26). The mutation site in these revertants was localized in the *ara*C regulatory gene. The loss of cAMP and CRP was found to be compensated by an alteration in the regulatory function of the *ara*C gene product allowing for transcription in the absence of cAMP and CRP. It was also shown by DeRoberts *et al.* (17) that the *cya*$^-$ deletion mutant, referred to above (CA8306), required exogenous addition of cAMP for optimal growth. Higher concentrations of cAMP, however, inhibited its growth. These results suggest a basic involvement of cAMP in the regulation of bacterial growth, and that its requirement for growth, under certain conditions, may be compensated by specific mutations in the bacterial genome.

IN VITRO REGULATION OF THE ADENYLATE CYCLASE ACTIVITY BY DIFFERENT CARBOHYDRATES

Work in our laboratory on the *in vitro* regulation of the adenylate cyclase activity by different carbon sources in *E. coli* B/r strains ABDR01 and EE1001, showed that D-glucose, D-ribose, D-mannose, D-galactose, D-xylose and succinate produced varying degrees of stimulation or inhibition of the adenylate cyclase activity in both strains (Table 6). It is interesting to note that the sugar effects on the adenylate cyclase activity were, once again, inversely related to the growth response of the cultures on each specific carbon source. For example, ABDR01 grows slowly on glucose in comparison to EE1001, while on ribose it shows normal growth in comparison to the residual

TABLE 6. *The effect of different carbohydrates on the adenylate cyclase activity in vitro*

Carbohydrate Addition (0.2 *M*)	Adenylate Cyclase Activity*			
	EE1001		ABDR01	
	Activity	% of Control Activity**	Activity	% of Control Activity
-	1.48	100	2.12	100.0
Glucose	1.36	92	6.54	308.0
Ribose	2.73	185	2.30	108.0
Mannose	2.56	174	3.69	174.0
Galactose	2.83	191	4.56	215.1
Xylose	6.30	426	5.95	280.0
Succinate	5.68	384	3.50	165.0
Glucagon ($5 \cdot 10^{-5}$ *M*)	3.58	242	4.23	200.0

**Adenylate cyclase activity determined as nmol cyclic AMP produced/30 min/mg membrane protein.*

***The regulative effect of the various carbon sources on the adenylate cyclase activity is reflected by the increase or decrease in the enzyme activity (as % of control activity) in the presence of the carbon source in the reaction mixture.*

From Abou-Sabé et al. (5).

growth of EE1001 (1). The effects of glucose and ribose on the adenylate cyclase activity shows glucose stimulation in ABDR01 and ribose stimulation in EE1001 (slow growth), whereas glucose and ribose effects were only marginal on EE1001 and ABDR01, respectively (fast growth). Carbon sources which support poor growth in both strains, such as xylose and succinate, stimulated the adenylate cyclase activity in each strain.

As noted above, regulation of cAMP synthesis in bacteria is, therefore, not limited to glucose, as carbon sources other than glucose can regulate the adenylate cyclase activity. The speculation being that carbon sources may regulate the adenylate cyclase activity via interactions with their specific membrane bound carrier proteins, simulating hormonal regulation of mammalian adenylate cyclase activity via their specific receptor sites.

The analogy presented thus far remains limited to the conception of these carbon sources acting as hormones. The question is, therefore, asked, whether mammalian hormones can regulate the bacterial adenylate cyclase. If so, then the analogy of the regulatory mechanism of bacterial and mammalian adenylate cyclases is further strengthened, providing for a possible detailed investigation of the mechanism of hormone action using the bacterial adenylate cyclase system in a much simpler and straight forward manner, with the use of the important facility of bacterial mutants.

EFFECT OF INSULIN ON THE ADENYLATE CYCLASE ACTIVITY OF ESCHERICHIA COLI B/r

Insulin is a mammalian hormone which has been shown to modulate both the adenylate cyclase activity and glucose uptake in mammalian cell lines (11,15,22,30). Due to the dual function of insulin, and the fact that glucose regulation of bacterial adenylate cyclase activity does involve the glucose transport molecules (see above), the action of insulin on glucose transport and on the adenylate cyclase activity in *E. coli* B/r became a logical choice for study. To test the action of insulin on *E. coli* B/r, its effect on glucose transport was first determined. As shown (Table 7 and Figure 3), insulin stimulated glucose transport (using C^{14}-α-methyl glucoside) at insulin concentrations generally comparable to those used with mammalian cell cultures (6).

TABLE 7. *Effect of zinc-free insulin on C^{14}-α-MG uptake in Escherichia coli B/r*

Insulin Concentration (μunits/ml)	Rate of α-MG Uptake* (10^{-1})	α-MG Uptake as % of Control
0 (control)	3.0	100.0
30	3.2	106.6
60	4.5	150.0
180	5.0	166.6
240	4.75	158.3

**Rate of α-MG uptake calculated as nanomoles of C^{14}-α-MG/min/mg protein.*

Glucose uptake was determined using C^{14}-α-MG glucoside (a non-metabolized glucose analogue α-MG) at 30 μM concentration as previously described (5). Cultures were incubated in the presence of insulin for 15 min prior to C^{14}-α-MG uptake assay (37°C). From Abou-Sabê and Reilly (6).

Determination of insulin's effect on the adenylate cyclase activity showed a marked inhibition of its activity at different concentrations (Table 8). It was also found that at high concentrations, insulin stimulated the adenylate cyclase activity (last two columns of Table 8). These combined effects of insulin (stimulation of glucose transport and inhibition of adenylate cyclase activity) suggests a common mediator for insulin action on both activities. It is conceivable that if insulin interacts with the glucose Pts E_{II} - increasing its activity and/or affinity for glucose - the resulting conformation of the Pts E_{II} molecule could stimulate glucose uptake, and also

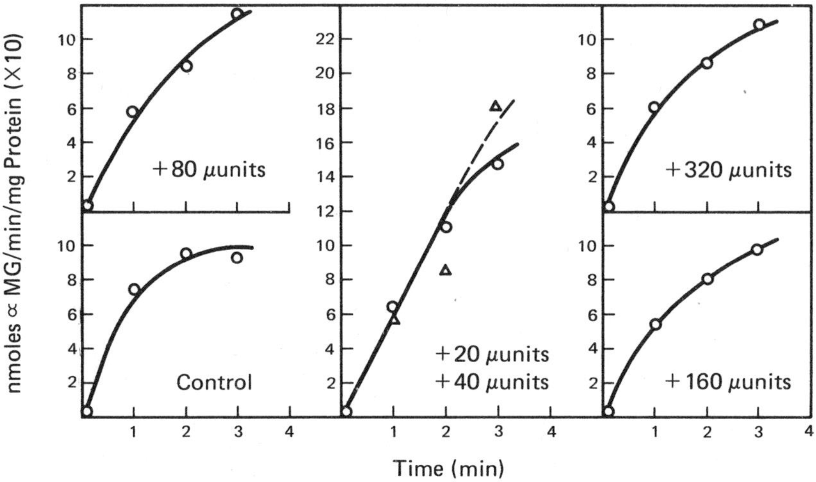

FIGURE 3. *Determination of the effect of insulin on C^{14}-α-MG uptake in Escherichia coli B/r (ABDR01) was performed on cultures grown on minimal glycerol (0.4%) media. Cultures were centrifuged at room temperature, washed with saline and resuspended in fresh minimal glycerol media, incubated with insulin for 15 min (at 37°C) in a water bath shaker, the C^{14}-α-MG was added (30 μM) and its uptake followed.*

inhibit the adenylate cyclase activity. A mutant of ABDR01 specifically lacking the phosphorylating activity for glucose (Pts E_{II} *gpt*$^-$) was isolated. Determination of the effect of insulin on the adenylate cyclase activity of this mutant showed continued insulin inhibition despite the loss of the phosphorylating activity for glucose (Table 9), thus making it unlikely that the glucose Pts E_{II} is involved in insulin's action on the adenylate cyclase activity. The possibility, however, remains that insulin's action on glucose transport and adenylate cyclase activity may utilize a common mediator at an earlier step in insulin's interaction with the cell membrane.

It is thus indicated that insulin, while a specific mammalian hormone, does recognize *E. coli* B/r cells, stimulates its uptake of glucose and regulate the adenylate cyclase activity in both directions (as has been observed in mammalian cells and the eukaryote *Neurospora* (22)).

A basic difficulty in accepting these implications comes from the fact that insulin is a specific mammalian hormone which requires, as many other hormones do, specific receptor sites to interact with. Insulin interaction with these receptors is envisioned to elicity its effects through a signal carrier or through general perturbation of the fluidity of the mammalian cell membrane (11). The complexity of such interactions would tend to reduce the likelihood of insulin action on prokaryotic cells such as

TABLE 8. *Effect of insulin on the adenylate cyclase activity* of Escherichia coli B/r***

Protein Concentration per Reaction (µg)	Control Activity	Additions to Reaction Mixture									
		Glucose (20 m*M*)	Insulin Concentration (µunits)								
			0.25	1.0	2.0	3.75	7.5	22.5	37.5	225	375
30.0	2.55	1.97 (77)***	2.78 (109)	2.57 (101)	2.49 97.6	0.93 (36.47)	1.57 (61.6)	2.06 (80.1)	-	7.0 (274)	9.6 (376)
.33.5	4.43	3.07 (69.3)	-	-	-	0.31 (6.9)	1.01 (22.8)	1.46 (32.96)	3.24 (73.14)	-	-
40.0	3.68	-	-	-	-		1.1 (29.9)	-	-	9.37 (255)	11.40 (310)
50.0	3.19	-	-	-	-	1.47 (46.0)	1.76 (55.2)	2.0 (62.7)	2.87 (90.0)	-	-

**Activity determined as nonomoles cAMP produced per 10 minutes per mg protein.*
***Toluenized membrane preparations were from a 200 Klett culture of ABDR01.*
****Percent of control activity.*

Adenylate cyclase assays were performed on toluenized cell membranes prepared as described (24) and modified (7). From Abou-Sabé and Reilly (6).

TABLE 9. *Effect of insulin on the adenylate cyclase activity in a phosphotransferase E_{II} (gpt^-) mutant of Escherichia coli B/r*

Bacterial Strain	Pts E_{II} Activity[1]	Adenylate Cyclase Activity[2] Insulin Concentration (μunits) 0	3.75	22.50	37.50	225.00	375
E. coli B/r (ABDR01)	17.86	2.50 (100)[3]	0.58 (23.2)	0.84 (33.6)	1.86 (74.4)	6.49 (259.6)	7.90 (316)
E. coli B/r (PR134)	1.87	1.85 (100)	0.50 (27.0)	0.57 (31.0)	0.40 (22.0)	1.96 (106.0)	2.09 (113)[4]

[1]*Pts E_{II} activity determined as nanomoles C^{14}-α-MG phosphorylated per mg protein in 10 min reactions, as previously described (25).*

[2]*Adenylate cyclase activity determined as nanomoles cAMP produced per 10 min/mg protein.*

[3]*Percent of control activity.*

[4]*Note the loss of the stimulation of the adenylate cyclase activity, by high insulin concentrations, in the gpt^- mutant. Analysis (in progress) of insulin action on this mutant might lead to the characterization of the nature of the stimulatory vs inhibitory effects of insulin on the adenylate cyclase activity.*

E. coli. Consider, however, the nature of the insulin receptor which is envisioned as a glycoprotein molecule with oligosaccharide chains tipped with sialic acid residues. Although sialic acid does not participate in insulin binding, it is considered a part of the recognition portion of the receptor (12,42), and may be involved in the transmission of the insulin binding signal from the recognition site to the glucose transport site. Consider, if you will, the analogy in the structure of the sialic acid residues with those of the N-acetyl muramic acids which are important building blocks in the murein layer of the cell wall. These negatively charged molecules, in conjunction with the periplasmic lipoproteins and/or cytoplasmic membrane proteins, may provide for a rudimentary binding and recognition structure capable of interacting with insulin.

CHOLERA ENTEROTOXIN REGULATION OF THE ADENYLATE CYCLASE ACTIVITY IN ESCHERICHIA COLI B/r

Cholera enterotoxin (CT) is a heat and acid labile hormone-like protein of a molecular weight 84,000 daltons, produced by the organism *Vibrio cholerae* (20). It consists of two subunits of molecular weights 28,000 and 14,500 daltons (subunits A and B, respectively). Previous investigations in mammalian cells have shown that the A subunit is responsible for the biological effects of the toxin, while the B subunit is responsible for the binding of the toxin to the mammalian cell membrane. The toxin effect on the adenylate cyclase activity in mammalian cells has been interpreted to be mediated by the toxin's interaction with ganglioside GM_1 (13), where the toxin binds to the cell membrane (via GM_1) followed by a 30-90 min lag before the toxin's effect on the adenylate cyclase activity occurs (13,14,19,29,42). It is not clear if conformational changes occur in the cell membrane during the lag period. It is known, however, that the lag is not a reflection of the toxin concentration or the number of toxin receptor sites on the cell membrane (13,14,19).

In an effort to investigate the effect of cholera enterotoxin on the bacterial system, the toxin effect was determined on two levels: (a) *in vivo* effect on the inducible synthesis of TPase, which was used as an indirect measure of the toxin's effect on cAMP synthesis; and (b) its effect on the adenylate cyclase activity *in vitro* in membrane preparations, without prior incubation of the live cells with the toxin. Results (7) showed an immediate stimulation of TPase synthesis in *E. coli* B/r (Table 10). The extent of stimulation of TPase synthesis was found to be dependent on two factors: one, the toxin concentration used, and two, incubation of the bacterial culture with the toxin prior to induction of TPase synthesis. Use of high toxin concentrations (above 500 ng/ml culture) or prolonged incubation with low concentrations of the toxin (e.g., with 160 ng/ml, 120 min) resulted in a reduction of the stimulatory effect of the toxin and/or

TABLE 10. *Effect of incubation time with cholera enterotoxin on TPase synthesis by Escherichia coli B/r**

Incubation Period (min)	Rate of TPase Synthesis ($x10^{-3}$)** Control Culture (No Toxin)	Toxin Added (160 ng/ml)	% of Control Activity
0	1.93	3.02	156.5
30	1.62	3.51	216.6
60	1.76	3.98	226.1
90	1.82	4.86	267.0
120	-	3.29	170.5

**Cultures of Escherichia coli B/r (EE1001) were grown to 200 Kletts in minimal glycerol media (0.4%) at 37°C and 350 rpm shaking speed. Cells were collected by centrifugation at room temperature, washed with saline and resuspended in equal volume of minimal medium (without the carbon source). Toxin was added and cultures kept in stationary incubator at 37°C for the desired incubation interval. To each culture, glycerol (0.4%) was added and tryptophanase was induced and its differential rate determined.*

***Differential rate determined as units of tryptophanase per mg dry wt of bacterial cells. (Data collected from (7).)*

TABLE 11. *Effects of cholera enterotoxin concentration and incubation period on TPase synthesis in Escherichia coli B/r**

Toxin Concentration ng/ml culture	Differential Rate of TPase Synthesis** Incubation Period with Toxin (min): 0	30	% of Control Activity	
0	1.93	1.96	100.0	100.0
20	2.30	-	119.2	-
40	2.54	-	131.6	-
70	-	2.90	-	148.0
80	3.03	-	154.9	-
140	-	3.60	-	183.6
160	3.02	-	156.5	-
280	-	2.30	-	118.0
320	2.54	-	131.6	-
500	-	1.70	-	86.7
750	-	1.10	-	59.2

**Conditions of experiment were as described in Table 10.*

***Differential rate determined as units of tryptophanase per mg dry wt of bacterial cells. (Data collected from (7).)*

inhibition of TPase synthesis (Tables 10 and 11). These results thus indirectly reflect a toxin effect on cAMP synthesis in *E. coli*. To substantiate these results, the effect of the toxin on the adenylate cyclase activity *in vitro* was determined using membrane preparation of cells without prior exposure to the toxin. It was found that the adenylate cyclase activity responded to the toxin's modulation in a manner parallel to that observed *in vivo* with TPase synthesis (Figure 4). As low toxin concentrations (0.2-2.0 nanograms/reaction) stimulated the adenylate cyclase activity (85%), higher toxin concentrations (> 2.0 nanograms/reaction) inhibited the enzyme activity (44%).

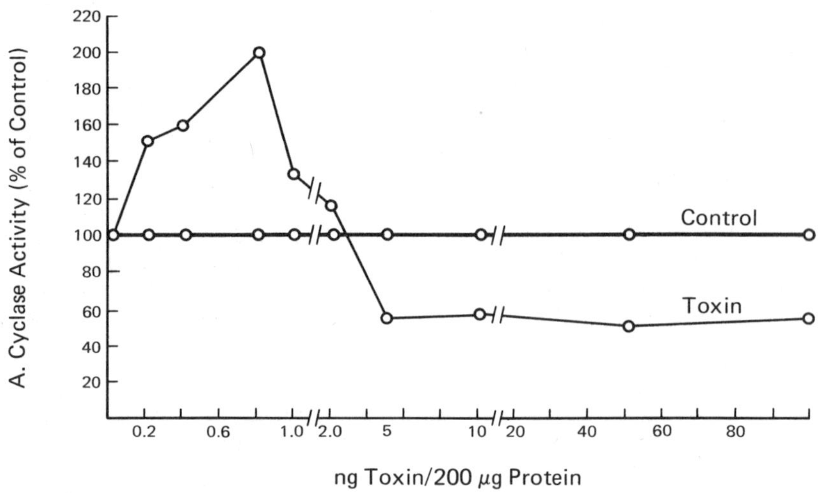

FIGURE 4. *Effect of cholera enterotoxin on the adenylate cyclase activity in toluenized cell membrane preparations. Toxin concentrations were added to the adenylate cyclase reaction mixture and the activity determined in 10 min reactions. Each point on the graph represents an average of triplicate assays.*

Once again, the question arises whether *E. coli* B/r has a recognition or receptor site for cholera enterotoxin.

In conclusion, irrespective of the mechanism by which cholera enterotoxin and insulin interact with the adenylate cyclase of *E. coli* B/r, it appears that the membrane bound bacterial adenylate cyclase is subject to regulation by a variety of molecules which interact with the cell membrane. That, this kind of interaction, in addition to regulating the rate of cAMP synthesis, also alters the growth rates of the cultures. That interactions with the cell membrane are not limited to growth factors, but extend to other environmental effectors capable of regulating the cell's adenylate cyclase. It is further indicated that for an understanding of the mechanism of growth control by cyclic nucleotides, an examination of the cell's interaction with the immediate environment surrounding it, is necessary to delineate the initial steps which can ultimately regulate cell growth.

ACKNOWLEDGEMENTS

This work was supported in part by grants from the Charles and Johanna Busch Memorial Fund and the Rutgers Research Council.

REFERENCES

1. Abou-Sabé, M. 1971. The isolation and characterization of d-ribose positive revertants of *E. coli* B/r. J. Gen. Microbiol. *65*:375-377.
2. Abou-Sabé, M. 1973. Stimulation of cAMP synthesis by glucose in a mutant of *E. coli* B/r. Nature New Biol. *243*:182-185.
3. Abou-Sabé, M. and R. Nardi. 1973. On the regulation of cAMP in bacteria. Stimulation of adenylate cyclase activity *in vitro* by glucose in a mutant of *E. coli* B/r. Biochem. Biophys. Res. Commun. *51*:551-557.
4. Abou-Sabé, M. and S. Mento. 1975. On the regulation of cAMP level in bacteria. II. *In vitro* regulation of the adenylate cyclase activity. Solubilization and reconstitution of a functional membrane bound adenylate cyclase system responsive to regulation by glucose. Biochim. Biophys. Acta *385*:294-307.
5. Abou-Sabé, M., M. Burday and J. Gentsch. 1975. On the regulation of cAMP synthesis in bacteria. I. Effect of carbon source variation on cAMP synthesis on *E.coli* B/r. Biochim. Biophys. Acta *385*: 281-293.
6. Abou-Sabé, M. and T. Reilly. 1976. Insulin action on *E. coli* B/r. Submitted for publication.
7. Abou-Sabé, M., L. Sussman and T. Unnash. 1976. Cholera enterotoxin regulation of bacterial adenylate cyclase. Submitted for publication.
8. Brickman, E., L. Soll and J. Beckwith. 1973. Genetic characterization of mutations which affect catabolite sensitive operons in *E. coli* including deletions of the gene for adenylate cyclase. J. Bacteriol. *116*:582-587.
9. Buettner, M., E. Spitz and H. Rickenberg. 1973. cAMP in *E. coli*. J. Bacteriol. *114*:1068-1073.
10. Cuatrecasas, P. 1971. Insulin receptor interactions in adipose tissue cells: direct measure net and properties. Proc. Nat. Acad. Sci. USA *68*:1264-1268.
11. Cuatrecasas, P. 1973. Insulin receptors of liver and fat cell membranes. Fed. Proc. *32*:1838-1846.
12. Cuatrecasas, P. and G. Illiano. 1971. Membrane sialic acid and the mechanism of insulin action in adipose tissue cells: effects of digestion with neuramidase. J. Biol. Chem. *246*:4938-4946.

13. Cuatrecasas, P. 1973. Interactions of *Vibrio cholerae* enterotoxin with cell membranes. Biochemistry *12*:3547-3557.
14. Cuatrecasas, P. and J. Bennett. 1975. Mechanism of activation of adenylate cyclase by *Vibrio cholerae* enterotoxin. J. Membr. Biol. *22*:29-52.
15. DeAsua, L., E. Surian, M. Flawia and H. Torres. 1973. Effect of insulin on the growth pattern and adenylate cyclase activity of BHK fibroblasts. Proc. Nat. Acad. Sci. USA *70*:1388-1392.
16. DeCrombrugghe, B., B. Chen, W. Anderson, M. Gottesman, R. Perlman and I. Pastan. 1971. Role of cyclic adenosine 3',5'-monophosphate and cyclic adenosine 3',5'-monophosphate receptor protein in the initiation of *lac* transcription. J. Biol. Chem. *246*: 7343-7348.
17. DeRobertis, E., Jr., N. Judewicz and H. Torres. 1973. On the control mechanism of bacterial growth by cAMP. Biochem. Biophys. Res. Commun. *55*:758-764.
18. Epstein, W., L. Rothman-Denes and J. Hesse. 1975. Adenosine 3',5' cyclic monophosphate as mediator of catabolite repression in *E. coli*. Proc. Nat. Acad. Sci. USA *72*:2300-2304.
19. Field, M. 1974. Mode of action of cholera toxin: stabilization of catecholamine-sensitive adenylate cyclase in turkey erythrocytes. Proc. Nat. Acad. Sci. USA *71*:3299-3303
20. Finkelstein, R., M. Boesman, S. Neoh, M. LaRue and R. Delaney. 1974. Dissociation and recombination of the subunits of the cholera enterotoxin (choleragen). J. Immunol. *113*:145-150.
21. Flawia, M. and H. Torres. 1973. Response to trypsin treated brown and white fat cells to hormones: preferential inhibition of insulin action. J. Biol. Chem. *244*:3500-3506.
22. Flawia, M. and H. Torres. 1973. Insulin inhibition of membrane bound adenylate cyclase in *Neurospora crassa*. FEBS Lett. *30*:74-78.
23. Gilman, A. 1970. A protein binding assay for adenosine 3',5' monophosphate. Proc. Nat. Acad Sci. USA *67*:305-312.
24. Harwood, J. and A. Peterkofsky. 1975. Glucose sensitive adenylate cyclase in toluene-treated cells of *E. coli* B. J. Biol. Chem. *250*:4656-4662.
25. Harwood, J. P., C. Gazdar, C. Prasad and A. Peterkofsky. 1976. Involvement of the glucose enzymes II of the sugar phosphotransferase system in the regulation of adenylate cyclase by glucose in *Escherichia coli*. J. Biol. Chem. *251*:2462-2468.
26. Heffernan, L., R. Bass and E. Englesberg. 1976. Mutations affecting catabolite repression of the l-arabinose regulation in *Escherichia coli* B/r. J. Bacteriol. *126*:1119-1131.
27. Ide, M. 1969. Adenyl cyclase of *E. coli*. Biochem. Biophys. Res. Commun. *36*:42-46.

28. Judewicz, N., E. DeRobertis and H. Torres. 1974. Control of uracil transport by cAMP in *E. coli*. FEBS Lett. *45*:155-158.
29. Kimberg, D., M. Field, J. Johnson, A. Henderson and E. Gershon. 1971. Stimulation of intestinal mucosal adenyl cyclase by cholera enterotoxin and prostaglandins. J. Clin. Invest. *50*:1218-1230.
30. Kons, T. and F. Barham. 1971. The relationship between the insulin binding capacity of fat cells and the cellular response to insulin. Studies with intact and trypsin treated cells. J. Biol. Chem. *246*:6210-6216.
31. Krishna, G., Wise, B. and Brodie, B. 1968. A simple sensitive method for the assay of adenyl cyclase. J. Pharmacol. Expt. Ther. *163*:379-385.
32. Kundig, W. and S. Roseman. 1971. Sugar transport. II. Characterization of constitutive membrane bound enzyme II of the *Escherichia coli* phosphotransferase system. J. Biol. Chem. *246*:1407-1418.
33. Makman, R. and E. Sutherland. 1965. cAMP in *E. coli*. J. Biol. Chem. *240*:1309-1314.
34. Mawe, R., B. Doore, M. McCaman, B. Feucht and M. Saier. 1974. Regulation of cAMP excretion in bacteria and cultured animal cells. J. Cell. Biol. *63*:211-216.
35. Nielsen, L., D. Monard and H. Rickenberg. 1973. Cyclic AMP phosphodiesterase of *E. coli*. J. Bacteriol. *116*:857-866.
36. Pastan, I. and R. Perlman. 1969. Stimulation of tryptophanase synthesis in *Escherichia coli* by cyclic 3',5'-adenosine monophosphate. J. Biol. Chem. *244*:22226-2232.
37. Peterkofsky, A. and C. Gazdar. 1971. Glucose and the metabolism of cAMP in *E. coli*. Proc. Nat. Acad. Sci. USA *68*:2794-2798.
38. Peterkofsky, A. and C. Gazdar. 1974. Glucose inhibition of adenylate cyclase in intact cells of *E. coli* B. Proc. Nat. Acad. Sci. USA *71*:2324-2328.
39. Peterkofsky, A., J. Harwood and C. Gazdar. 1975. Inducibility of sugar sensitivity of adenylate cyclase in *E. coli* B. J. Cyclic Nucleotide Res. *1*:11-20.
40. Peterkofsky, A. 1976. Cyclic nucletodies in bacteria. In: *Advances in Cyclic Nucleotide Research*. In press.
41. Rickenberg, H. 1974. Cyclic AMP in prokaryotes. Ann. Rev. Microbiol. *28*:353-369.
42. Rosenthal, J. and J. Fain. 1971. Insulin-like effect of clostridial phospholipase C, neuraminidase and other bacterial factors on brown fat cells. J. Biol. Chem. *246*:5888-5895.
43. Shafer, D., W. Lust, B. Sircar and N. Goldberg. 1970. Elevated concentrations of adenosine 3',5'-cyclic monophosphate in intestinal mucosa after treatment with cholera toxin. Proc. Nat. Acad. Sci. USA *67*:

851-856.

44. Tao, M. and F. Lipmann. 1969. Isolation of adenyl cyclase from *E. coli*. Proc. Nat. Acad. Sci. *63*: 86-92.

45. Tao, M. and A. Huberman. 1970. Some properties of *E. coli* adenyl cyclase. Arch. Biochem. Biophys. *141*:236-240.

46 Wicks, W. 1974. Regulation of protein synthesis by cAMP. Adv. Cyclic Nucleotide Res. *4*:335-438.

The Mechanism of Regulation of *Escherichia coli* Adenylate Cyclase

Alan Peterkofsky

Mammalian cells correspond with one another via hormonal stimuli. Interaction of hormones with cell surface receptors stimulate changes in the intracellular level of the cyclic nucleotides, cyclic AMP and cyclic GMP. The cellular levels of cAMP in the bacterium, *Escherichia coli*, are also controlled via communication with the outside world. This discussion will center on the mechanism by which carbohydrates present in the environment in which *E. coli* cells are suspended regulate adenylate cyclase, the enzyme that makes cAMP. It was shown many years ago that the addition of glucose to a culture growing on a number of different catabolites led to the inhibition of synthesis of many catabolic enzymes (1). This glucose effect (termed transient or catabolite repression) could be overcome by cAMP addition. That the inhibition by glucose of adenylate cyclase provides the molecular explanation for the phenomenon of catabolite repression should become clear from this presentation.

Using either intact (2) or permeabilized cells (3) of *E. coli*, it has been shown that adenylate cyclase can be inhibited by glucose. The data shown in Figure 1 suggest that the glucose inhibition observed is not due simply to an inhibition of the enzyme catalytic center. While intact and toluene-treated cells show a strong inhibition resulting from the addition of glucose, broken-cell extracts prepared by passage of cells through a French pressure cell have adenylate cyclase activity which is insensitive to inhibition by glucose. These studies

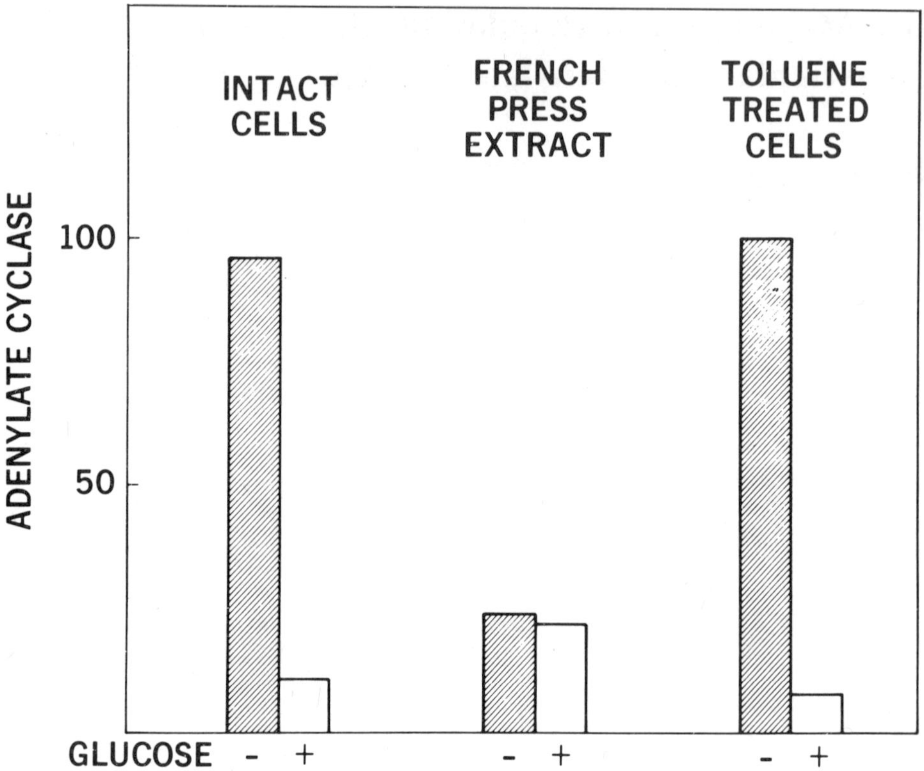

FIGURE 1. *The effect of glucose on adenylate cyclase activity in intact cells, French press extracts and toluene-treated cells. For intact cell assays (4), E.coli B was cultured in minimal medium (5) supplemented with 0.3% glucose. Stationary phase cells were suspended in salts medium and tested at 37°C during an assay period of 2 min. For assays in French press extracts or toluene-treated cells (3), E. coli B was grown on medium supplemented with 0.1% glucose to late logarithmic phase. The assays were performed at 30°C. Where indicated, glucose (1 mM) was included in the assays.*

led us to consider the possibility that demonstration of adenylate cyclase activity that is subject to inhibition by glucose (the repressible enzyme) required the presence not only of the catalytic unit but also some other factors, perhaps structural.

We undertook a survey of the sugars that would replace glucose for inhibition of the repressible adenylate cyclase. The results of this study indicated that the glucose inhibition was relatively specific; most of the other stereoisomeric hexoses were inactive. Glucose-6-phosphate, which is the first product of glucose metabolism in *E. coli*, was also inactive, suggesting that the observed inhibition was due to free glucose itself. The absence of a requirement for glucose metabolism was also supported by the inhibition effects of the glucose analogs 2-deoxyglucose and methyl-α-glucoside, neither of which support growth or are metabolized, although they are taken up by cells and phosphorylated.

An important observation we made during the course of studies on the specificity of the inhibition of adenylate cyclase by various sugars was that the profile of specificity varied depending on the carbon source that the cells were cultured (Table 1) (6). The data we accumulated led to several conclusions. No matter what the carbon source, the adenylate cyclase was always repressible by glucose. However, if the cells were grown on a variety of other sugars, like fructose, mannitol or lactose, the adenylate cyclase from such cells was also inhibited by the growth compound. We concluded that induction of the metabolic pathways for utilization of these sugars endowed cells with the new property of adenylate cyclase repressible by the metabolized sugar.

TABLE 1. *Changes in sugar specificity for inhibition of adenylate cyclase depending on growth conditions.*

Sugar in Growth Medium	Sugar in Assay	Adenylate Cyclase Activity
Glucose	None	10,020
	Glucose	540
	Fructose	10,320
	Mannitol	12,120
Fructose	None	2,580
	Glucose	300
	Fructose	360
	Mannitol	4,260
Mannitol	None	4,140
	Glucose	1,140
	Fructose	4,320
	Mannitol	1,380
Lactose	None	3,780
	Glucose	720
	Lactose	360

E. coli B was grown on salts medium (5) supplemented with the indicated sugars at a concentration of 0.3%. Stationary phase cultures were assayed for adenylate cyclase activity in intact cells (4) at 37°C for 2 min. When sugars were added to the assay, the final concentrations were 3 mM. Data adapted from (6).

The observation that growth of cells on a carbohydrate to which they had not been previously exposed led to inhibition of adenylate cyclase by that carbohydrate had some interesting implications for cellular regulation in the context of the phenomenon of catabolite repression (Figure 2). Cells grown on a mixture of glucose and lactose grow sequentially on the glucose followed by the lactose. An examination of the cellular cAMP levels during diauxic growth on glucose plus lactose (Figure 2) indicated that cAMP levels rise abruptly during the transition period when cells are adapting from growth on glucose to growth on lactose. Just prior to the initiation of growth on lactose, the cAMP levels drop again and remain low throughout the period of lactose growth. When growth stops due to the depletion of lactose, cAMP levels rise again. This wave-like fluctuation of cAMP levels during repeated cycles of growth shows the tendency for cAMP levels to be low during periods of growth and high during periods of stasis or growth transition. The physiological significance of these alterations in cAMP levels may be that elevated cAMP levels overcome the barrier to effective transcription of operons for catabolic enzymes when only low levels of inducer penetrate the cells. After induction of the transport system for a catabolite, effective transcription can persist with lowered cAMP levels.

Our survey of the sugars that could be induced to inhibit adenylate cyclase indicated that most of these sugars were either transported into cells via the phosphoenolpyruvate-dependent phosphotransferase system (pts) (for example, mannitol or fructose) (9) or in some way regulated by the pts system (for example, lactose) (9). This correlation prompted us to examine the properties of adenylate cyclase in some strains of *E. coli* with lesions in various components of the pts system.

The phosphoenolpyruvate:sugar phosphotransferase system is made up of the following series of reactions:

1. PEP + Enzyme I → Enzyme I-P + pyruvate.
2. Enzyme I-P + HPr → HPr-P + Enzyme I.
3. HPr-P + Enzyme II (sugar-specific) → Enzyme II-P + HPr.
4. Enzyme II-P + sugar → sugar-P + Enzyme II.

The two cytoplasmic cellular components, Enzyme I and HPr are utilized for the transport of all pts sugars. The Enzyme II components are membrane-bound, sugar specific, induced by growth on the respective sugars and fulfill the function of the classical permeases.

The studies shown in Figure 3 represent an examination of adenylate cyclase activity in some *E. coli* strains that have mutations in the sugar-specific Enzymes II for glucose and mannose. Activities were measured in the presence of glucose (a substrate for both the glucose (GPT) and mannose (MPT) permease), methyl-α-glucoside (a specific substrate for the glucose permease (GPT)) and 2-deoxyglucose (a specific substrate for the mannose permease (MPT)). All three sugar substrates inhibit adenylate cyclase in the wild-type, but none of these compounds

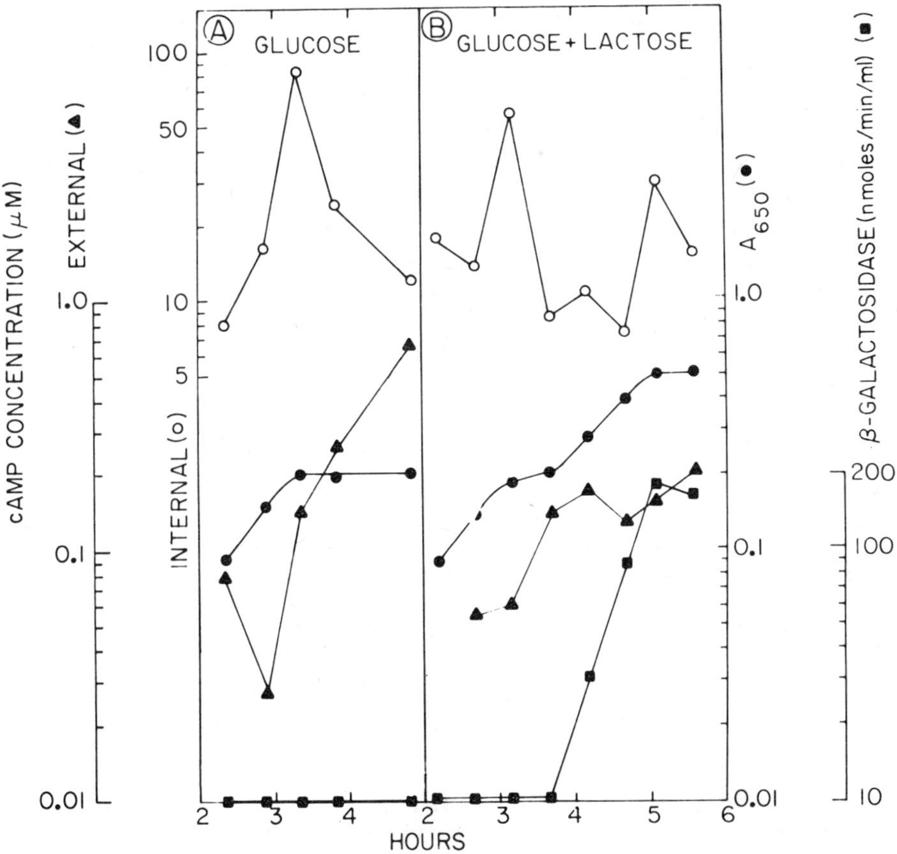

FIGURE 2. *Variation in extracellular and intracellular cyclic AMP accompanying diauxic growth on glucose and lactose. E. coli B were grown on minimal salts medium supplemented either with glucose (.015%) (panel A) or glucose (.015%) plus lactose (.03%) (panel B) in a gyrotory shaker at 37°C. At the indicated times, the following measurements were made: optical density (●); β-galactosidase (7) (■); extracellular cAMP (8) (▲); intracellular cAMP (2) (O).*

produce inhibition of the enzyme when both permeases are inactive (GPT^-, MPT^- strain). The GPT^- mutant contains adenylate cyclase activity that is inhibited by glucose and 2-deoxyglucose but not by methyl-α-glucoside. These data provide strong evidence that the conversion of adenylate cyclase that is nonrepressible by a particular sugar to a sugar-repressible form by exposure of cells to that sugar is a result of the induction of the permease (the sugar specific Enzyme II) for that sugar. While the absence of Enzyme II does not influence the activity of the adenylate cyclase, it does control the repressibility

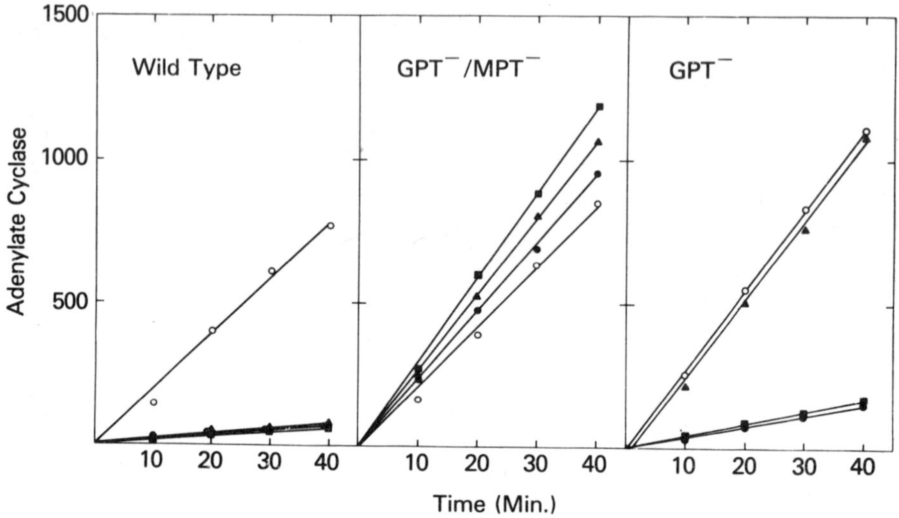

FIGURE 3. *Effect of glucose, methyl-α-glucoside and 2-deoxyglucose on adenylate cyclase activity in E. coli strains with mutation in Enzymes II for glucose. The three strains indicated: wild-type (left panel), GPT^-, MPT^- mutant deficient in both glucose Enzymes II (center panel) and the GPT^- mutant, deficient in the glucose Enzyme II specific for glucose and methyl-α-glucoside (right panel) were grown on salts medium (5) supplemented with 1% nutrient broth and thiamine (5 µg/ml). Washed cells were assayed for adenylate cyclase activity in toluene-treated cells as previously described (3). Some assays were performed in the presence of 1 mM concentrations of the indicated sugars O——O, control; ■——■, glucose; ▲——▲, methyl-α-glucoside; ●——●, 2-deoxyglucose. Adenylate cyclase activity is expressed as pmoles/mg protein.*

of the enzyme.

It is important to note that cells can contain numerous sugar-specific Enzymes II (for example, for glucose and mannose), interaction of any one with its sugar substrate leading to *complete* inhibition of adenylate cyclase. This observation suggested that if adenylate cyclase interacts directly with sugar specific Enzyme II's to produce repression, the nature of the interactions might be highly complex. We, therefore, considered an alternative possibility, that an *effect* of sugar interaction with a sugar specific Enzyme II on a general pts factor might influence adenylate cyclase action. We, therefore, measured adenylate cyclase activity in a set of *E. coli* strains that had mutations in the general cytoplasmic pts proteins, Enzyme I or HPr (Table 2). The important finding that we made was that, while a mutation in the HPr protein did not diminish the adenylate cyclase activity, a leaky mutation in the Enzyme I protein led to essentially complete loss of the adenylate cyclase activity. It became clear that the depressed levels of adenylate cyclase in the Enzyme I mutant were the result of a regulation defect, since the addition of PEP which had no effect on enzyme activity in the wild-type or HPr mutant, fully restored the activity in the leaky Enzyme I mutant. We concluded from this study that the general pts protein, Enzyme I, plays a role in regulating the adenylate cyclase activity.

TABLE 2. *Effect of glucose and phosphoenolpyruvate on adenylate cyclase activity in E. coli strains with mutations in phosphotransferase proteins.*

	Adenylate Cyclase Activity		
Strain	Control	Glucose Added	PEP Added
Wild-type	814	71	1108
HPr^-	2055	138	2305
Enzyme I^-	52	26	2500

Cultures of the indicated strains were grown on salts medium (5) supplemented with 1% nutrient broth. Adenylate cyclase activity was determined in toluene-treated cells (3), in some cases adding glucose or PEP (final concentration = 1 mM). Adenylate cyclase activity is expressed as pmoles cAMP formed/h/mg protein. Reproduced from (10).

Our current working hypothesis to explain the interaction of the pts system with adenylate cyclase involves the idea that adenylate cyclase is normally complexed with Enzyme I (Figure 4). When Enzyme I exists in a phosphoryl-

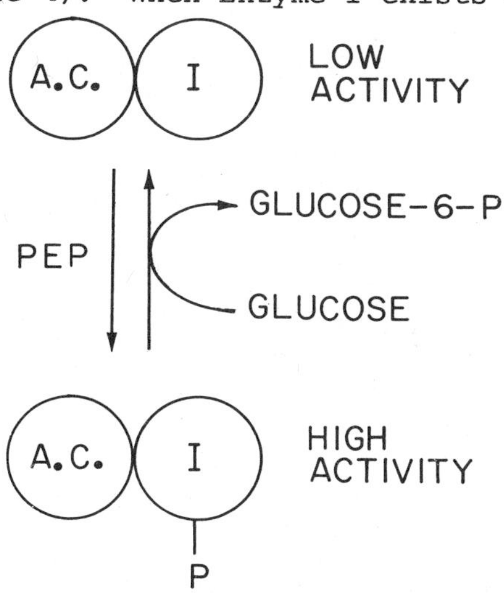

FIGURE 4. *Interconversion of adenylate cyclase between low and high activity forms dependent on the state of phosphorylation of Enzyme I. The model suggests an oppos-*

ing action of PEP and glucose on the activity of adenylate cyclase. The PEP-dependent phosphorylation of Enzyme I (I), complexed to adenylate cyclase (A.C.) results in enzyme activation. Glucose transport through the phosphotransferase system results in dephosphorylation of Enzyme I, converting adenylate cyclase to the low activity site.

ated form (10), adenylate cyclase can express a high level of activity; when Enzyme I is dephosphorylated, adenylate cyclase shows low activity. The high activity state is favored by the presence of PEP and the absence of glucose; the low activity state is favored by the presence of glucose and/or the absence of PEP.

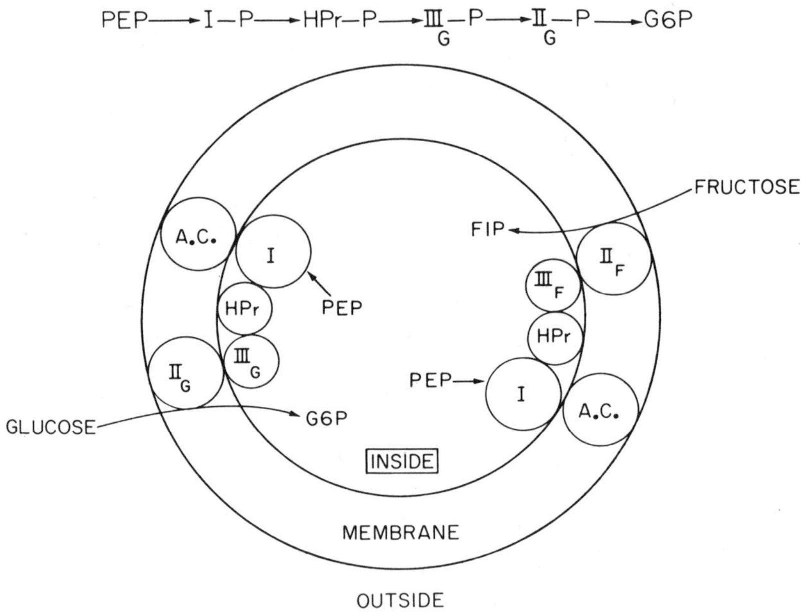

FIGURE 5. *The interaction of sugar transport proteins with membrane-bound adenylate cyclase. The model suggests that membrane-bound adenylate cyclase functions as a multi-enzyme complex regulated by the sugar transport proteins. It is suggested that the interaction of adenylate cyclase with the Enzyme I protein is direct. The inhibition of adenylate cyclase activity coupled to sugar transport is represented to be indirect and dependent on the presence of all the components of the transport system.*

Figure 5 presents a model indicating the location of adenylate cyclase and the various pts components in the cell. It emphasizes the direct interaction of adenylate cyclase with Enzyme I, a complex that can directly affect the catalytic activity. What also becomes clear from this model is that the inhibition of adenylate cyclase by glucose requires the function of all the pts components, but that inhibition by glucose is indirect--the direct regulation depends on the phosphorylation state of Enzyme I. Our current experiments are aimed at testing this model.

REFERENCES

1. Magasanik, B. Glucose effects: inducer exclusion and repression. In: *The Lactose Operon*, Beckwith, J. and Zipser, D. (eds.). Cold Spring Harbor Laboratory, Cold Spring Harbor, 1970, pp. 189-210.
2. Peterkofsky, A. and Gazdar, C. 1974. Glucose inhibition of adenylate cyclase in intact cells of *Escherichia coli* B. Proc. Nat. Acad. Sci. USA *71*:2324-2328.
3. Harwood, J. P. and Peterkofsky, A. 1975. Glucose-sensitive adenylate cyclase in toluene-treated cells of *Escherichia coli* B. J. Biol. Chem. *250*: 4656-4662.
4. Peterkofsky, A. and Gazdar, C. 1973. Measurements of rates of adenosine 3':5'-cyclic monophosphate synthesis in intact *Escherichia coli* B. Proc. Nat. Acad. Sci. USA *70*:2149-2152.
5. Vogel, H. J. and Bonner, D. M. 1956. Acetylornithinase of *Escherichia coli*: partial purification and some properties. J. Biol. Chem. *218*:97-106.
6. Peterkofsky, A., Harwood, J. and Gazdar, C. 1975. Inducibility of sugar sensitivity of adenylate cyclase of *Escherichia coli* B. J. Cycle Nucleotide Res. *1*:11-20.
7. Miller, J. H. (ed.). Assay of β-galactosidase. 1972. *Experiments in Molecular Genetics*. Cold Spring Harbor Laboratory, Cold Spring Harbor, pp. 352-355.
8. Peterkofsky, A. and Gazdar, C. 1971. Glucose and the metabolism of adenosine 3':5'-cyclic monophosphate in *Escherichia coli*. Proc. Nat. Acad. Sci. USA *68*:2794-2798.
9. Roseman, S. A bacterial phosphotransferase system and its role in sugar transport. In: *The Molecular Basis of Biological Transport*, Woessner, J. F., Jr. and Huijing, F. (eds.). Academic Press, New York, pp. 181-215.
10. Peterkofsky, A. and Gazdar, C. 1975. Interation of Enzyme I of the phosphoenolpyruvate: sugar phosphotransferase system with adenylate cyclase of *Escherichia coli*. Proc. Nat. Acad. Sci. USA *72*: 2920-2924.

Cyclic Nucleotides and the *Caulobacter* Cell Cycle

Nurith Kurn and Lucille Shapiro

The formation of polar surface structures during the cell cycle of the unicellular gram-negative bacterium, *Caulobacter*, renders it a useful model for the study of programmed cellular differentiation in prokaryotes (16,23,27,47). The execution of these events is superimposed on the common prokaryotic sequence of cell cycle events, i.e., initiation of DNA synthesis, DNA replication, septation and cell division (14,22). The coordinate formation of polar flagellum, pili which serve as RNA phage ϕCb5 attachment sites, and DNA phage ϕCbK receptor sites, occurs at the pole opposite the stalk of the predivisional cells. The formation of these surface structures occurs upon the completion of chromosomal replication and prior to cell division. Equatorial division of the "fully differentiated" cell then yields two distinct cell types: a nonmotile, phage-resistant, stalked cell and a motile, phage-sensitive swarmer cell (Figure 1). These daughter cells differ with respect to polar surface structures as well as the ability to replicate the chromosome. DNA replication is initiated immediately following cell division in the nonmotile stalked cell, but is delayed in the swarmer cell. DNA replication is initiated in the daughter swarmer cell only after the next differentiation event, stalk formation (7). At a fixed time in the cell cycle sequence stalk formation takes place at the flagellar pole and is accompanied by coordinate loss of polar flagellum, pili and DNA phage receptor sites. Stalk formation was shown to involve temporal and site-restricted

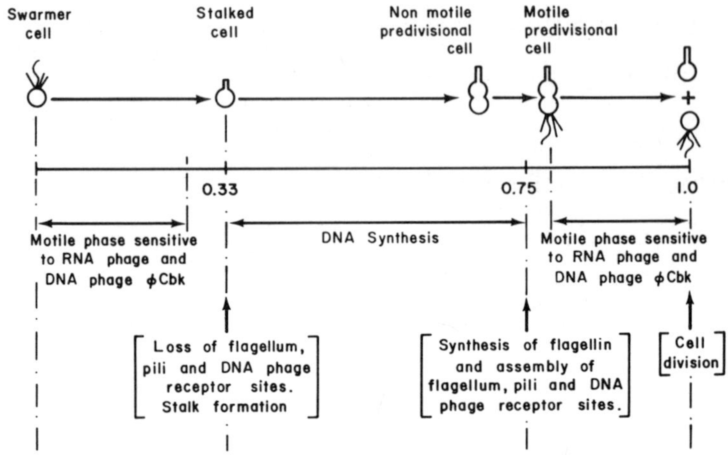

FIGURE 1. *Schematic representation of C. crescentus cell cycle events. Data have been compiled from Shapiro and Agabian-Keshishian (4), Newton (20) and Poindexter (27).*

synthesis of cell wall, distinct from cell wall synthesis for equatorial septation and division (34). Further development of the resultant stalked cell includes the initiation of DNA replication as well as cell and stalk elongation.

The strict surface structure polarity of the *Caulobacter* cell is maintained throughout the cell cycle and is not altered in mutant strains impaired in either of the polar differentiation events. Moreover, the coordinate formation of flagella, pili, and DNA phage receptor sites, although temporally regulated within the cell cycle, are not obligatory for the programmed progression of other cell cycle events (15,21).

The coordination of growth, cell division and cellular differentiation within each *Caulobacter* cell cycle suggests that a general regulatory mechanism functions in maintaining the temporal expression of these various functions. One of the most extensively studied general regulatory mechanisms of gene expression in prokaryotes is the glucose repression of both inducible operons for synthesis of catabolic enzymes (20) and operons controlling inducible morphogenesis, such as flagella formation and bacteriophage λ receptor synthesis in *Escherichia coli* (13,24,28,43,48). In eukaryotic cells a general regulatory system was shown to function in the coordination of cell cycle progression and the metabolic state of the cell culture (12,22,46).

Our studies of the regulation of programmed cell

cycle events in *Caulobacter* have concentrated on the coordinately controlled formation of the transient polar structures, flagellum, pili, and DNA phage φCbK receptor sites. This polar differentiation event is particularly suitable for study, since (a) the formation of these structures can be easily monitored for phage attachment sites and the onset of motility (35); (b) the flagella subunit, flagellin, has been purified and flagellin antibody prepared; thus control at the level of gene expression can be studied (38; and (c) mutants defective in expression of these structures, either at the level of gene expression or assembly of these structures, can be readily selected (16,23).

COORDINATE FORMATION OF TRANSIENT POLAR SURFACE STRUCTURES

Flagellin has been obtained from intact flagella released into the culture fluid during the swarmer-to-stalk cell transition, and its molecular weight, 25,000, was determined by electrophoresis on SDS-polyacrylamide gels (38). Using a *C. crescentus* flagellin marker it was shown in synchronized cultures that flagellin synthesis occurs at a fixed time in the cell cycle, 0.75 to 0.83 division units, just prior to and during flagellum assembly. Flagellum assembly can be monitored by the appearance of motility, as well as by electron microscopy.

Pili and DNA phage receptor site formation was shown to coincide with flagellin synthesis and flagellum assembly at 0.75 to 0.83 division units. Pili formation in synchronized cultures was monitored both by electron microscopy and adsorption of radio-labeled RNA phage φCb5 (4, 32,37). DNA phage φCbK attachment was shown to be at the flagellated pole of the predivisional and swarmer cells (2,3). The time of DNA phage receptor site formation was determined in synchronized cultures by measuring the time of injection of labeled phage DNA.

The coordinate control of assembly of these polar surface structures was suggested by the isolation of a pleiotropic mutant (AE20; Table 1) simultaneously defective in the assembly of these polar structures (10,15). AE20, selected as a nonmotile strain on soft agar plates (15), was found upon DNA phage φCbK infection to yield cloudy plaques and decreased efficiency of plating, suggesting altered phage receptor sites. The mutant was also shown to lack pili and to be resistant to RNA phage φCb5 infection. The single mutation was shown to specifically affect the coordinate assembly of these structures, since precursor synthesis was not affected. The mutant strain was shown to synthesize flagellin immunologically identical to that of wild-type. Growth rate, polarity of stalk formation, and equatorial cell division were not affected in this mutant strain. We thus conclude that assembly of the polar structures, although temporally regulated within the progression of cell cycle events, is not an obligatory function. Moreover, mutant strains defective in specific polar structures, but normal with respect to others, have been isolated. Flagella-less mutants, both temperature-sensitive and nonconditional (21,38), were isolated, which exhibited normal polar pili and DNA phage receptor site formation. The defect in both types of mutants was shown to be at the level of flagellin synthesis.

TABLE 1. *List of bacterial strains*

	Genotype	Phenotype	Source
C. crescentus *CB13*	Wild-type		J. Schmidt
ZE7	*lac*^{c}-505	β-gal constitutive	H. Stern, NTG mutagenesis of CB13 and selection on x-gal media
AE9	*lac*$^{-}$-507	Pleiotropic C-source block, not rescuable by dibutyryl cyclic AMP	NTG mutagenesis of CB13
AE16	cGMP^{r}-50	cGMP-resistant non-conditional growth on lactose	NTG mutagenesis of CB13
AE20	Morph-501	Pleiotropic surface structure defective	NTG mutagenesis of CB13

CONTROL OF CELL CYCLE EVENTS AND CARBON SOURCE UTILIZATION

Sugar metabolism in *C. crescentus*, like *Pseudomonas*, is carried out *via* a combination of the Entner-Doudoroff pathway and the pentose phosphate shunt, to the exclusion of the Embden-Meyerhoff pathway (18,27,39). In addition, *C. crescentus* was shown to lack the components of phosphotransferase systems (M. Saier, personal communication) responsible for sugar transport in many bacteria (29).

Caulobacter crescentus CB13 was shown to be able to grow in minimal medium supplemented with a variety of carbohydrates as sole carbon source. Growth rates on a number of carbohydrates at both optimal (30°C) and elevated (37°C) temperatures are given in Table 2. Some of the enzymes required for the utilization of lactose, galactose and mannose have been assayed in cultures grown either on glucose or the corresponding sugars in order to determine their inducible or constitutive nature (Table 3). β-Gal-

TABLE 2. *Growth rate of C. crescentus CB13 on various carbon sources*[a] *and the effect of dibutyryl cyclic AMP*

	Doubling Time (h)			
	30°C		37°C	
Carbon Source	- Dibutyryl Cyclic AMP	+ Dibutyryl Cyclic AMP[b]	- Dibutyryl Cyclic AMP	+ Dibutyryl Cyclic AMP
Glucose	4	4	4	4
Maltose	4	4	4	4
Sucrose	4	4	4	4
Lactose	10	10	ng	16
Galactose	ng[c]	26	ng	-
Mannose	32	18	42	12
Ribose	28	18	35	12
Fructose	42	24	58	18

[a] *Cultures were grown in minimal salt media containing 0.2% of either glucose, maltose or sucrose and 0.5% of the other sugars listed.*

[b] *Addition of* 3×10^{-3} *M dibutyryl cyclic AMP.*

[c] *No growth observed.*

TABLE 3. *Enzyme activities and transport constants during growth on various carbon sources*[a]

Strain	Carbon Source	β-Galactosidase	Lactose Transport	Galactokinase	Phosphomannose Isomerase	Mannose Transport
		(nm/mg/h)	(C = V_{max}/K_m [min^{-1}])	(nm/mg/h)	(μm/mg/h)	(C = V_{max}/K_m [min^{-1}])
Wild-type	Glucose	1.5	0.30	2.15	0.42	-
AE7 *lac*[c]	Glucose	28.8	0.22	-	-	-
Wild-type	Lactose	33.0	3.08	125.0	-	-
Wild-type	Mannose	-	-	-	0.45	95
Wild-type	Mannose + dib cAMP[b]	-	-	-	0.47	95
Wild-type	Galactose + dib cAMP[b]	-	-	45.0	-	-

[a] *Enzyme activities and transport constants were assayed as described (18).*
[b] *3 mM dibutyryl cyclic AMP.*

tosidase, lactose transport, and galactokinase were all shown to be inducible, whereas mannokinase and phosphomannose isomerase were constitutively expressed. Although the enzymes for ribose and fructose utilization were not characterized in *C. crescentus*, growth on these sugars paralleled growth on mannose, as will be discussed below.

Catabolite repression of enzymes required for lactose utilization has been demonstrated in *C. crescentus* cultures grown in the presence of glucose (36) and this repression could be obviated by the addition of dibutyryl cyclic AMP. In addition, the normal cell cycle of *C. crescentus* CB13 can be transiently arrested by shifting cultures from glucose minimal medium to minimal media supplemented with either lactose or galactose, both of which require the synthesis of inducible enzymes. (Unlike a shift to lactose, the eventual growth on galactose is possible only in the presence of dibutyryl cyclic AMP.) Following this carbon source shift, the cell culture accumulated at the nonmotile predivisional stage, prior to the formation of the flagellum, pili and DNA phage receptor sites (Figure 2). Cell viability was maintained during the period of arrested cellular development. The rate of synthesis of RNA, protein and DNA, as measured by incorporation of radio-labeled precursor into macromolecules, was shown to drop sharply upon glucose to lactose shift and was maintained at a low level during the lag period (Fukuda, unpublished results). The growth lag following glucose to lactose shift coincided with delayed induction of β-galactosidase (36). Induction of metabolic enzymes and onset of growth on lactose occurred either spontaneously after a period of 20 to 30 h, or could be stimulated earlier by the addition of 3 *mM* dibutyryl cyclic AMP. The onset of growth was initiated by coordinate formation of polar surface structures followed by cell division.

The cell cycle block following the carbon source shift appears to be the result of nutritional deprivation, since:

(a) The cell cycle block which followed glucose to lactose shift coincided with delayed induction of β-galactosidase required for lactose utilization.

(b) A shift of glucose-grown cultures to mannose, which was shown to be constitutively utilized, did not result in cell cycle arrest, although there was a growth lag. This lag period most likely reflects the adjustment of the culture to balanced growth (19). A similar growth pattern was obtained following a shift from glucose to either ribose or fructose.

(c) A shift of β-galactosidase-constitutive mutant strain (AE7; Table 1) from glucose to lactose did not result in cell cycle block, and onset of growth on lactose was delayed by a short lag period. This is in agreement with the observed growth pattern following a shift to other constitutively utilized carbon sources.

(d) Starvation for carbon source, following transfer of glucose-grown cultures to minimal salt medium, resulted

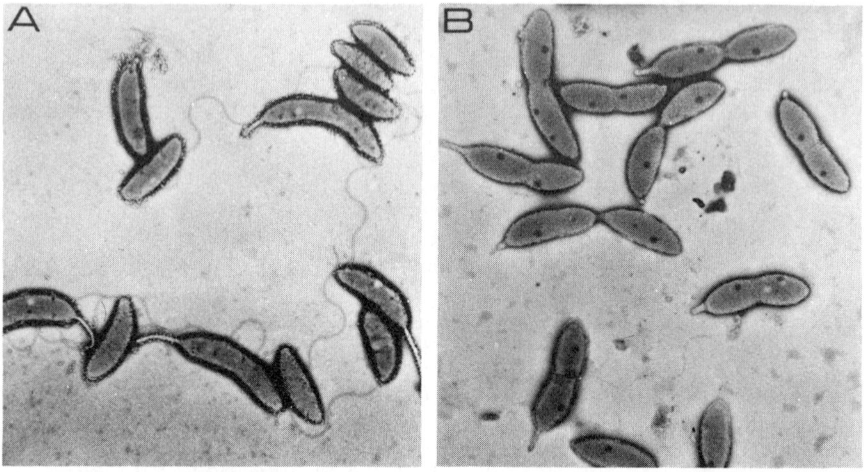

FIGURE 2. *Electron micrographs of C. crescentus CB13. (A) Culture fully induced for growth on lactose and (B) culture during the diauxic arrest period in lactose showing accumulation of cells at the nonmotile predivisional stage.*

in accumulation of cells at the nonmotile predivisional cell stage. Similarly, late stationary phase cultures grown on glucose accumulate mainly at this cell stage.

The nonmotile predivisional cell appears to represent a stable hypometabolic stage in the cell cycle. This stage occurs at the end of cell elongation and chromosomal replication, and just prior to flagellin synthesis. Pulse-labeling of blocked cultures with ^{14}C amino acids, followed by resolution of total cellular protein by electrophoresis on SDS-polyacrylamide gradient slab gels and autoradiography, indeed showed that cell cycle block occurred prior to synthesis of structural protein for polar differentiation (18). The synthesis of flagellin, the precursor for flagellar assembly, as well as three protein species identified as protein components of the outer membrane (1), were shown to be blocked during the cell cycle arrest period. In order for a new cell cycle to initiate, active metabolism must occur. Indeed, stimulation of β-galactosidase synthesis and onset of growth were shown to be dependent on *de novo* RNA synthesis. Onset of growth was accompanied by a sharp increase in RNA synthesis prior to increased protein and DNA synthesis. Onset of β-galactosidase synthesis and growth could be blocked by addition of rifampicin (60 μg/ml). Cell cycle initiation coincided with the expression of polar surface structures and cell division.

EFFECT OF DIBUTYRYL CYCLIC AMP ON CARBON SOURCE UTILIZATION

The availability of cyclic AMP and its specific binding protein has been shown to regulate induction of catabolite-sensitive operons in gram-negative bacteria (6,7,25, 26,44,49). This cyclic nucleotide was also shown to have a regulatory function in surface structure formation, such as flagellar formation in *E. coli* and *Salmonella typhimurium*, as well as phage receptor site formation (13,24,28, 43,48). The mechanism of cyclic AMP function in *Caulobacter* has not been ascertained, although both adenylate cyclase (40) and cyclic AMP-binding protein (40) were demonstrated in *C. crescentus* cell extracts. The intracellular concentration of this cyclic nucleotide was shown not to vary with carbon source and growth rate (18,36). However, the stimulatory effect of dibutyryl cyclic AMP on growth with sugars utilized by either inducible or constitutive enzymes indicates that it exerts a general control on sugar utilization. Dibutyryl cyclic AMP was shown to relieve catabolite repression and stimulate synthesis of β-galactosidase in the presence of lactose (36). In addition, the transient cell cycle block following a shift of glucose-grown cultures to either lactose or galactose was obviated by dibutyryl cyclic AMP.

Growth on mannose was also stimulated by exogenous dibutyryl cyclic AMP. Enzymes for initial metabolism of mannose, phosphomannose isomerase and mannokinase, were shown to be constitutively expressed. Moreover, the specific activity of the mannose uptake system and phosphomannose isomerase was not enhanced by growth in the presence of dibutyryl cyclic AMP. It thus appears that stimulation of growth on this carbon source was not at the level of increased catabolic enzyme activity. Stimulation of growth by dibutyryl cyclic AMP was also observed in cultures grown on ribose and fructose. Growth on all these sugars was found to be temperature-sensitive and stimulation of growth by dibutyryl cyclic AMP was observed at both optimal and elevated temperatures. Growth on lactose was also temperature-sensitive and growth at the restrictive temperature (37°C) was possible only in the presence of dibutyryl cyclic AMP.

Shifting glucose-grown cultures to growth on mannose, ribose or fructose resulted in a short lag period which could be eliminated by the addition of dibutyryl cyclic AMP. Similar stimulation was observed following a glucose-to-lactose shift of β-galactosidase-constitutive mutant strain (AE7; Figure 3). Growth stimulation on lactose, however, did not result in shortening of generation time, which was observed for growth on the constitutively utilized sugars.

It thus appears that dibutyryl cyclic AMP mediates a general mechanism for regulating growth both on inducible and constitutively utilized carbon sources. However, the effect of dibutyryl cyclic AMP on growth of a *lac*$^-$ mutant strain (AE9) suggested that the two components of a

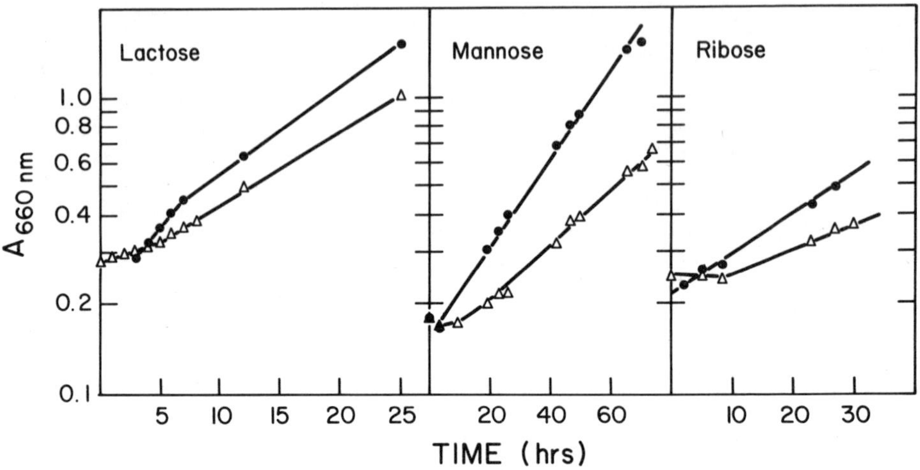

FIGURE 3. *Growth of the β-galactosidase-constitutive strain C. crescentus AE7 on lactose (0.5%) and CB13 on mannose (0.2%) and ribose (0.2%) after a shift from glucose (0.2%) (Δ-Δ-Δ) and stimulation of growth by exogenous dibutyryl cyclic AMP (3 × 10^{-3} M) (●-●-●). The generation time of AE7 on lactose was 12 h, both in the absence and presence of dibutyryl cyclic AMP.*

general regulatory system, mediated by dibutyryl cyclic AMP, can be dissociated. The mutant strain was isolated as a *lac*⁻ mutant and exhibited pleiotropic defects for growth on sugars: it was incapable of growth on either lactose or galactose, even in the presence of dibutyryl cyclic AMP, and growth on mannose was slower than that of wild-type cultures. Growth rate on glucose was not affected. Although growth on lactose and galactose was not stimulated by dibutyryl cyclic AMP, parallel to mutants of enteric bacteria lacking cyclic AMP-binding protein (CRP), growth on mannose was stimulated by this cyclic nucleotide (Figure 4). Thus, although the mutation resulted in a defective mechanism for induction of both lactose and galactose enzymes, the control mechanism that operates in the case of growth on mannose still functioned and was sensitive to exogenous dibutyryl cyclic AMP. Similar stimulation of growth of this mutant strain on ribose and fructose was observed.

THE EFFECT OF EXOGENOUS CYCLIC GMP DERIVATIVES ON POLAR DIFFERENTIATION

Although it has been proposed that cyclic GMP functions as a regulatory molecule in both eukaryotes and prokaryotes (5,11,31,45), little is known about the role of cyclic GMP in bacteria. Studies on the function of this cyclic nucleotide in *E. coli* have concentrated on its relation to cyclic AMP regulation of inducible operons

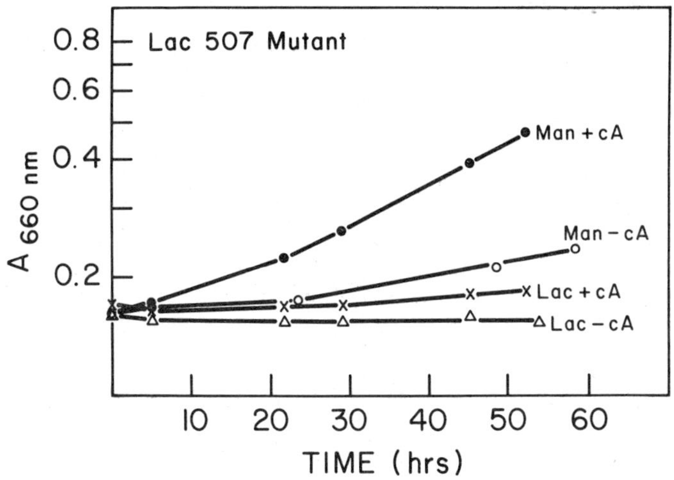

FIGURE 4. *Differential effect of exogenous dibutyryl cyclic AMP (3 × 10^{-3} M) on growth on lactose and mannose of C. crescentus AE9.*

sensitive to catabolite repression (9,30,49,50). The presence of cyclic GMP (18), guanylate cyclase (41), and a specific receptor protein for both cyclic GMP and dibutyryl cyclic GMP (42) in *C. crescentus* has been previously demonstrated. Cyclic GMP has also been reported to regulate stalk elongation in a mutant of *C. crescentus* (33). In addition, the intracellular concentration of this cyclic nucleotide was shown to vary with carbon source (18; Table 4). Intracellular concentrations of cyclic GMP in mid-log phase cultures grown on glucose were 3- to 5-fold higher than that of cultures grown on mannose or lactose. Cyclic GMP concentrations also varied with the stage of growth, since intracellular concentrations are lower in stationary phase cultures grown on glucose (Table 4). These observations suggest that the guanine cyclic nucleotide has a regulatory function in *C. crescentus*.

Addition of exogenous 8-bromo or dibutyryl cyclic GMP to cultures grown in minimal media supplemented with either lactose or mannose, but not glucose, resulted in repression of polar structure formation (17). Loss of motility was observed within 6 to 8 h after addition of the cyclic nucleotide derivatives. Electron micrographs of the nonmotile cultures showed the absence of polar flagellum and pili. No attachment of DNA phage ϕCbK was observed in electron micrographs of cultures grown in the presence of these derivatives. Furthermore, these cultures were shown to be resistant to both RNA phage ϕCb5 and DNA phage ϕCbK infection in the presence of cyclic GMP derivatives (3 *mM*). The repression of polar structure formation did not affect subsequent polar stalk formation, equatorial septation, or growth rate (Figure 5).

TABLE 4. *Intracellular concentrations of cyclic GMP*[a]

Carbon Source	Growth Stage	Cyclic GMP ($\times 10^{-8}$ *M*)
Glucose (0.2%)	Stationary phase	0.230
		0.268
		0.186
Glucose (0.2%)	Mid-log phase	0.975
Mannose (0.2%)	Mid-log phase	0.232
Lactose (0.5%)	Mid-log phase	0.330
Glucose → lactose	1 h block	1.05
	6 h block	0.46
		0.41
	14 h block	0.58
	18 h block	-
	22 h induced[c]	0.26
		0.29

[a]*Intracellular concentrations of cyclic nucleotides were determined by radioimmune assay (18).*
[b]*A culture was grown to late-log phase (OD-1.2) in minimal media containing 0.2% glucose. Cells were collected by centrifugation, washed with minimal salt media and resuspended in lactose (0.5%) minimal medium at OD = 0.340.*
[c]*Self-induced growth on lactose.*

The target of cyclic GMP repression was shown to be synthesis of structural gene products (17). Cultures grown in the presence and absence of cyclic GMP derivatives were pulse-labeled with ^{14}C amino acids. The profile of *de novo* protein synthesis was determined by electrophoresis of total cell proteins on SDS-polyacrylamide gradient slab gels, followed by autoradiography. The synthesis of the flagellar precursor molecule, flagellin, as well as two proteins identified as components of the outer membrane (1), was shown to be repressed by cyclic GMP derivatives. It could be argued that the repressed synthesis of structural gene products resulted from an altered assembly mechanism which, in turn, represses precursor synthesis *via* a feedback control. This possibility, however, seems unlikely, since a pleiotropic mutation which affected the coordinate assembly of the polar surface structures did not cause repression of precursor synthesis (15).

The repression of polar differentiation by cyclic GMP derivatives was shown to be reversed by simultaneous addition of dibutyryl cyclic AMP (17). The protein profiles of pulse-labeled cultures grown in the presence of both di-

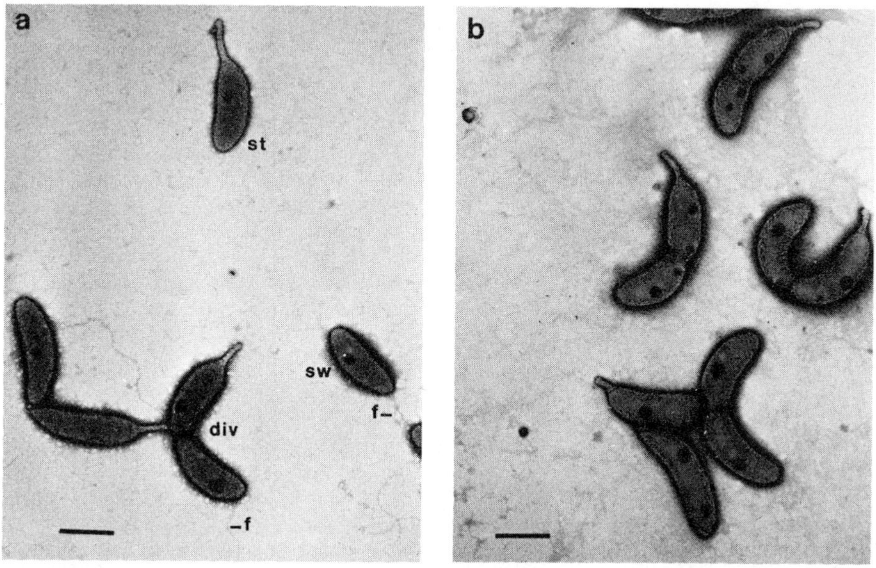

FIGURE 5. *Electron micrographs of C. crescentus CB13 grown in lactose minimal medium in the absence (a) and presence (b) of 3 mM 8-bromo cyclic GMP. Div = predivisional cell; st = stalked cell; sw = swarmer; f = flagellum*

butyryl cyclic AMP and cyclic GMP derivatives were identical with those of control cultures. Cyclic GMP derivatives, on the other hand, did not interfere with the stimulation of synthesis of the inducible lactose and galactose enzymes by dibutyryl cyclic GMP, nor was the stimulation of growth on the constitutively utilized mannose affected by cyclic GMP derivatives. Moreover, the induction of lactose enzymes following a glucose-to-lactose shift and the consequent onset of growth on lactose was not affected by the presence of cyclic GMP derivatives. Polar differentiation, however, was repressed in these cultures. It thus appears that dibutyryl cyclic AMP affects the specific regulatory mechanism mediated by cyclic GMP derivatives. This conclusion is further supported by the isolation of mutants resistant to the effect of exogenous cyclic GMP derivatives, by selection for the ability to grow on lactose at the restrictive temperature (37°C) in the absence of dibutyryl cyclic AMP. As was discussed earlier, wild-type cultures were found to be temperature-sensitive with respect to growth on lactose, and growth at 37°C was possible only in the presence of dibutyryl cyclic AMP. These pleiotropic mutants exhibited the identical phenotype to that of mutant strains isolated by direct selection for motility maintenance in the presence of cyclic GMP derivatives. The phenotype of both mutant classes is summarized in Table 5.

TABLE 5. *Phenotype of wild-type and mutant strains of C. crescentus in the presence of exogenous 8-bromo cyclic GMP*

Phenotype	Wild-Type – 8-BrcGMP	Wild-Type + 8-BrcGMP	Mutant Class A[e] + 8-BrcGMP	Mutant Class B[e] + 8-BrcGMP
Generation time[a]	11 h	11 h	11 h	11 h
Growth on lactose at 37°C	–	–	+	+
Motility[b]	+	–	+	+
Flagella[c]	+	–	+	+
Intracellular flagellin[d]	+	–	+	+
Pili[c]	+	–	+	+
RNA phage ϕCb5 adsorption[c]	+	–	+	+
RNA phage ϕCb5 infection[e]	+	–	–	–
DNA phage ϕCbK infection[e]	+	–	+	+

[a] *In minimal medium with lactose as the sole carbon source.*
[b] *Phase contrast microscopy.*
[c] *Electron microscopy.*
[d] *Autoradiograms of SDS-polyacrylamide gels of pulse-labeled cultures.*
[e] *Class A mutants were selected by their ability to maintain motility in the presence of cyclic GMP derivatives. Class B mutants were selected by their ability to grow on lactose at 37°C in the absence of exogenous dibutyryl cyclic AMP.*

Electron micrographs of the cyclic GMP-resistant mutant cultures grown in the presence of cyclic GMP derivatives showed the presence of polar pili, and RNA phage ϕCb5 was seen to attach to pili. However, all the mutant clones were found to be resistant to RNA phage infection. This observation strongly suggests that cyclic GMP derivatives directly interfere with RNA phage development in *Caulobacter*. Further studies on the mechanism of inhibition of RNA phage development will help to elucidate the mechanism of repression of the specific structural gene expression by these cyclic GMP derivatives.

REFERENCES

1. Agabian, N. and Unger, B. 1976. J. Bacteriol, submitted.
2. Agabian-Keshishian, N. and Shapiro, L. 1970. Stalked bacteria: properties of deoxyribonucleic acid bacteriophage ϕCbk. J. Virol. *5*:795-800.
3. Agabian-Keshishian, N. and Shaprio, L. 1971. Bacterial differentiation and phage infection. Virology *44*:46-53.
4. Bendis, I. K. and Shapiro, L. 1970. Properties of *Caulobacter* ribonucleic acid bacteriophage ϕCbk. J. Virol. *6*:847-854.
5. Bernlohr, R. W., Haddox, M. K. and Goldberg, N. D. 1974. Cyclic guanosine 3':5'-monophosphate in *Escherichia coli* and *Bacillus lichenformis*. J. Biol. Chem. *249*:4329-4331.
6. deCrombrugghe, B., Chen, B., Anderson, W., Nissley, P., Gottesman, M., Pastan, I. and Perlman, R. 1971. *Lac* DNA, RNA polymerase and cyclic AMP receptor protein, cyclic AMP, *lac* repressor and inducer are the essential elements for controlled *lac* transcription. Nature New Biol. *231*:139-142.
7. deCrombrugghe, B., Perlman, R. L., Varmus, H. and Pastan, I. 1969. Regulation of inducible enzyme synthesis in *Escherichia coli* by cyclic adenosine 3',5'-monophosphate. J. Biol. Chem. *244*:5828-5835.
8. Degnan, S. T. and Newton, A. J. 1972. Chromosome replication during development in *Caulobacter crescentus*. J. Mol. Biol. *64*:671-680.
9. Emmer, M., deCrombrugghe, B., Pastan, I. and Perlman, R. 1970. Cyclic AMP receptor protein of *E. coli*: its role in the synthesis of inducible enzymes. Proc. Nat. Acad. Sci. USA *66*:480-487.
10. Fukuda, A., Miyakawa, K. and Yoshimi, O. 1974. A pleotropic mutation in *Caulobacter crescentus* that affects the flaggelar formation and susceptibility to phage infection. Proc. Japan. Acad. *50*:839-842.
11. Goldberg, N. D., O'Dea, R. E. and Haddox, M. K. 1973. Cyclic GMP. Adv. Cyclic Nucleotide Res. *3*:155-223.
12. Hartwell, L. H. 1974. *Saccharomyces cerevisiae* cell cycle. Bacteriol. Rev. *38*:164-198.

13. Hazelbauer, B. L. 1975. Role of the receptor for bacteriophage lambda in the functioning of the maltose chemoreceptor of *Escherichia coli*. J. Bacteriol. *124*:119-126.
14. Koch, A. L. and Blumberg, I. 1976. Microbiol. Physiol., in press.
15. Kurn, N., Ammer, S. and Shapiro, L. 1974. A pleiotrophic mutation affecting expression of polar development events in *Caulobacter crescentus*. Proc. Nat. Acad. Sci. USA *71*:3157-3161.
16. Kurn, N. and Shapiro, L. Regulation of the *Caulobacter* cell cycle. In: *Current Topics in Cellular Regulation*, Horecker, B. L. and Stadtman, E. R. (eds.), Vol 9. Academic Press, New York, 1975, pp. 41-63.
17. Kurn, N. and Shapiro, L. 1976. Proc. Nat. Acad. Sci. USA, in press.
18. Kurn, N., Shapiro, L. and Agabian, N. 1976. J. Bacteriol., submitted.
19. Maaløe, O. and Kjeldgaard, N. O. *Control of Macromolecular Synthesis*. W. A. Benjamin, New York-Amsterdam, 1966.
20. Magasanik, B. Glucose effects: inducer exclusion and repression. In: *The Lactose Operon*, Beckwith, J. and Zipser, D. (eds.). Cold Spring Harbor, New York, 1970, pp. 189-219.
21. Marino, W., Ammer, S. and Shapiro, L. 1976. J. Mol. Biol., submitted.
22. Mitchison, J. M. *The Biology of Cell Cycle*. Cambridge University Press, 1971.
23. Newton, A., Osley, M. A. and Terrana, B. *Caulobacter crescentus*: a model for the temporal and spatial control of development. *Microbiology*, Schlessinger, D. (ed.). Amer. Soc. for Microbiol., Washington, D. C., 1975.
24. Pearson, M. L. 1972. The role of adenosine 3',5'-cyclic monophosphate in the growth of bacteriophage lambda. Virology *49*:605-609.
25. Perlman, R. and Pastan, I. 1968. Cyclic 3',5'-AMP: stimulation of β-galactosidase and tryptophanase induction in *E. coli*. Biochem. Biophys. Res. Commun. *30*:656-664.
26. Perlman, R. and Pastan, I. The role of cyclic AMP in bacteria. In: *Current Topics in Cellular Regulation*, Horecker, B. L. and Stadtman, E. R. (eds.). Academic Press, New York, 1971, pp. 117-133.
27. Poindexter, J. S. 1964. Biological properties and classification of the *Caulobacter* group. Bacteriol Rev. *28*:231-295.
28. Randall-Hazelbauer, L. L. and Schwartz, M. 1973. Isolation of the bacteriophage lambda receptor from *Escherichia coli*. J. Bacteriol. *116*:1436-1446.
29. Roseman, S. 1969. The transport of carbohydrates by a bacterial phosphotransferase system. J. Gen. Physiol. *54*:138s-180s.

30. Sanders, R. and McGeoch, D. 1973. A mutant transcription factor that is activated by 3':5'-cyclic guanosine monophosphate. Proc. Nat. Acad. Sci. USA *70*:1017-1021.
31. Sarkar, N. and Pandus, H. 1974. A guanosine 3':5'-monophosphate-sensitive nuclease from *Bacillus brevis*. J. Biol. Chem. *250*:684-690.
32. Schmidt, J. M. 1966. Observations on the adsorption of *Caulobacter* bacteriophages containing ribonucleic acid. J. Gen. Microbiol. *45*:347-353.
33. Schmidt, J. M. and Samuelson, G. M. 1972. Effects of cyclic nucleotides and nucleoside triphosphates on stalk formation in *Caulobacter crescentus*. J. Bacteriol. *112*:593-601.
34. Schmidt, J. M. and Stanier, R. Y. 1966. The development of cellular stalks in bacteria. J. Cell. Biol. *28*:423-436.
35. Shapiro, L., Agabian-Keshishian, N. and Bendis, I. 1971. Bacterial differentiation. Science *173*: 884-893.
36. Shapiro, L., Agabian-Keshishian, N., Hirsch, A. and Rosen, O. M. 1972. Effect of dibutyryladenosine 3':5'-cyclic monophosphate on growth and differentiation in *Caulobacter crescentus*. Proc. Nat. Acad. Sci. USA *69*:1225-1229.
37. Shapiro, L. and Bendis, I. K. RNA phages of bacteria other than *E. coli*. In: *RNA Phage*, Zinder, N. (ed.). Cold Spring Harbor, New York, 1975, pp. 397-410.
38. Shapiro, L. and Maizel, J. V., Jr. 1973. Synthesis and structure of *Caulobacter crescentus* flagella. J. Bacteriol. *113*:478-485.
39. Shedlarski, J. G. 1974. Glucose-6-phosphate dehydrogenase from *Caulobacter crescentus*. Biochim. Biophys. Acta *358*:33-43.
40. Sun, I. 1976. Cyclic nucleotide metabolism in *Caulobacter crescentus*. Ph.D. Thesis, Albert Einstein School of Medicine.
41. Sun, I., Shapiro, L. and Rosen, O. M. 1974. Purification and characterization of guanylate cyclase from *Caulobacter crescentus*. Biochem. Biophys. Res. Commun. *61*:193-203.
42. Sun, I., Shapiro, L. and Rosen, O. M. 1975. A specific cyclic guanosine 3':5'-monophosphate-binding protein in *Caulobacter crescentus*. J. Biol. Chem. *250*:6181-6187.
43. Szmelcman, S. and Hofnung, M. 1975. Maltose transport in *Escherichia coli* K-12: involvement of the bacteriophage lambda receptor. J. Bacteriol. *124*:112-118.
44. Tyler, B., Deleo, A. and Magasanik, B. 1974. Activation of transcription of *hut* DNA by glutamine synthetase. Proc. Nat. Acad. Sci. USA *71*:225-229.
45. Vincenzo, M., Stelio, V., Weissbach, H., Miller, D. L. and Pastan, I. 1975. Guanylate cyclase in *Escherichia coli*. J. Biol. Chem. *250*:6214-6217.

46. Wolfner, M. D., Yep, F., Messenguy, and Fink, G. R. 1975. Integration of amino acid biosynthesis into the cell cycle of *Saccharomyces cerevisiae*. J. Mol. Biol. *96*:273-290.
47. Wood, N. B. and Shapiro, L. Morphogenesis during the cell cycle of the prokaryote *Caulobacter crescentus*. In: *Results and Problems in Cell Differentiation*, Reinart, J. and Holtzer, H. (eds.), Vol. 7. Springer Verlag, Berlin-New York, 1975, pp. 133-149.
48. Yokota, T. and Kosuga, T. 1974. Requirement of adenosine 3',5'-cyclic phosphate for formation of the phage lambda receptor in *Escherichia coli*. J. Bacteriol. *109*:13-4-1306.
49. Zubay, G., Chambers, D. A. and Cheony, L. B. Cell-free studies on the regulation of the *lac* operon. In: *The Lactose Operon*, Beckwith, J. R. and Zipser, D. (eds.). Cold Spring Harbor, New York, 1970, pp. 375-391.
50. Zubay, G., Schwartz, D. and Beckwith, J. 1970. Mechanism of activation of catabolite-sensitive genes: a positive control system. Proc. Nat. Acad. Sci. USA *66*:104-110.

Part II

ROLE OF CYCLIC NUCLEOTIDES IN THE REGULATION OF THE CELL CYCLE

FRIEDMAN et al.
Role of Cyclic AMP in the Regulation of the HeLa Cell Cycle

BYRON
Cyclic Nucleotides and the Cell Cycle of the Hematopoietic Stem Cell

Role of Cyclic AMP in the Regulation of the HeLa Cell Cycle

Daniel L. Friedman, Roger A. Johnson, Charles E. Zeilig, James B. Kurz, Kusum V. Kumar, and Patrick Gray

Widely varying effects of adenosine 3',5'-monophosphate (cAMP) on the cell cycle have been claimed by different laboratories in recent years. cAMP analogs or agents thought to elevate intracellular cAMP have been reported to inhibit cell cycle traverse in G_1 in normal fibroblasts (10,11,20), S49 mouse lymphosarcoma cells (6), and neuroblastoma cells (24); to slow traverse through G_2 in skin (34), normal and transformed fibroblasts (21,28, 30,37), and several other cell types (18,19,25); to stimulate passage through late G_1 in fibroblasts (37), thymocytes, regenerating liver [for review see (36)] and hemopoietic stem cells (5); and to inhibit DNA synthesis in cultured hepatoma cells (33).

While multiple cell cycle actions of cAMP are possible, both the validity and the interpretation of many of the reported results must be questioned. Several important pitfalls should be considered. First is the common problem that agents thought to elevate intracellular cAMP often have other actions as well. Nonspecificity has been observed with the phosphodiesterase inhibitors (3,31), with hormones (9,16), and as the result of the degradation products of cAMP and its analogs (e.g., 14,17,32,38). A second pitfall is the inhibition of thymidine transport by some of these agents (15,29), leading to misleading results with regard to effects on DNA synthesis. A third problem, often difficult to circumvent, is that the methods used to synchronize the cell cycle may induce unbalanced growth or other changes in cells resulting in abnormal responses to agents. Finally, with regard to the interpretation of results it should be apparent that the observation that cAMP inhibits cells at a specific point in the cell cycle does not constitute proof of a physiological role. To come to such a conclusion requires, as a minimum, the additional observation of consistency between the given effect and the natural fluctuations in cAMP content at the particular point during the cell cycle.

In the present studies we have attempted to determine if cAMP plays a role in the cell cycle using HeLa cells as a model system. The use of this cell line was dictated in part by the fact that it has been well characterized with regard to the cell cycle. Furthermore, HeLa cells have apparently lost the ability to enter a quiescent (G_0) state and, therefore, allow "pure" cell cycle studies, uncomplicated by events associated with the interconversion between growing and resting states. We have attempted to correlate the effects of chemical agents which modify internal cAMP levels with fluctuations in cAMP during the cell cycle. To ascertain the specificity of cAMP analogs and of hormones, we have used the criterion of potentiation between these agents and phosphodiesterase inhibitors. In addition, we have attempted to establish the observed effects in both synchronized and unsynchronized cells. Finally, we have begun attempts to determine if changes in cAMP content during the cell cycle are reflected in changes in protein kinase activity.

LEVELS OF INTRACELLULAR cAMP DURING A SYNCHRONIZED CELL CYCLE

Following release from a double high-thymidine block, cAMP levels decay progressively from the beginning of S-phase until mitosis, interrupted by transient elevations at 4 and 6 h (Figure 1) (40,41). Changes in cAMP are not observed if high thymidine is readded to cells (dashed line, Figure 1), supporting the idea that the fluctuations are cell-cycle-related and not due to resuspension of cells in fresh growth medium. Moreover, the same fluctuations are observed if cells are synchronized by a hydroxyurea-amethopterin double block (data not shown). The elevations in cAMP at 4 and 6 h are highly reproducible. However, only the peak at 4 h is statistically significant.

Following mitotic shake-off, cAMP levels rise rapidly as cells exit from mitosis (Figure 2). Levels then continue to rise more slowly until 40 percent of the cells in the population have entered S-phase, at which time levels fall. This fall is prevented if DNA synthesis is blocked by addition of high levels of thymidine.

Taken together these results suggest that cAMP levels are at a maximum at, or near the G_1-S border, while the lowest levels are seen in late G_2 and M. In addition, reproducible fluctuations occur in S-phase.

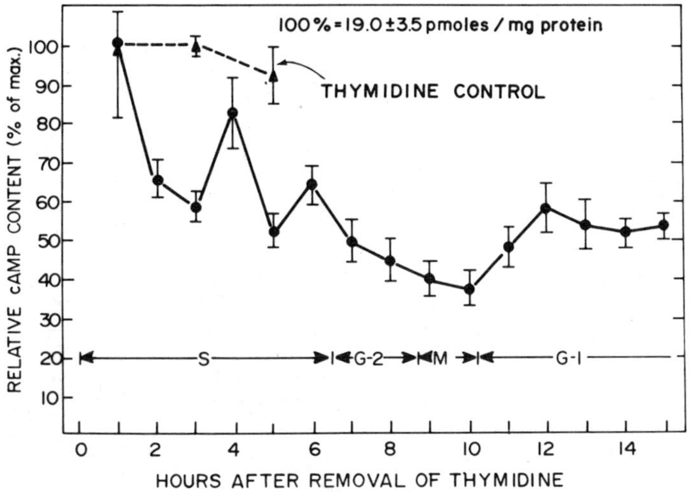

FIGURE 1. *Cyclic AMP content in cells synchronized by the double thymidine block method. Cells were synchronized by addition of 2 mM thymidine for 17 h followed by fresh thymidine-free medium for 11 h and a second thymidine treatment for 13 h. At 0-time, the cells were resuspended in fresh thymidine-free growth medium and allowed to proceed through the cell cycle. Thymidine (2 mM, final concentration) was immediately readded to a portion of the cultures to prevent further DNA synthesis. Cyclic AMP content was determined on purified samples by the method of Gilman (13). —●—, no thymidine; —▲—, + thymidine. Values represent the mean and standard errors of three experiments with thymidine and seven without. Within each experiment, points represent average values from duplicate samples from a single culture, each sample assayed for cAMP in duplicate.*

EFFECT OF AGENTS ON S AND G_2

Addition of the phosphodiesterase inhibitor, 1-methyl, 3-isobutyl-xanthine (MIX) (1 m*M*) to cells synchronized in S-phase by a double thymidine block resulted in about a three- to five-fold increase in cAMP levels. Markedly elevated levels were maintained throughout S-phase and G_2. In spite of the high cAMP levels, the length of S-phase was unaffected by MIX (Figure 3) as determined by the ^{3}H-thymidine labeling index. In contrast, G_2 phase was markedly prolonged, from 2.3 h in control cells to 10.7 h in the MIX-treated cells. In addition, mitosis was also prolonged (Figure 3). MIX appeared to be acting by increasing cAMP levels since isoproterenol and MIX acted synergistically to inhibit G_2 traverse (Figure 4).

Because of the possibility that the observed G_2 inhibition resulted artifactually from the synchronization procedure, two additional methods were used to determine if high levels of cAMP blocked cells in G_2 phase. The

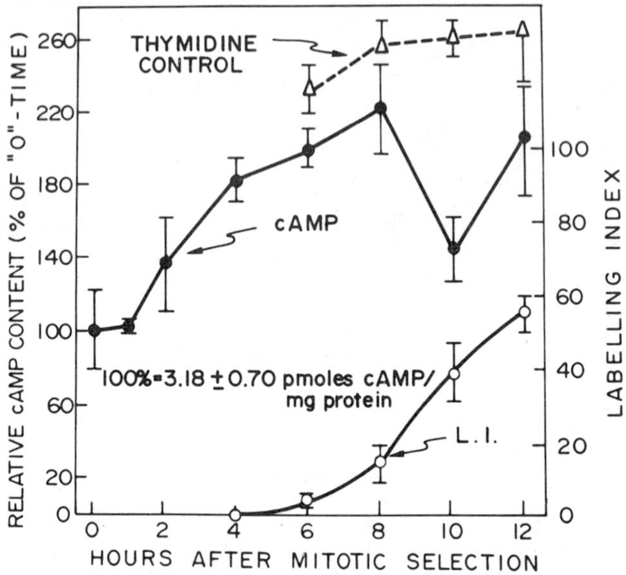

FIGURE 2. *Cyclic AMP levels in HeLa cells synchronized by mitotic selection. At 0-time, mitotic cells were selectively dislodged from monolayer cultures, previously synchronized by a double thymidine block, and placed in suspension culture. Cyclic AMP, mitotic index (not shown), and labeling index ([5-^{3}H]-deoxycytidine) were monitored at various times thereafter. At 5 h, the cultures were divided and thymidine (2 mM, final concentration) was added to one group of cells. Cyclic AMP content: —•—, control; --Δ--, + thymidine. Labeling index: —o—, control. Values are the mean and standard errors of three experiments.*

first procedure (23) was to simply follow the cell density with time following addition of agents. A G_2 block should result in a very early decrease in the rate of cell multiplication. In contrast, a G_1 block would not affect the rate for at least 10 h and an S-phase block for 4-5 h since the cells beyond the block point would be unaffected until the subsequent cell cyle. Hence, a marked inhibition of cell density increase at early times is indicative of inhibition of traverse through G_2 or M.

A control cell density curve is shown in Figure 5. The doubling time is 25 h. The dashed line represents a calculated curve which would result if the rate of cell accumulation were inhibited 25 percent. For comparison, these two curves are superimposed on the data obtained after addition of various agents. It can be seen (Figure 6a,c,e,g) that increasing concentrations of MIX, from 0.2 m*M* to 1 m*M*, leads to increasing inhibition, from 12 percent (panel a) to 61 percent (panel g). Furthermore, these inhibitions were all potentiated by dibutyryl cAMP (panels b,d,f), which by itself had little or no effect (see

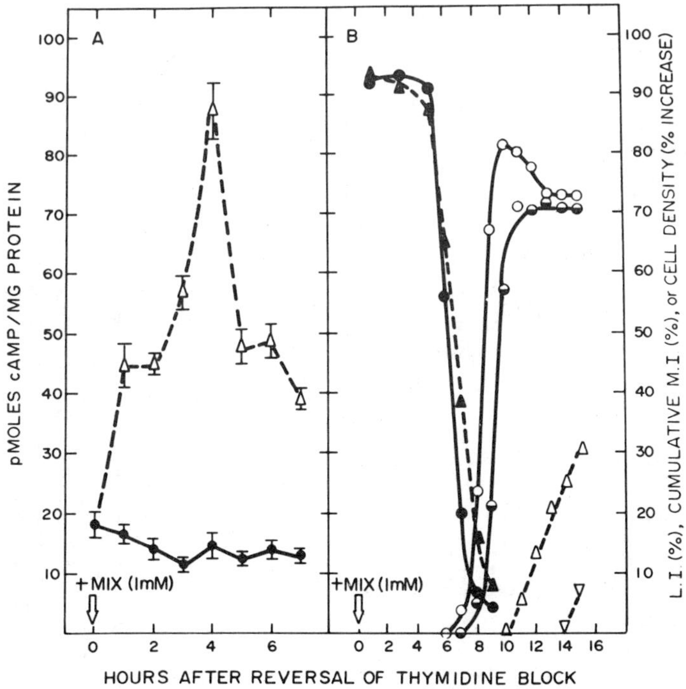

FIGURE 3. *Effect of MIX on cAMP levels during S-phase and on the lengths of S-phase, G_2 and M. Cells were synchronized in early S-phase by the double thymidine block technique. After resuspension of the cells in thymidine-free medium, MIX was added (arrow) to the indicated cultures to give a final concentration of 1 mM.* (*A*) *cAMP content: —●—, control; --Δ--, MIX. Values represent the mean ± standard errors of three samples.* (*B*) *Labelling index: —o—, control; --Δ--, MIX. Percent cumulative mitotic index: —o—, control; --Δ--, MIX. Percent increase in cell density: —θ—, control; --∇--, MIX. Values shown are the average of two experiments. Each point in the curve of cumulative mitotic index is obtained by summing the mitotic index (expressed as % and corrected for increasing cell density) and the % increase in cell density.*

Figure 8). Essentially, the same results were observed with theophylline (Figure 7) and caffeine (data not shown) although in both cases about five-fold greater levels of the phosphodiesterase inhibitor were required to obtain the same effects as those seen with MIX. Similar potentiation was also observed between 0.2 m*M* MIX and either monobutyryl cAMP or 8-Br-cAMP (Figure 8), but not with cAMP, 5'AMP (Figure 8), 8-Br-5'AMP, sodium butyrate, cGMP or 8-Br-cGMP (not shown). The finding that MIX and cAMP analogs act synergistically supports the conclusion that cAMP mediates the inhibition of cell growth. The magnitude of the inhibition at these short times can only be explained

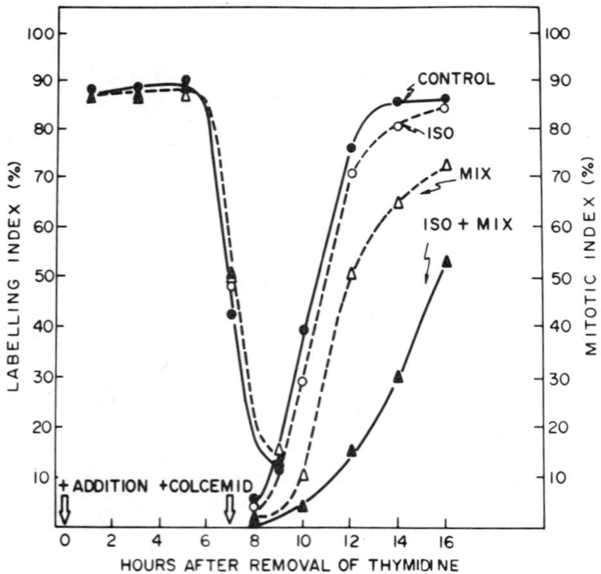

FIGURE 4. *Effects of isoproterenol and MIX on the duration of S-phase and* G_2. *The various agents were added immediately after reversal of the second thymidine block (arrow). Colcemid (0.064 µg/ml) was added at 7 h (as indicated) to accumulate cells in metaphase. Labeling index (0-9 h) or mitotic index (8-16 h): —●—, control; --o--, 10 µM L-isoproterenol·HCl; --Δ--, 0.5 mM MIX: —▲—, isoproterenol + MIX.*

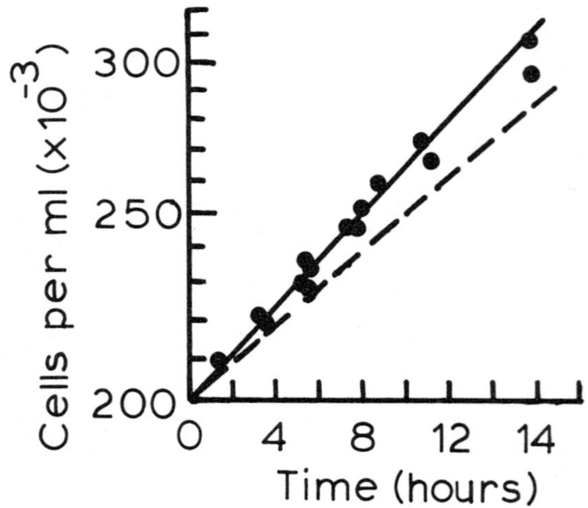

FIGURE 5. *The data points from two experiments were normalized to a 0 time cell concentration of 200,000/ml. The solid line drawn through these points if the curve expected for asynchronous cells with a doubling time of 25 h. The broken line represents the curve that would be obtained if cell growth were inhibited 25% at 0 time. In all of the experiments to be described, growth in the control bottles fell close to the solid line and in no case deviated by greater than the broken line.*

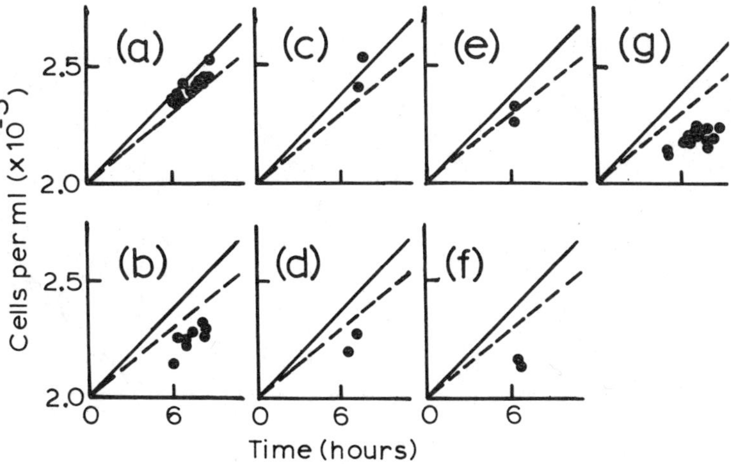

FIGURE 6. *Inhibition of division by MIX, with and without dibutyryl cAMP. For reference, the data points are shown with a control curve (solid line) and a curve representing 25% inhibition of the control curve (broken line). (a) 0.2 mM MIX, average inhibition 12.5%; (b) 0.2 mM MIX + 0.2 mM dibutyryl cAMP, 50%; (c) 0.33 mM MIX, 4%; (d) 0.33 mM MIX + 0.2 mM dibutyryl cAMP, 48%; (e) 0.5 mM MIX, 33%; (f) 0.5 mM MIX + 0.2 mM dibutyryl cAMP, 65%; (g) 1 mM MIX, 61%.*

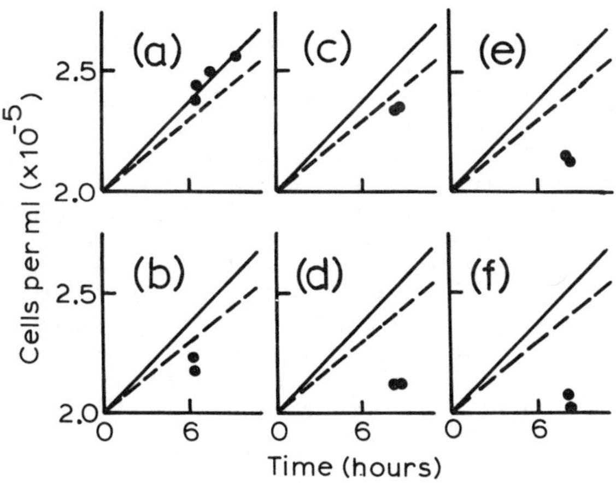

FIGURE 7. *Inhibition of division by theophylline, with and without dibutyryl cAMP. (a) 1 mM theophylline, average inhibition, 2%; (b) 1 mM theophylline + 0.2 mM dibutyryl cAMP, 48%; (c) 2 mM theophylline, 37%; (d) 2 mM theophylline + 0.2 mM dibutyryl cAMP, 79%; (e) 5 mM theophylline, 76%; (f) 5 mM theophylline + 0.2 mM dibutyryl cAMP, 92%.*

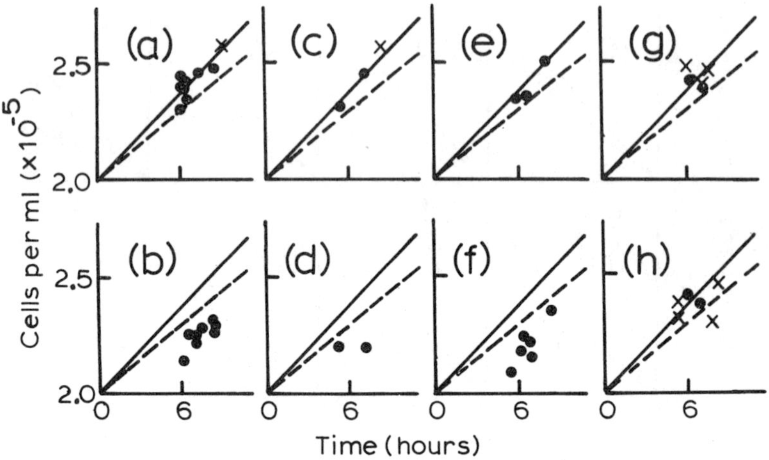

FIGURE 8. *Inhibition of division by cAMP and its analogs, with and without 0.2 mM MIX. (a) dibutyryl cAMP: •, 0.2 mM, average inhibition = 4%, ×, 1 mM; (b) 0.2 mM dibutyryl cAMP + 0.2 mM MIX, 50%; (c) monobutyryl cAMP: •, 0.2 mM, 5%; ×, 1 mM; (d) 0.2 mM monobutyryl cAMP + 0.2 mM MIX, 53%; (e) 0.2 mM 8-Br-cAMP, 6%; (f) 0.2 mM 8-Br-cAMP + 0.2 mM MIX, 54%; (g) cAMP: •, 0.33 mM, 3%; ×, 1 mM, -6%; (h) cAMP + 0.2 mM MIX: •, 0.33 mM cAMP, 7%; 1 mM cAMP,15%.*

by inhibition in G_2 or M. Since mitotic cells did not accumulate during the block, an inhibition in G_2 is indicated.

An additional method for detecting inhibition of traverse through G_2 in unsynchronized cultures has been developed in our laboratory recently (Kurz and Friedman, manuscript in preparation). The method entails a light pulse (low level of radioactivity) with ^{3}H-thymidine, followed, after a short interval without label, by a heavy pulse. This protocol results in the specific tagging of early G_2 cells with a light load of ^{3}H-thymidine. This G_2 "cohort" can be identified autoradiographically and followed through mitosis. The curve (from two experiments) of such cells entering mitosis is shown in Figure 9. Essentially, none of the agents at the concentrations used affected the rate at which the cohort entered mitosis except the combination of 8-Br-cAMP and MIX (0.2 m*M*). Thus, by this technique also, a clear potentiation is observed, indicating G_2 delay by cAMP.

EFFECT OF AGENTS ON MITOSIS

While it was shown above that elevations in cAMP during S or G_2 phase cause a prolongation of G_2 and mitosis, once cells have entered mitosis, addition of MIX leads to an acceleration of the remainder of mitosis (39,41). This is evidenced by a more rapid decay of the mitotic index and the more rapid rise in the cell density as cells enter

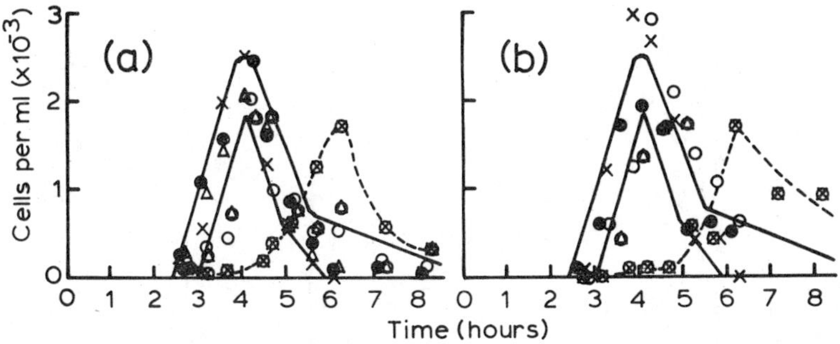

FIGURE 9. *The effect of chemical agents upon the flow of "tagged" G_2 cells through mitosis. The ordinate is the number of lightly labeled mitotic cells/ml. "Tagging" of cells with a light label was carried out as described in the text and was completed at 0-time. The agents were added at 1.3 h. Aliquots were then taken at intervals and analyzed for cell number and, by autoradiography, for the fraction of lightly labeled mitotic cells. From these determinations the number of lightly labeled cells/ml were calculated. (a) and (b) are duplicate experiments carried out on different days. The space enclosed by heavy lines in (a) roughly defines reasonable error limits of a control curve. For reference, the same space is reproduced in (b). ●, control; o, 0.2 mM MIX; ×, 0.2 mM 8-Br-cAMP; Δ, 0.2 mM 8-Br-5'AMP; --⊗--, 0.2 mM MIX + 0.2 mM 8-Br-cAMP; ◐, 0.2 mM MIX + 0.2 mM 8-Br-5'AMP.*

G_1 (Figure 10). In early experiments this effect of MIX on mitosis was potentiated by isoproterenol and mimicked by Ca^{++} (3 m*M*). However, the effect was variable from one experiment to the next and it was observed that the magnitude of the effect was directly related to the duration of mitosis in control cells. We, therefore, undertook a study of various conditions which might slow mitosis, to determine if there was any specificity to the accelerating actions of MIX or Ca^{++}.

Three conditions which slow mitosis have now been shown to be reversed by MIX or Ca^{++}: heat (Figure 11), D_2O (Figure 12), and low levels of colcemid (Figure 13). The accelerating effects of MIX on mitosis are potentiated by 8-Br-cAMP (Figure 14), but not 8-Br-5'AMP or 8-Br-cGMP indicating that the action of MIX is mediated through cAMP. Several other conditions which slow mitosis are not reversed by either MIX or Ca^{++}. These include cold shock, dinitrophenol and nitrous oxide.

The fact that MIX and Ca^{++} reverse the same inhibitory conditions suggests that they act on the same process. This is further supported by the finding that when maximal levels of MIX and Ca^{++} are added together they produce less than additive effects (Figure 11). Suboptimal concentrations, on the other hand, act synergistically (Figure 15).

The relationship between MIX and Ca^{++} has been

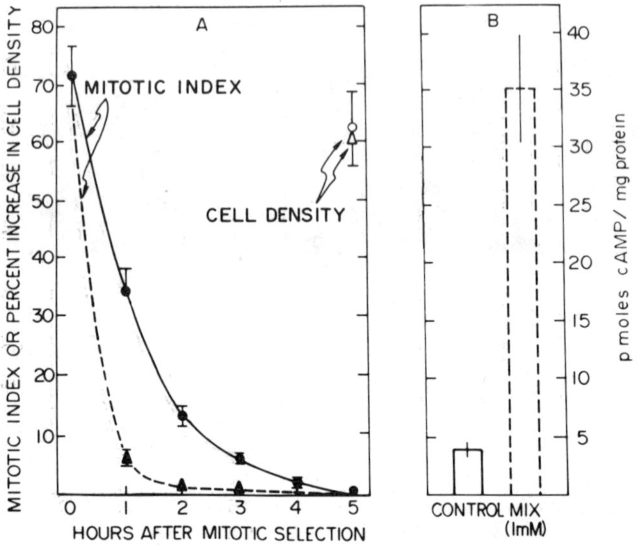

FIGURE 10. *Effects of MIX on cAMP levels and on completion of mitosis. MIX was added to synchronized cells immediately after isolation in metaphase by mitotic selection, to give a final concentration of 1 mM.* (A) *Effect of MIX on completion of mitosis. Mitotic index:* —●—, *control;* --▲--, *+ MIX. Percent increase in cell density:* o, *control;* Δ, *+ MIX. Values represent the mean ± standard errors from four separate experiments.* (B) *Effect of MIX on cAMP content of mitotic cells. cAMP levels were measured 60 min after the addition of either 0.9% saline or 1 mM MIX in 0.9% saline.* ——, *control;* ---, *MIX. Vertical lines represent the range of cAMP levels in duplicate experiments.*

investigated further. Ca^{++} does not raise intracellular cAMP levels in mitotic cells. MIX, on the other hand, does stimulate $^{45}Ca^{++}$ uptake (data not shown), thus suggesting an explanation for the synergism observed with suboptimal levels of these agents. However, the MIX effect is also observed in the presence of EGTA, suggesting that its action is not entirely due to effects on Ca^{++} influx. The ionophore, A-23187 (26), also stimulates mitosis (following heat treatment) in the presence of EGTA. These findings are consistent with the hypothesis that MIX (and A-23187) act by releasing Ca^{++} from intracellular storage sites. According to this hypothesis, Ca^{++} would be the final effector of the mitotic stimulation.

EFFECT OF AGENTS ON G_1

While we have observed a number of effects of chemical agents on G_1 traverse, none of these effects can be clearly attributed to cyclic nucleotides at the present

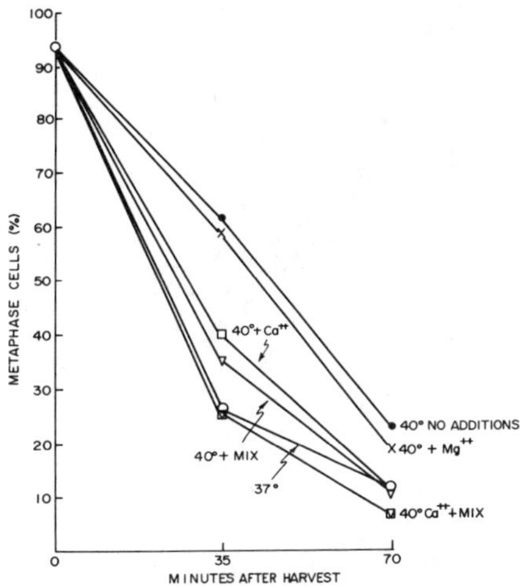

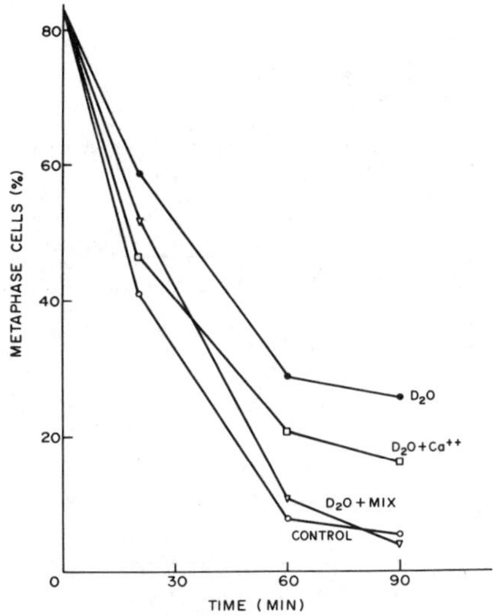

FIGURE 11. *MIX or Ca^{++} reverse the inhibitory effect of elevated temperature on mitosis. At 0-time, mitotic cells were selectively dislodged from monolayer cultures previously synchronized by a single thymidine block. Cells were placed in suspension culture at 37°C, o; or at 40°C: ●, no additions; ×, 4 mM $MgCl_2$; □, 4 mM $CaCl_2$; ∇, 1 mM MIX; 1 mM MIX + 4 mM $CaCl_2$. At the indicated times, cells were pelleted and fixed in 50% acetic acid. The metaphase index (metaphase cells/total cells × 100) was determined under phase contrast microscopy. The data points are the average of the results from duplicate flasks.*

FIGURE 12. *MIX or Ca^{++} reverse the inhibitory effect of D_2O on mitosis. At 0-time, dislodged mitotic cells were diluted into media in the presence or absence of 40% D_2O: o, control without D_2O; ●, media contained D_2O; □, media contained D_2O and 3 mM $CaCl_2$; ∇, media contained D_2O and 1 mM MIX. Values are the average of determinations on duplicate flasks.*

time. Thus, MIX inhibits the entrance into S-phase (Figure 16), but isoproterenol stimulates G_1 slightly and tends to reverse the inhibition by MIX (Figure 17). 8-Br-cAMP stimulates G_1 at low concentrations and inhibits at high concentrations. However, both of these effects are also observed with 8-Br-5'AMP at about ten-fold lower concentrations than 8-Br-cAMP (not shown).

HELA CELLS PROTEIN KINASES

HeLa cell protein kinases have been characterized in 27,000 x g supernatants of sonicates from randomly growing cells. To differentiate between cAMP-dependent and independent activities, the specific inhibitor of cAMP-dependent protein kinase described by Walsh *et al.* (35) was found

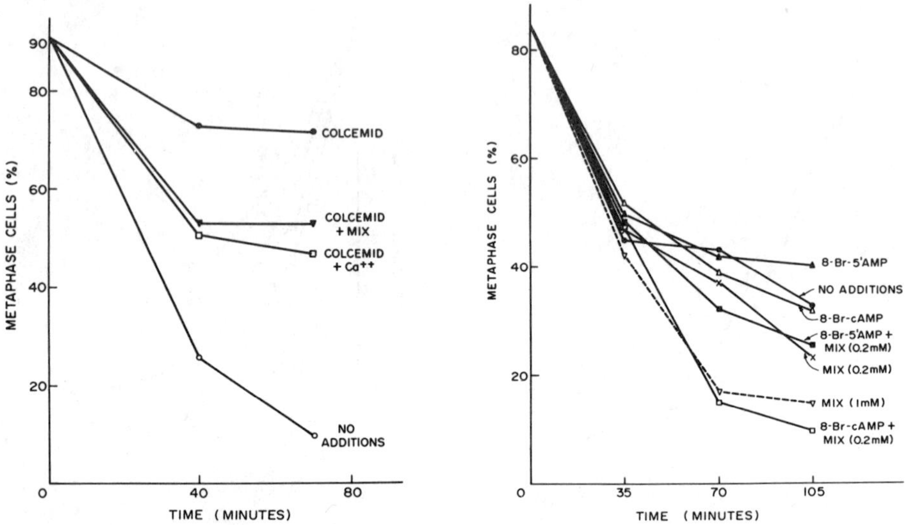

FIGURE 13. *MIX or Ca^{++} reverse the inhibitory effect of colcemid on mitosis. At 0-time, dislodged mitotic cells were placed in suspension culture in the absence, o, or presence, •, of colcemid (0.016 µg/ml). Other flasks with colcemid also contained 1 mM MIX, ▼, or 4 mM $CaCl_2$,* □ *. Values are the average of determinations on duplicate flasks.*

FIGURE 14. *MIX and 8-Br-cAMP act synergistically to reverse inhibition of mitosis by elevated temperature. At 0-time, dislodged mitotic cells were placed in suspension culture at 40°C with the additions (in duplicate flasks) described in the figure. 8-Br-cAMP and 8-Br-5'AMP, where present, were at a concentration of 0.2 mM. Synergism is shown between 0.2 mM MIX and 0.2 mM 8-Br-cAMP, but not between 0.2 mM MIX and 8-Br-5'AMP.*

useful. Studies suggest that the cAMP-dependent protein kinases of HeLa cells are completely and specifically inhibited by this protein. However, unlike partially purified peak II protein kinase from pig heart, when HeLa cell extracts were treated with the inhibitor much residual cAMP-independent activity remained. Furthermore, at very high concentrations of the crude "Walsh" inhibitor (> 500 µg/ml) the inhibition was changed to stimulation (Figure 18). A similar pattern was seen with BHK cell extracts though the reversal was not nearly so pronounced. The reversal was apparently due to the phosphorylation of a protein in the "Walsh inhibitor" preparation as suggested by the finding that the high concentrations stimulated incroporation in the absence of added histone and in a time-dependent manner. After further purification of the "Walsh inhibitor" by DEAE cellulose chromatography (35) the stimulatory effect of the Walsh inhibitor was no longer observed.

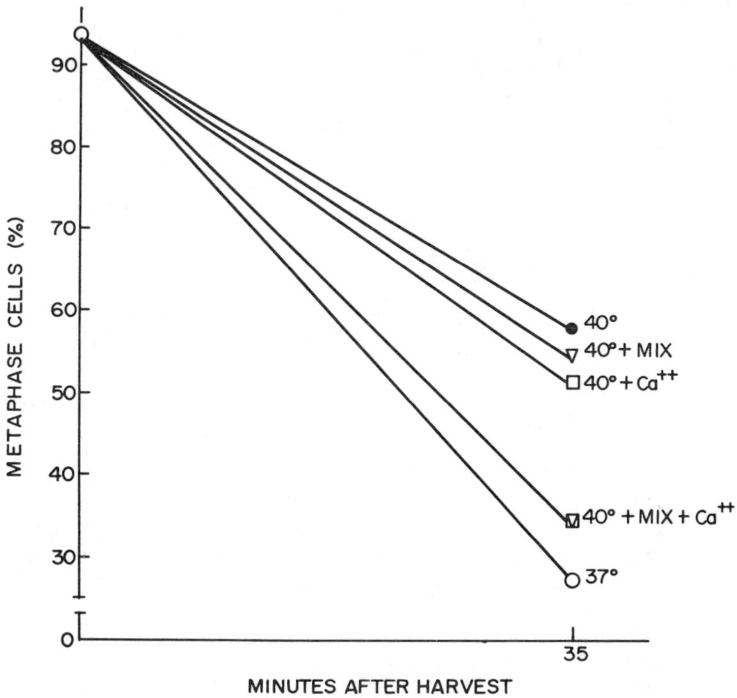

FIGURE 15. *Suboptimal concentrations of MIX and Ca^{++} act synergistically to reverse the inhibitory effects of elevated temperature. At 0-time, dislodged mitotic cells were placed in suspension culture at 37°C, o, or at 40°C: •, no additions; ∇, 0.2 mM MIX; □, 0.5 mM $CaCl_2$; ⊠, 0.2 mM MIX + 0.05 mM $CaCl_2$. Values are the average of determinations on duplicate flasks.*

The results obtained with "Walsh inhibitor" suggested that HeLa cell extracts contained cAMP-dependent protein kinase as well as other kinase activity(ies). The results of a comparison of the protein kinase activity with different substrates were consistent with this interpretation (Table 1). Considerable activity was observed with endogenous substrate. This activity was not inhibited by "Walsh inhibitor," nor was it activated by cAMP. Addition of casein as substrate led to a doubling of ^{32}P incorporation over that seen with endogenous substrate. This activity was also unaffected by "Walsh inhibitor" or by cAMP (Table 1). When mixed histone was used as substrate it was observed that the incorporation into endogenous substrate was suppressed. The activity observed with histone was partly inhibited by the "Walsh inhibitor" and was stimulated four- to eight-fold by cAMP. This cAMP-dependent activity was completely suppressed by the "Walsh inhibitor."

Further characterization of cAMP-dependent activity toward histone indicated that it was inhibited by salt,

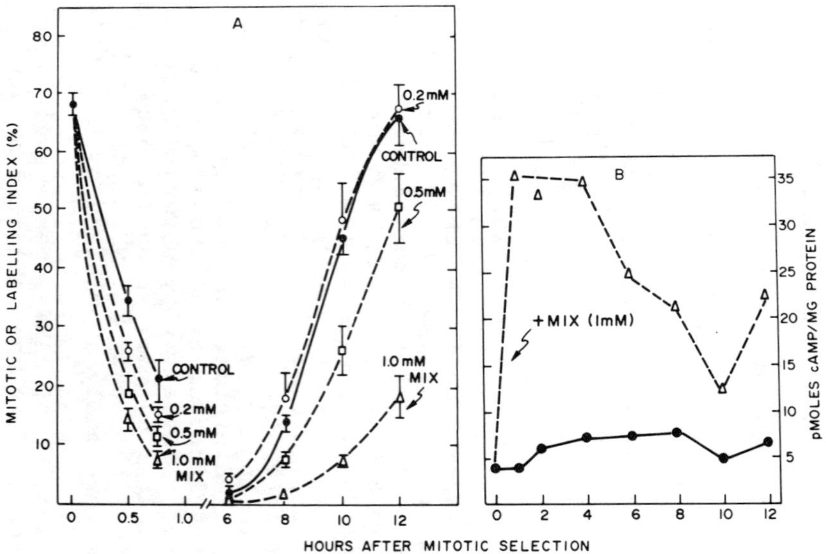

FIGURE 16. *Effects of MIX on G1. (A), the effect of varying concentrations of MIX on the duration of M and G1. All additions were made immediately after selection of mitotic cells (0-time). Mitotic index (0-0.75 h) or labeling index (6-12 h): ●, control; o, MIX (0.2 mM); □, MIX (0.5 mM); Δ, MIX (1.0 mM). Values represent the mean ± standard errors from four experiments. (B), the effect of MIX on cAMP levels during G1: ●, control; Δ, MIX (1.0 mM). Values are the average of two experiments.*

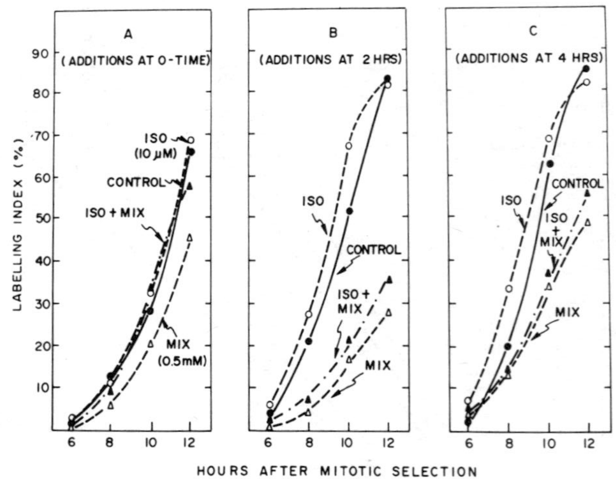

FIGURE 17. *Effects of isoproterenol and MIX on the duration of G1. Labeling index: ●, control; Δ, MIX (0.5 mM); o, isoproterenol (10 μM); ▲, isoproterenol + MIX. Values are the average of two experiments. (A), additions at 0-time; (B), additions at 2 h; (C), additions at 4 h.*

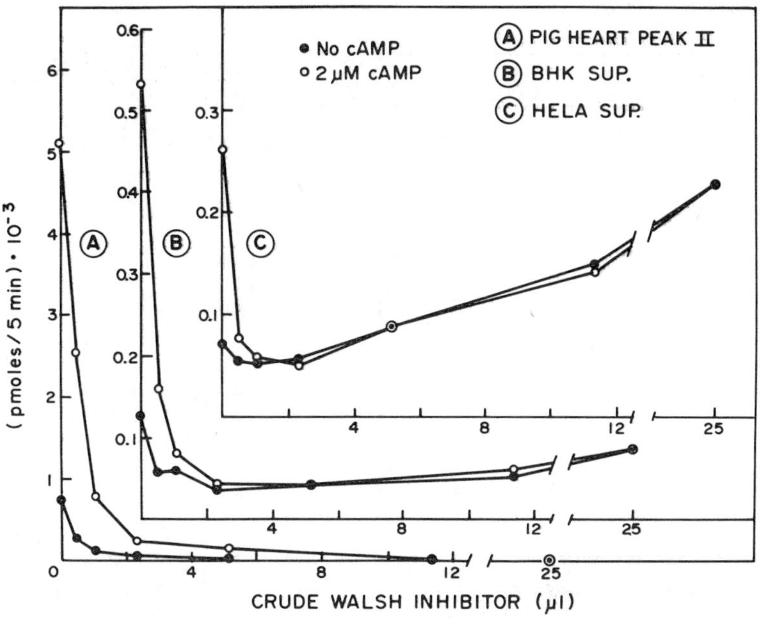

FIGURE 18. *Effects of crude "Walsh inhibitor" on protein kinase activities. The ordinate represents protein kinase activity using mixed histone as substrate. Protein kinase assay mixtures contained 30 mM MOPS buffer (pH 6.8), 7.5 mM Mg acetate, 2.5 mg/ml mixed histones, 200 μM γ [^{32}P] ATP (0.2-1.0 μ ci), and either 1 μg (pig heart), 43 μg (BHK) or 9 μg (HeLa) of protein. Acid insoluble radioactivity was determined by the method of Reimann et al. (27). The pig heart enzyme was the peak II cAMP-dependent protein kinase partially purified on DEAE cellulose (26). The HeLa and BHK kinase preparations were 27,000 x g (30 min) supernates of sonicates of exponentially growing cells. Walsh inhibitor was prepared as the acid precipitate from a boiled supernate of rabbit skeletal muscle (35).*

inorganic phosphate and high levels of $MgCl_2$. The half-maximal concentration of cAMP required for activation was about 5 x 10^{-8} *M*. The activity was greatest against histone F_1 and F_{2b} and was weak against F_{2a} and F_3. No cAMP-dependent activity was detectable with casein or protamine as substrate under our assay conditions.

Soluble extracts from randomly growing cells were chromatographed on DEAE cellulose, eluting the column with a linear NaCl gradient. The results (Figure 19) indicated that there were two types of cAMP dependent activity, one eluting at 0.05 *M* NaCl and the other at 0.15 *M* NaCl. This finding has been previously described for other tissues and studied in detail by Corbin and co-workers (7,8). Variable proportions of these two activities exist in different tissues but their physiological significance is

TABLE 1. *HeLa supernate protein kinase activities*

Protein Substrate[a]	Walsh Inhibitor[b]	cAMP (2 μ*M*)	Protein Kinase Activity Exp.1	Exp.2	Exp.3
			pmoles/min/10^6 cells		
None	-	-	710	874	878
	+	-	643	-	-
Casein	-	-	1374	1503	1472
	+	-	1377	-	-
	-	+	1360	-	-
Mixed histone	-	-	338	257	199
	+	-	223	171	162
	-	+	1579	1576	1574
Crude Walsh inhibitor	+	-	1263	-	-

[a]*Casein or mixed histones present at 2.5 mg/ml; crude Walsh inhibitor at 6.7 mg/ml.*

[b]*Partially purified with Walsh inhibitor (fractionated by DEAE-cellulose chromatography (35)) present at 4.3 μg/ml.*

Crude supernates (without NaCl in the homogenizing medium) of exponentially-growing HeLa cells (∿ 10 μg protein) were incubated 5 min at 30°C as described.

unknown (7).

The cAMP-independent protein kinase that phosphorylates casein was eluted at 0.20 to 0.30 *M* NaCl. This activity was apparently responsible for the phosphorylation of the protein in crude Walsh inhibitor preparation since it chromatographed with this activity. The endogenous substrate phosphorylating activity was found exclusively in the fractions which were not retained by the column. These fractions also contained some casein kinase activity and cAMP-dependent protein (histone) kinase. However, in subsequent columns, using increased ratios of DEAE cellulose:protein, the complete retention of casein kinase and cAMP-dependent kinase was achieved. In these columns the endogenous activity was lost, suggesting the separation of endogenous substrate from its kinase.

To further characterize the two cAMP-dependent activities, the two DEAE cellulose peaks were separately pooled and concentrated. They were then characterized by their ability to dissociate (RC → R + C) and reassociate (R + C → RC) in the presence or absence of 0.5 *M* NaCl as described by Corbin *et al.* (7). The results of these studies (not shown) indicated that the two peaks had properties similar to those previously described in other tissues (7). The Peak I (0.25 *M* NaCl peak) enzyme was readily dissociable

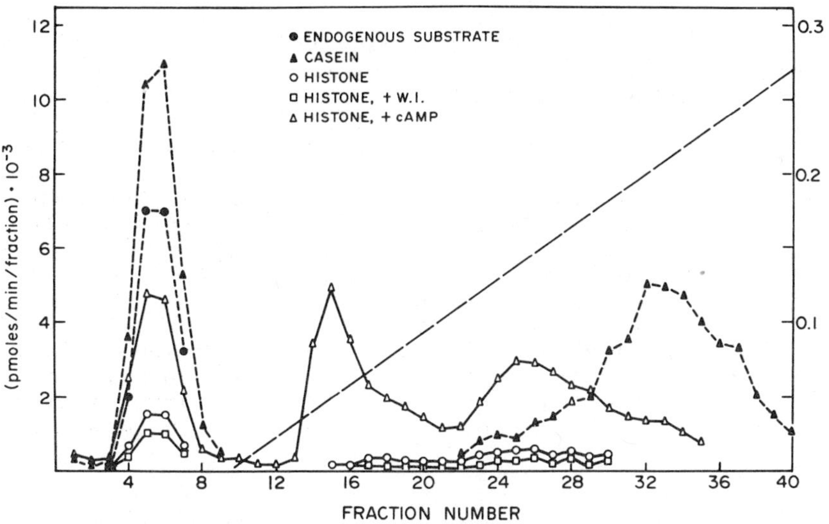

FIGURE 19. *Separation of HeLa cell protein kinases on DEAE cellulose column. Seven ml of HeLa cell supernate, containing 14 mg protein (derived from 5.6 × 10^7 exponentially growing cells) were passed over a 0.7 × 6.3 cm column of DE-11, pre-equilibrated with 5 mM Tris (pH 7.5), 1 mM EDTA. After washing with 20 ml of the same buffer, the column was eluted with a linear gradient (0 to 0.35 M) of NaCl. Three ml fractions were collected and assayed as described. The recovery of cAMP-dependent histone kinase was 68%. W. I. = Walsh inhibitor.*

in high salt while the peak II enzyme was not. The subunits of the Peak II activity rapidly reassociated in the absence of NaCl.

The results of these studies may be summarized as follows:

(1) HeLa cell extracts contain, in about equal proportions, the two previously described types of cAMP-dependent protein kinase. These enzymes utilize histone as substrate, but not casein or endogenous soluble proteins.

(2) Extracts contain a cAMP-independent casein kinase, which can apparently also phosphorylate some protein in the crude Walsh inhibitor preparation, but which has little or no activity with histone.

(3) Extracts contain endogenous cAMP-independent activity which is inhibited by histone. This activity is apparently not due to the cAMP-dependent protein kinase, but may be due to the casein kinase.

These three types of activity as well as activity ratios

$$\left(\frac{\text{activity} - \text{cAMP}}{\text{activity} + \text{cAMP}}\right)$$

of the cAMP-dependent protein kinase can be separately

determined in crude extracts by varying the type of substrate and by using the "Walsh inhibitor" to identify cAMP-dependent protein kinase. The presence of the two peaks of cAMP-dependent protein kinase makes it difficult to establish homogenization conditions conducive to maintaining *in vivo* activity ratios. However, the results suggested that the optimal conditions would be the use of low salt concentrations. Under these conditions, the peak I enzyme is stable, while the peak II enzyme reassociates. However, reassociation was very slow at 0°C in crude extracts and would be negligible during the 10-15 min required to prepare and assay homogenates.

Protein kinase determinations during the cell cycle have thus far only been applied to S, G_2, M and early G_1 following the double thymidine block. The results of these studies are still considered preliminary. A typical pattern is seen in Figure 20. All activities remained

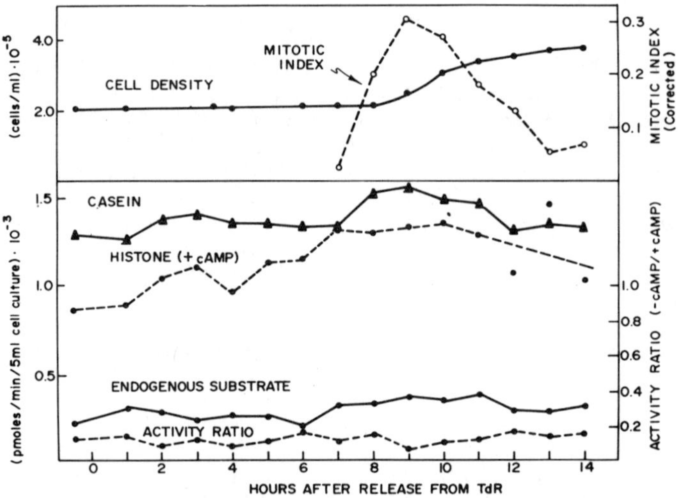

FIGURE 20. *Protein kinase activities in synchronized HeLa cells. Cells were synchronized at the beginning of S-phase by the double thymidine block technique, then resuspended at 200,000 cells/ml in thymidine-free medium at 0-time. At the indicated times, cells were harvested from 25 ml of suspension and sonicated in 5.0 ml of 5 mM Tris (pH 7.5), 1 mM EDTA, 0.5 mM MIX, 150 mM NaCl. Homogenates (25 µl, about 6 µg protein) were assayed (10 min at 30°C) as described, within 15 min of cell collection. Cyclic AMP-dependent histone kinase activities were corrected by using the Walsh inhibitor.*

relatively constant during the cell cycle. There appeared to be a minor elevation of casein kinase activity and of activity with endogenous substrate during mitosis. The total (+ cAMP) cAMP-dependent protein kinase activity increased slowly until about 7 h and then remained relatively stable thereafter. The most surprising finding was that the cAMP-dependent protein kinase activity ratios were extremely low throughout the experiment and no consistent changes, corresponding to changes in cAMP (Figure 1) were observed. However, at these low ratios, small changes in activity ratio would be missed. The low activity ratios were observed if homogenates were prepared in 0.5 *M* NaCl, 0.15 *M* NaCl or in the absence of NaCl, and increased ratios (0.3-0.4) were only observed if homogenates were allowed to stand on ice for 30 min to 1 h in the presence of 0.5 *M* NaCl, a finding consistent with the expected dissociation of the subunits of the Peak I enzyme.

DISCUSSION

The results suggest an inhibitory role for cAMP during G_2 phase. This is supported by the natural depression in cAMP levels that occurs as cells enter G_2 and M, and by the inhibitory effects of agents which elevate or mimic cAMP in both synchronized and unsynchronized cells. That these agents act through cAMP is supported by their potentiation by MIX. Evidence for G_2 inhibition also derives from studies by other workers, the most extensive being those with epidermis (4,34). A possible physiological regulatory role for cAMP in G_2 in epidermis is suggested by the fact that a significant proportion of the population of cells is naturally arrested in G_2 (12). These cells are readily activated to divide in response to wounding of the skin (12).

The stimulatory action of cAMP during mitosis is of questionable physiological significance, since the levels of cAMP are normally minimal during mitosis. However, a highly transient rise (e.g., one lasting but a few minutes) in cAMP cannot be ruled out. To detect such a rise a high degree of synchrony of cells during their progress through mitosis would be required. Such synchrony is not possible with present techniques in mammalian cells. The results suggest the possibility that the effect of cAMP at mitosis may be secondary to the stimulation of Ca^{++} release from an internal store. Ca^{++} would appear to stimulate mitosis in certain situations where mitosis is slowed. The diversity of inhibitory conditions which are reversed by Ca^{++} and our lack of information as to how these conditions affect mitosis does not as yet allow a plausible model of Ca^{++} action. The effects of Ca^{++} could reflect a role for this ion in chromosome movement.

The studies on G_1 illustrate the nonspecificity of chemical agents used to elevate cyclic nucleotide levels. All of the agents tested either shortened or prolonged G_1. However, none of the results could clearly be attributed to cAMP. It remains to be determined whether the high level of cAMP observed in G_1 phase is an important regulatory mechanism in the traverse of HeLa cells through G_1. A stimulatory role for cAMP in late G_1 is supported by additional studies reported in this volume (5,22,36).

Specific phosphorylation of histone during mitosis

has been reported by several laboratories (2). Increased phosphorylation could result either from an increase in a histone kinase, a decrease in protein phosphatase or an increased availability of the substrate. Our results showed no significant change in histone kinase activity during mitosis, although the data suggested a small elevation in kinase activity toward casein around mitosis.

Extremely low activity ratios

$$\left(\frac{\text{activity - cAMP}}{\text{activity + cAMP}} = \text{ca } 0.1\right)$$

of cAMP-dependent protein kinase were observed throughout S-phase, G_2 and M. Despite efforts to retain the ratios of intact cells by establishing the optimal salt concentration for homogenization and by assaying samples within minutes of homogenization, the results could reflect rapid changes in ratio occurring during preparation. Alternatively, the ratios may actually be very low in HeLa cells despite rather high (ca 2 μ*M*) cellular cAMP levels and a relatively low Ka for cAMP (ca 5×10^{-8} *M*). There are several possible explanations for low ratios. The Ka for cAMP may be much greater in intact cells then in dilute lysates (1). This may be related to the high level of cAMP dependent protein kinase observed in these cells (4 to 8 nmoles/mg protein/min). Other conditions or factors may act to prevent cAMP from dissociating the subunits of protein kinase *in vivo*. A further possibility is that internal compartmentation exists such that the kinase is not affected by cAMP. Cyclic AMP might influence cell cycle progress by means other than activation of protein kinase. However, because of the low activity ratios and inprecision in the assay,it is not yet possible to determine if fluctuations in cAMP levels are reflected in changes in kinase activity.

While these studies and others to be reported in this volume clearly suggest roles for cAMP in the cell cycle, the mechanisms by which it exerts its effects are poorly understood. The ability to unravel these mechanisms is limited by our primitive understanding of the basic mechanisms of cell cycle traverse.

SUMMARY

Measurements of cAMP during the synchronized cell cycle of HeLa cells demonstrate that levels are at a minimum during late G_2 and mitosis and at a maximum near the G_1-S border. If cAMP levels are raised during G_2 in either synchronized or unsynchronized cells, traverse through this stage is slowed. If, on the other hand, cAMP levels are elevated during mitosis in cells which have been slowed in this stage by several treatments, then the exit from mitosis is accelerated. Evidence suggests the possibility that this latter effect of cAMP is related to the release of calcium from internal stores. No effect of elevated cAMP is observed in S-phase and, although various effects have been observed in G_1, none of them as yet can be attributed to cAMP.

Using the specific inhibitor of cAMP-dependent protein kinase, we have characterized the protein kinases of HeLa cells. Three activities are described. With regard to the cAMP-dependent protein kinase, two peaks of activity are characterized in order to obtain conditions for determining *in vivo* activity ratios. Preliminary data is presented suggesting that activity ratios are extremely low *in vivo* and that they are relatively unresponsive to changes in cAMP.

ACKNOWLEDGEMENTS

This work was supported by grants AM-07462, HE-08332, and CA-13638 from the National Institutes of Health and IN-25P-1 from the American Cancer Society.

REFERENCES

1. Beavo, J. A., Bechtel, P. J. and Krebs, E. G. 1974. Activation of protein kinase by physiological concentrations of cyclic AMP. Proc. Nat. Acad. Sci. USA *71*:3580-3583.
2. Bradbury, E. M. 1975. Histones, chromatin structure, and control of cell division. Current Topics in Dev. Biol. *9*:1-13.
3. Browning, E. T., Groppi, V. E., Jr. and Kon, C. 1974. Papaverine, a potent inhibitor of respiration in C-6 astrocytoma cells. Mol. Pharmacol. *10*:175-181.
4. Bullough, W. S. 1962. The control of mitotic activity in adult mammalian tissues. Biol. Rev. *37*: 307-342.
5. Byron, J. W. 1971. Effect of steroids and dibutyryl cyclic AMP on the sensitivity of haemopoietic stem cells to ^{3}H-thymidine *in vitro*. Nature *234*:39-40.
6. Coffino, P., Gray, J. W. and Tomkins, G. M. 1975. Cyclic AMP, nonessential regulator of the cell cycle. Proc. Nat. Acad. Sci. USA *72*:878-882.
7. Corbin, J. D., Keely, S. L. and Park, C. R. 1975. The distribution and dissociation of cyclic adenosine 3':5'-monophosphate-dependent protein kinases in adipose, cardiac, and other tissues. J. Biol. Chem. *250*:218-225.
8. Corbin, J. D., Soderling, T. R. and Park, C. R. 1973. Regulation of adenosine 3',5'-monophosphate-dependent protein kinase. I. Preliminary characterization of the adipose tissue enzyme in crude extracts. J. Biol. Chem. *248*:1813-1824.
9. Durham, J. P., Baserga, R. and Butcher, F. R. Lack of correlation between catacholamine analog effects on cyclic AMP levels and adenylate cyclase activity and the stimulation of DNA synthesis in mouse parotid gland. In: *Control of Proliferation in Animal Cells*, Clarkson, B. and Baserga, R. (eds.).

Cold Spring Harbor Laboratory, Cold Spring Harbor, New York, 1974, pp. 595-607.
10. Frank, W. 1972. Cyclic 3',5'-AMP and cell proliferation of embryonic rat cells. Exp. Cell Res. *71*: 238-239.
11. Froehlich, J. E. and Rachmeler, M. 1972. Effect of adenosine 3',5'-cyclic monophosphate on cell proliferation. J. Cell Biol. *55*:19-31.
12. Gelfant, S. A new theory on the mechanism of cell division. In: *Cell Growth and Cell Division*, Harris, R. J. C. (ed.), Vol. 2. Academic Press, New York, 1963, pp. 229-260.
13. Gilman, A. G. 1970. A protein binding assay for adenosine 3',5'-cyclic monophosphate. Proc. Nat. Acad. Sci. USA *67*:305-312.
14. Ginsburg, E., Solomon, D., Sreevalsan, T. and Freese, E. 1973. Growth inhibition and morphological changes caused by lipophilic acids in mammalian cells. Proc. Nat. Acad. Sci. USA *70*:2457-2461.
15. Hauschka, P. V., Everhart, L. P. and Rubin, R. W. 1972. Alteration of nucleoside transport of Chinese hamster cells by dibutyryl adenosine 3',5'-cyclic monophosphate. Proc. Nat. Acad. Sci. USA *69*:3542-3546.
16. Johnson, M. and Romwell, P. W. In: *Advances in Bio-Sciences*, Bergstrom, B. and Bernhard, H. (eds.), Vol. 9. Pergamon Press, Vieweg, Braunschwerg, 1973, pp. 205-212.
17. Leder, A. and Leder, P. 1975. Butyric acid, a potent inducer of erythroid differentiation in cultured erythroleukemic cells. Cell *5*:319-322.
18. Millis, A. J. T., Forrest, G. and Pious, D. A. 1972. Cyclic AMP in cultured human lymphoid cells: relationship to mitosis. Biochem. Biophys. Res. Commun. *49*:1645-1649.
19. Millis, A. J. T., Forrest, G. A. and Pious, D. A. 1974. Cyclic AMP dependent regulation of mitosis in human lymphoid cells. Exp. Cell res. *83*:335-343.
20. Pardee, A. 1974. A restriction point for control of normal animal cell proliferation. Proc. Nat. Acad. Sci. USA *71*:1286-1290.
21. Paul, D. 1973. Quiescent SV 40 virus transformed 3T3 cells in culture. Biochem. Biophys. Res. Commun. *53*:745-753.
22. Pawelek, J., Halaban, R. and Christie, G. 1975. Melanoma cells which require cyclic AMP for growth. Nature *258*:539-540.
23. Petersen, D. F., Tobey, R. A. and Anderson, E. C. Essential biosynthetic activity in synchronized mammalian cells. In: *The Cell Cycle*, Padilla, G. M., Whitson, G. L. and Cameron, I. I. (eds.). Academic Press, New York, 1969, pp. 341-359.
24. Prasad, K. N. and Kumar, S. Cyclic AMP and differentiation of neuroblastoma cells in culture. In: *Control of Proliferation in Animal Cells*, Clarkson,

B. and Baserga, R. (eds.). Cold Spring Harbor Laboratory, Cold Spring Harbor, New York, 1974, pp. 581-594.
25. Radley, J. M. and Hodgson, G. S. 1971. Effect of isoprenaline on cells in different phases of the mitotic cycle. Exp. Cell Res. *69*:148-160.
26. Reed, P. W. and Lardy, H. A. 1972. A23187: a divalent cation ionophore. J. Biol. Chem. *247*:6980-6977.
27. Reimann, E. M., Walsh, D. A. and Krebs, E. G. 1971. Purification and properties of rabbit skeletal muscle adenosine 3',5'-monophosphate-dependent protein kinases. J. Biol. Chem. *246*:1986-1995.
28. Remington, J. A. and Klevecz, R. R. 1973. Hormone-treated CHO cells exit the cell cycle in the G_2 phase. Biochem. Biophys. Res. Commun. *50*:141-146.
29. Roller, B., Hirai, K. and Defendi, V. 1974. Effect of cAMP on nucleoside metabolism. I. Effect on thymidine transport and incorporation in monkey cells (CV-1). J. Cell Physiol. *83*:163-176.
30. Smets, L. A. 1972. Contact inhibition of transformed cells incompletely restored by dibutyryl cyclic AMP. Nature New Biol. *239*:123-124.
31. Stellwagen, R. H. 1974. The effect of theophylline and certain other purine derivatives on tyrosine aminotransferase activity in hepatoma cells in culture. Biochim. Biophys. Acta *338*:428-439.
32. Thomas, D. B., Medley, G. and Lingwood, C. A. 1973. Growth inhibition of murine tumor cells *in vitro*, by puromycin, $[^6N]O^{2'}$-dibutyryl 3',5'-adenosine monophosphate, or adenosine. J. Cell Biol. *57*: 397-405.
33. VanWijk, R., Wicks, W. D., Beevers, M. M. and VanRijn, J. 1973. Rapid arrest of DNA synthesis by $N^6,O^{2'}$-dibutyryl cyclic adenosine 3',5'-monophosphate in cultured hepatoma cells. Cancer Res. *33*:1331-1338.
34. Voorhees, J. J., Duell, E. A., Stawiski, M. and Harrell, E. R. 1974. Cyclic nucleotide metabolism in normal and proliferating epidermis. Adv. Cyclic Nucleotide Res. *4*:117-162.
35. Walsh, D. A., Ashby, C. D., Gonzalez, C., Calkins, D., Fischer, E. H. and Krebs, E. G. 1971. Purification and characterization of a protein inhibitor of adenosine 3',5'-monophosphate-dependent protein kinases. J. Biol. Chem. *246*:1977-1985.
36. Whitfield, J. F., MacManus, J. P., Rixon, R. H., Boynton, A. L., Youdale, T. and Swierenga, S. 1976. Control of cell proliferation. In Vitro *12*:1.
37. Willingham, M. C., Johnson, G. S. and Pastan, I. 1972. Control of DNA synthesis and mitosis in 3T3 cells by cyclic AMP. Biochem. Biophys. Res. Commun. *48*:743-749.
38. Wright, J. A. 1973. Morphology and growth rate changes in Chinese hamster cells cultured in

presence of sodium butyrate. Exp. Cell Res. *78*: 456-460.

39. Zeilig, C. E., Johnson, R. A., Friedman, D. L., Kumar, K. and Sutherland, E. W. 1974. Regulatory influences of cyclic AMP in synchronized HeLa cells. Second International Conference on Cyclic AMP, Vancouver, Canada (Abs. Tha-10).

40. Zeilig, C. E., Johnson, R. A., Friedman, D. L. and Sutherland, E. W. 1972. Cyclic AMP concentrations in synchronized HeLa cells. J. Cell Biol. *55*:296a.

41. Zeilig, C. E., Johnson, R. A., Sutherland, E. W. and Friedman, D. L. 1974. Cyclic AMP levels in synchronized HeLa cells and a dual effect on mitosis. Fed. Proc. *33*:948a.

Cyclic Nucleotides and the Cell Cycle of the Hematopoietic Stem Cell

J. W. Byron

Mouse bone marrow contains a primitive pluripotent cell (stem cell) which following transplantation produces macroscopic colonies in the spleen of heavily irradiated mice (1); each colony arises from a single cell (2, 3). These cells are called spleen colony forming units or CFU-S (4).

With respect to the cell cycle, CFU-S of normal adult bone marrow have generally been considered to be in a resting stage (G_o) or to have long cell cycles with a relatively small fraction occupied by DNA synthesis. Accordingly, they are insensitive to the cytocidal actions of high specific activity of H^3-thymidine (5) and of inhibitors of DNA synthesis; a more accurate description of the cell cycle state of resting CFU-S may be that they are held up in late G_1; i.e., close to the G_1-S boundary (6). Under conditions requiring increased numbers of differentiated bone marrow cells, CFU-S may enter into the cell cycle; hence their sensitivity to H^3-thymidine and to inhibitors of DNA synthesis will increase.

The hemopoietic stem cell possesses the property of self-renewal, and also gives rise to hemopoietic cells committed to differentiate towards a particular bone marrow end-cell. Accordingly, the erythropoietin sensitive cell (7), *in vitro* colony forming cells related to red cell formation (8), and the *in vitro* colony forming cell committed to granulocyte and macrophage formation (9), all ultimately derive from the pluripotent stem cell. The committed progenitor cells are thought to have a finite life span. Day-to-day growth and differentiation of bone marrow cells is maintained by these cells derived from CFU-S. In contrast to the stem cell in adult mouse bone marrow, the committed progenitor cells are in cell cycle, actively synthesize DNA and are, therefore, susceptible to killing by high specific activity H^3-thymidine or by other agents which kill cells in cycle. Increased demand for differentiated bone marrow cells can alter the cell cycle state of CFU-S such that these cells enter into active cell cycle and now become sensitive to H^3-thymidine killing (H^3-TdR).

Hematopoietic cells give rise to a series of proliferative and differentiation processes that eventually provide physiological functions essential to survival. Unfortunately, these cells also make up a tissue vulnerable to the development of a number of undesirable conditions including neoplasias, genetically based anemias, drug-induced bone marrow aplasia, etc. The extent to which cell-cycle state influences the complexity of proliferative and differentiation processes in the bone marrow is not completely understood. Knowledge of cellular mechanisms controlling the cell cycle, and the ability to manipulate this cycle at will both *in vitro* and *in vivo*, may allow for further clarification of normal biological events within the hematopoietic system and may uncover new facts underlying pathological changes in this tissue. Accordingly, normal tissue activity may be modified, and the quantitative response of tissues to both therapeutic and to toxicologic agents may be dependent upon the position of the cells within the cell cycle. Cell proliferation and differentiation, the response of tissues to certain antitumor agents and cellular responses to certain carcinogens represent some important events closely aligned with stages of the cell cycle.

Mechanisms regulating growth and differentiation processes are of basic importance to an understanding of cell kinetics in the bone marrow, and for understanding pathophysiological conditions connected with bone marrow. Molecular events influencing the hemopoietic system are far from understood.

These studies describe two extremely sensitive cellular mechanisms that serve to trigger the hematopoietic stem cell from out of cycle (G_0) into the S-phase of the cell cycle. Through either β-adrenergic receptors or through cholinergic receptors, specific agonists initiate DNA synthesis. It is very likely that cyclic nucleotides mediate in receptor-initiated G_0 to S transition. Furthermore, some evidence is presented to suggest that CFU-S cyclase systems may have a regulatory role in the late G_1 and early S regions of the cell cycle. The extreme sensitivity of the aforementioned drug-receptor mechanisms is clearly demonstrated since low concentrations of receptor stimulating agents giving on average 150 molecules per nucleated bone marrow cell or less are sufficient to initiate DNA synthesis.

EXPERIMENTAL METHODS

The technique, in its more recent modifications (10, 11), used to study the action of drugs on CFU-S of mouse bone marrow has been described, in detail, elsewhere. Briefly, bone marrow cells suspended in Fischer medium are exposed *in vitro* to drugs for 2.5 h. Cultures are then treated for an additional 0.5 h with 200 µCi/ml of high specific activity H^3-TdR. Control cultures receive a comparable amount of unlabeled thymidine. After exposure to thymidine the cultures are diluted. The diluted suspension is injected intravenously (i.v.) into irradiated recipients, and spleen colonies are counted 9 days after transplantation. A comparison is made between numbers of colonies per spleen after transplanting control (H^1-TdR) and H^3-TdR treated cultures. The percent loss of CFU-S is then calculated. For agents switching-on DNA synthesis the percent loss of spleen colonies due to H^3-TdR is expected to increase. The system is free from serum or plasma. This may allow for a more realistic assessment of the concentration of free drug molecules needed to act on cells since many extracellular factors that might bind or consume compounds are not present.

As an additional test as to whether cells are synthesizing DNA, 10^{-3} *M* hydroxyurea is substituted for high specific activity H^3-TdR. Hydroxyurea (HU), however, is added at the start of culturing and is generally present for 3 h.

Agents were selected for study having known mechanisms of action, at least in other tissues. Hence, such compounds served as analytical tools for defining properties of the hemopoietic stem cell.

RESULTS

The response of the pluripotent stem cells to isoproterenol and to isoproterenol preceded by the β-adrenergic blocking agent propranolol is shown in Figure 1. The response to isoproterenol is compared with that to the β-adrenergic agonists orciprenaline and salbutamol (Figure 2). When exposed to low concentrations of the β-adrenergic receptor stimulating agent, isoproterenol, CFU-S greatly increased their sensitivity to the cytocidal actions of H^3-TdR. At low concentrations of isoproterenol this effect was blocked by the β-adrenergic receptor blocking agent, propranolol. Both orciprenaline and salbutamol were less effective than isoproterenol. This reduced effect is indicated by both agents requiring higher concentrations than isoproterenol to increase sensitivity to H^3-TdR and by maximum responses lower than that for isoproterenol. Salbutamol appears to be the least effective compound. The differences in the chemical structures of the three β-adrenergic stimulating agents are also shown.

The response of CFU-S to the cholinergic agonist, carbamylcholine, and to carbamylcholine preceded by the cholinergic receptor blocking agent, d-tubocurarine is shown in Figure 3. The response to acetylcholine alone and to acetylcholine in the presence of the cholinesterase inhibitor neostigmine is shown in Figure 4.

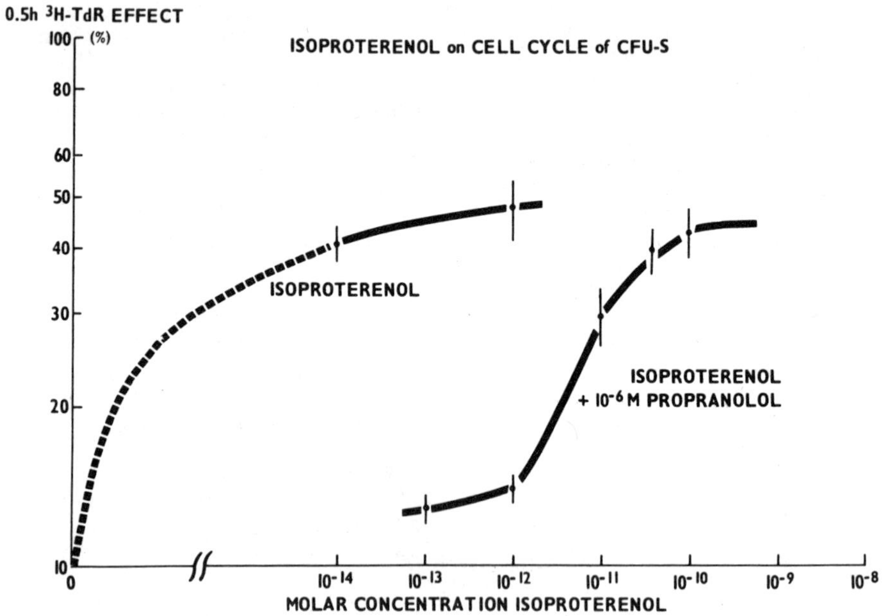

FIGURE 1. *Percent loss CFU-S due to H^3-thymidine (± SEM) in cultures treated with isoproterenol alone or in the presence of 10^{-6} M propranolol. Each point is made up from a minimum of 24-32 spleens. Propranolol was added 5 min before isoproterenol. The point at 10^{-4} M isoproterenol, however, is from three individual experiments giving a total of 81 spleens.*

Following chemical excitation of bone marrow with carbamylcholine or with acetylcholine in the presence of neostigmine, H^3-TdR decreased the colony forming ability of suspensions of mouse bone marrow. Low concentrations (10^{-14} to 10^{-12} *M*) of the cholinergic agents were effective. Alone, neither neostigmine nor acetylcholine stimulated DNA synthesis. In combination with neostigmine, acetylcholine increased the sensitivity of CFU-S to H^3-TdR. When added 5 min prior to carbamylcholine, the anticholinergic agent d-tubocurarine (10^{-6} *M*), inhibited the effect of low concentrations of carbamylcholine. This inhibition could be overcome by increasing the concentration of carbamylcholine.

The responses of CFU-S to the dibutyryl derivatives of adenosine or guanosine 3':5'-cyclic monophosphate are shown in Figure 5. The response to adenosine 5'-monophosphate is also shown. CFU-S responds to the dibutyryl derivatives of the cyclic monophosphates with increased sensitivity to H^3-TdR. Adenosine-5'-monophosphate (5'-AMP) was not active on CFU-S. The concentration-response curve to dibutyryl cyclic-AMP is bell-shaped giving a peak effect at 10^{-8} *M* but no effect at 10^{-6} *M*.

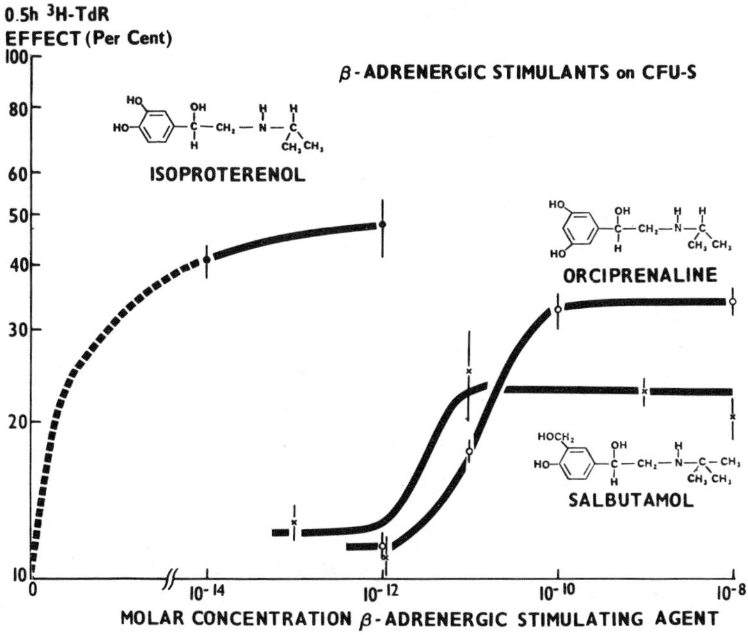

FIGURE 2. *Percent loss CFU-S due to H^3-thymidine (± SEM) in cultures treated with β-adrenergic stimulating agents. The data for orciprenaline and salbutamol are compared with that for isoproterenol (Fig. 1). Each point is made up from a minimum of 24-32 spleens.*

Figure 6 compares the responses of CFU-S to dibutyryl cyclic-AMP and 5'-AMP with their response to an inhibitor of cyclic nucleotide phosphodiesterase, 4-(3-butoxy-4-methoxybenzyl)-2-imidazolidinone; Ro 20-1724 (12). Given alone, Ro 20-1724 increases sensitivity to H^3-TdR.

Table 1 demonstrates that increased sensitivity to the cytocidal action of hydroxyurea is also induced by some of the agents tested with H^3-TdR. Hence carbamylcholine, isoproterenol, and dibutyryl cyclic AMP all increase the ability of hydroxyurea to inhibit spleen colony formation.

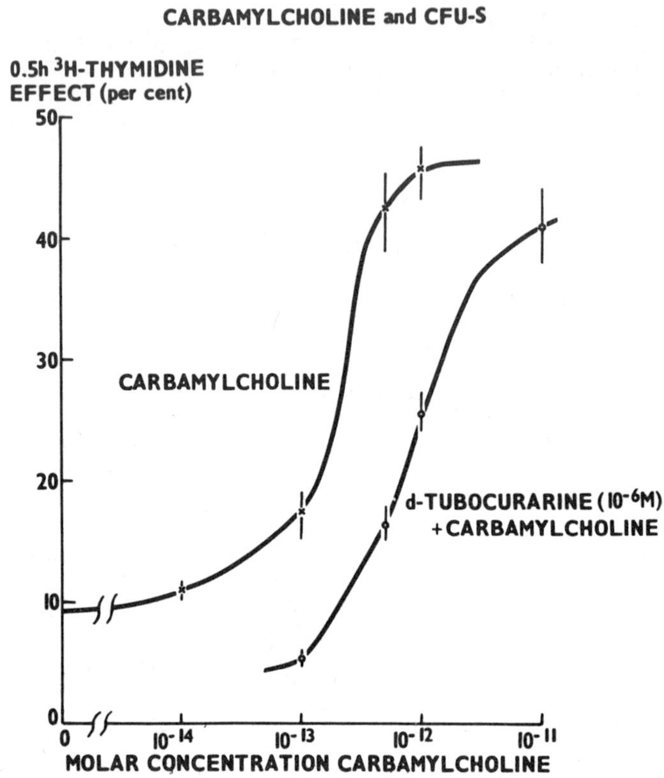

FIGURE 3. *Percent loss CFU-S due to H^3-thymidine (± SEM) in bone marrow cultures treated with carbamylcholine alone or in the presence of 10^{-6} M d-tubocurarine. d-Tubocuarine was added 5 min before carbamylcholine. Each point is made up from a minimum of 24-32 spleens. The point for 10^{-12} M carbamylcholine alone is from four experiments giving a total of 98 spleens.*

DISCUSSION

The studies described indicate the presence in bone marrow of drug receptor sites which, following chemical excitation, initiate DNA synthesis in the hematopoietic stem cell. The use of specific pharmacological agonists, suggests that β-adrenergic and cholinergic receptors may initiate the transition from G_0 to S-phase of the cell cycle. These sites act independently, however, to stimulate this cell cycle change. This is suggested since the β-adrenergic mechanism is not blocked by antagonists of cholinergic receptors and antagonists of β-adrenergic receptors do not block carbamycholine-induced DNA synthesis (13). From the studies comparing isoproterenol, orciprenaline, and salbutamol, it is suggested that the response to β-adrenergic agents is more β_1 than β_2 in character; salbutamol acts primarily at β_2-adrenergic receptor sites, while isoproterenol is active at both β_1 and β_2 receptor sites.

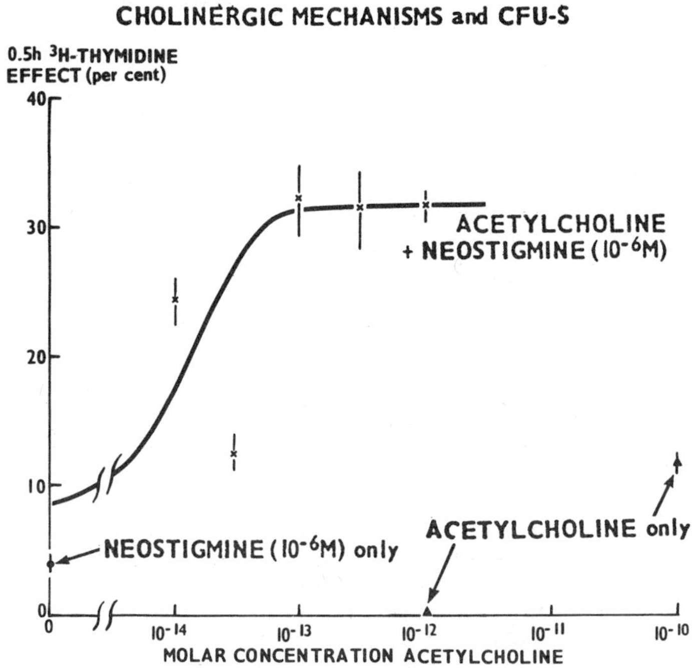

FIGURE 4. *Percent loss CFU-S due to H^3-thymidine (± SEM) in bone marrow cultures treated with acetylcholine, neostigmine, or with both neostigmine and acetylcholine. Each point is made up from a minimum of 24-32 spleens.*

A cyclic nucleotide may mediate in the cell-cycle actions of the β-adrenergic and cholinergic agonists. Cyclase systems could serve, therefore, to amplify the effects of small numbers of either β-adrenergic or cholinergic molecules on receptor sites. CFU-S respond to the dibutyryl derivative of 3':5'-cyclic monophosphate (cyclic-AMP) with increased sensitivity to H^3-TdR. Adenosine-5'-monophosphate (5'-AMP) is not active on CFU-S. The concentration-response curve to dibutyryl cyclic-AMP is bell-shaped giving a peak effect at 10^{-8} *M* but no effect at 10^{-6} *M*. Similarly, they respond to the dibutyryl derivative of guanosine 3':5'-cyclic monophosphage (cyclic-GMP) with increased sensitivity to the cytocidal effects of H^3-TdR. Cyclic-AMP might serve as a second messenger for the cell-cycle action of β-adrenergic agents, while cyclic-GMP might have a similar role following stimulation with cholinergic compounds.

Activation of phosphodiesterase with 4 *mM* imidazole suppressed the cell-cycle actions of both isoproterenol

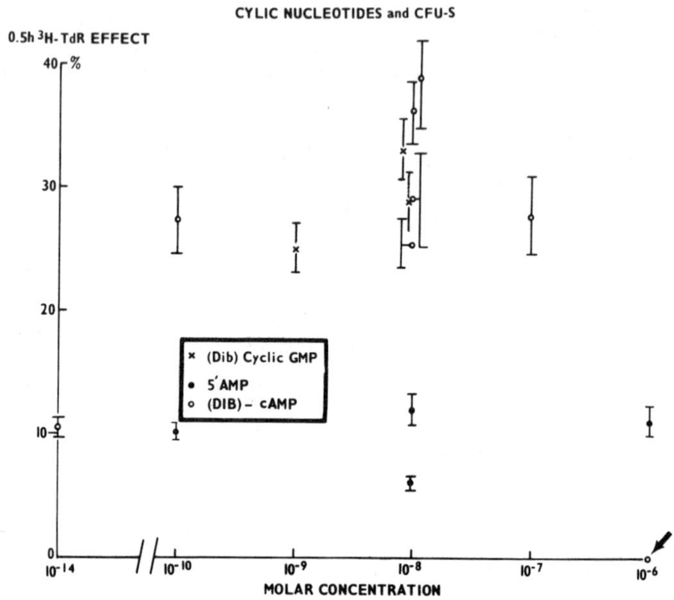

FIGURE 5. *Percent loss CFU-S due to H^3-thymidine (± SEM) in bone marrow cultures treated with either dibutyryl cyclic-AMP, dibutyryl cyclic-GMP or with adenosine-5'-monophosphate (5'-AMP). Each point is made up from a minimum of 24-32 spleens.*

and of dibutyryl c-AMP on CFU-S (10), and activation of phosphodiesterase with 4 *mM* imidazole suppressed the actions of both carbamylcholine and of dibutyryl cyclic-GMP (14). In contrast, phosphodiesterase activation does not suppress testosterone's ability to trigger CFU-S from G_0 into S-phase (10). Inhibition of phosphodiesterase initiates DNA synthesis in the hematopoietic stem cell. This finding suggests that CFU-S synthesize cyclic nucleotides, and that phosphodiesterase may act to maintain concentrations at sub-triggering levels in G_0 cells; hence it may be suggested that in the region of the G_1/S boundary, the enzyme phosphodiesterase may act to regulate the cell cycle state of the hemopoietic stem cell.

A question of importance is which, if any, of these mechanisms might serve, at the *physiological level*, to regulate the proliferation of the hemopoietic stem cell. The data presented only allow for speculation. These mechanisms must attract attention on the basis of the low concentrations of β-adrenergic and cholinergic agents

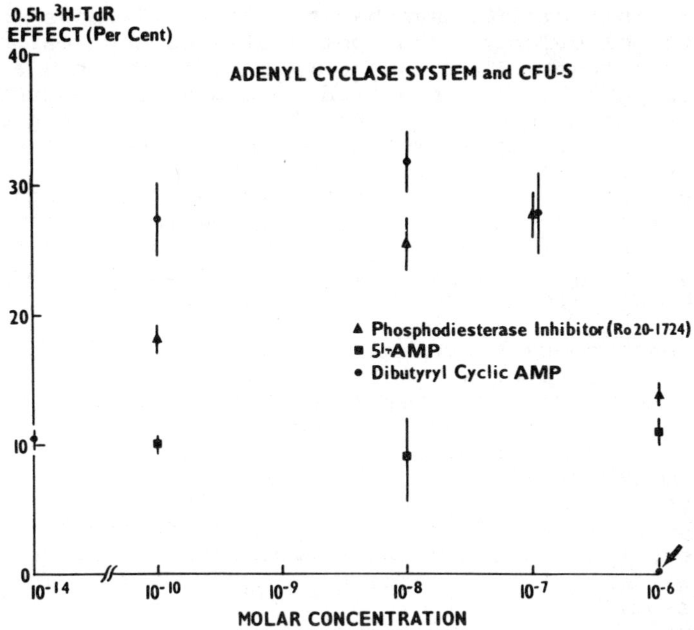

FIGURE 6. *Percent loss CFU-S due to H^3-thymidine (± SEM) in bone marrow cultures treated with dibutyryl cyclic-AMP, 5'-AMP or with Ro 20-1724. Each point is made up from a minimum of 24-32 spleens.*

TABLE 1. *10^{-3} M hydroxyurea (HU) on drug-induced DNA synthesis.*

	CFU-S/Spleen ± SEM		% Loss
Treatment	Without HU	HU	Due to HU
None	11.2 ±0.7 (13)	10.4 ±0.9 (13)	7.1
Carbamylcholine (10^{-12} *M*)	10.5 ±0.51 (16)	5.2 ±0.59 (16)	50.5
DL-Isoproterenol (10^{-14} *M*)	12.9 ±1.6 (12)	8.2 ±0.80 (10)	36.4
Dibutyryl cyclic-AMP (10^{-8} *M*)	11.1 ±0.5 (13)	7.5 ±0.5 (13)	32.4

( ) = Number of spleens counted.

needed to initiate DNA synthesis. Figure 7 shows the number of molecules of drug per nucleated bone marrow cell needed to trigger CFU-S into S. At 10^{-14} *M*, only a few molecules per cell are required. It may be, however, that CFU-S have an unusually strong attraction for isoproterenol or for cholinergic molecules and can concentrate molecules at these receptors.

The receptor sites, or possibly enzymes along the chain of reaction between receptors and initiation of DNA synthesis, may be points for the physiological triggering of DNA synthesis. As previously mentioned, the enzyme phosphodiesterase could act as a regulator of cell cycle state.

An endogenous source of agents capable of stimulating or blocking receptor sites must be located in order to consider the receptor sites as part of a regulatory mechanism. Nerve fibers, both myelinated and unmyelinated pass through the bone marrow and terminate in close proximity to bone marrow cells (15). This is no proof that neurotransmitters from these nerve endings influence or regulate CFU-S. It would, however, seem of interest to determine whether the stimulation of neurons leading to bone marrow leads to the transition of CFU-S from G_o into S. Future studies might also consider whether cells, other than nerve fibers, might serve as a source of compounds capable of acting on either adrenergic or cholinergic receptor sites; i.e., cell-cell communication via drug receptors. Given *in vitro*, the cholinesterase inhibitor neostigmine did not trigger CFU-S into DNA synthesis (Figure 4). This experiment would indicate that bone marrow cells, per se, may not serve as a source of agents capable of stimulating cholinergic receptors. Neostigmine given *in vivo* to mice very markedly increases the sensitivity of CFU-S to the cytocidal action of hydroxyurea (16). This effect has been demonstrated in two F_1 hybrids ($B6AF_1/J$; $C3D2F_1/J$). These findings suggest that endogenous acetylcholine may act, in some way, to eventually trigger CFU-S into DNA synthesis.

Receptor sites studied thus far serve to switch-on DNA synthesis in the hemopoietic stem cell. The possibility exists that receptor sites may be present which serve to switch-on DNA synthesis; i.e., receptors which return CFU-S to G_o. Future studies should, therefore, involve a search for such receptor sites. Preliminary studies to determine whether α-adrenergic receptor sites might serve to switch-off DNA synthesis in CFU-S have been negative.

To implicate the enzyme phosphodiesterase as a regulator of cell cycle state, requires the uncovering of endogenous mechanisms that could alter the activity of this enzyme. An endogenous inhibitor of phosphodiesterase, isolatable from rat adipocytes, has now been reported (17). The phosphodiesterase enzymes vary from tissue to tissue. This variation may provide the very basis for tissue specific inhibition or stimulation of cell proliferation. In time, when the much talked about tissue specific 'chalones' have been adequately characterized, it may be fruitful to study their effects on the activity of an enzyme such as phosphodiesterase.

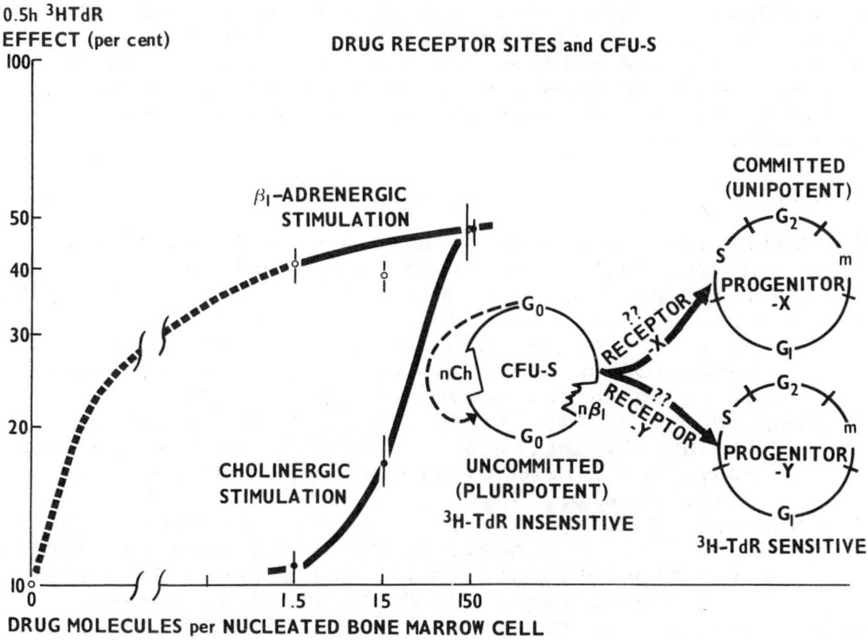

FIGURE 7. *The molar concentrations of isoproterenol and carbamylcholine are translated into molecules of drug per nucleated bone-marrow cell. For reference,* 10^{-14} *M is equal to 1.5 molecules per nucleated bone-marrow cell. CFU-S are pictured as having cholinergic receptors (nCh) and* β_1*-adrenergic receptors (n*β_1*) on their surfaces. The different shapes of the receptor sites represent their independence with respect to triggering CFU-S into S, and also serve to raise the question as to whether a particular receptor might determine the direction in which CFU-S differentiates.*

CFU-S may differentiate into the various committed bone marrow progenitor cells. The mechanism by which this occurs is not clear. Figure 7 also raises the question, for future study, as to whether receptor sites through a particular cyclic nucleotide, might influence the direction in which the stem cell differentiates. On the other hand, the presence of these two mechanisms in bone marrow may be speculated to represent a safety measure against the loss by the stem cell of its replication and/or differentiation characteristics.

ACKNOWLEDGEMENTS

Some of the experimental findings reported in this study were obtained when the author was a staff member at the Paterson Laboratories, Christie Hospital and Holt Radium Institute, Manchester, U.K. To -20-1724 was a donation from the Hoffman-LaRoche Co., U.S.A. The *in vivo* studies with neostigmine referred to (16) were supported by Grant No. CA 16147 from the National Cancer Institute, DHEW

REFERENCES

1. Till, J. E. and McCulloch,E. A. 1961. A direct measurement of the radiation sensitivity of normal mouse bone marrow cells. Radiat. Res. *14*:213.
2. Becker, A. J., McCulloch, E. A., and Till, J. E. 1962. Cytocidal demonstration of the clonal nature of spleen colonies derived from transplanted mouse marrow cells. Nature *197*:452.
3. Fowler, J. H., Wu, A. M., Till, N. E., McCulloch, E. A., and Siminovitch, L. 1967. The cellular composition of hemopoietic spleen colonies. J. Cell. Physiol. *69*:65.
4. Sutherland, D. J. A., Till, J. E., and McCulloch, E. A. 1970. A kinetic study of the genetic control of hemopoietic progenitor cells assayed in culture and *in vivo*. J. Cell. Physiol. *75*:267.
5. Becker, A. J., McCulloch, E. A., Siminovitch, L., and Till, J. E. 1965. The effect of differing demands for blood cell production on DNA synthesis by hemopoietic colony-forming cells of mice. Blood *26*:296.
6. Byron, J. W. 1972. Comparison of the action of ^{3}H-thymidine and hydroxyurea on testosterone-treated haemopoietic stem cells. Blood *40*:2.
7. Gurney, C. W. 1963. Effect of radiation on the mouse stem cell compartment *in vivo*. Persp. Biol. Med. *6*:233.
8. Axelrad, A. A., McLeod, D. L., Schreeve, M. M., and Heath, D. S. 1974. Properties of cells that produce erythrocytic colonies *in vitro*. *Hemopoiesis in Culture*, W. A. Robinson (ed.), p. 226. DHEW Publication No. (NIH) 74-205.
9. Bradley, T. R. and Metcalf, D. 1966. The growth of mouse bone marrow cells *in vitro*. Aus. J. Exp. Biol. Med. *44*:287.
10. Byron, J. W. 1972. Evidence for a β-adrenergic receptor initiating DNA synthesis in haemopoietic stem cells. Exp. Cell Res. *71*:228.
11. Byron, J. W. and Punchard, J. K. *In vitro* method for studying agents that trigger CFUs from G_O into S. In: *Hemopoiesis in Culture*, Second International Workshop, Airlie House, Virginia, DHEW Publication No. (NIH) 74-205, p. 436. Robinson, W. A. (ed.), 1974.

12. Sheppard, H. and Wiggan, G. 1971. Different sensitivities of the phosphodiesterases (adenosine-3', 5'-cyclic phosphate 3'-phosphohydrolase) of dog and cerebral cortex and erythrocytes to inhibition by synthetic agents and cold. Biochem. Pharmacol. *20*:2128.

13. Byron, J. W. Molecular basis for triggering haemopoietic stem cells into DNA synthesis. In: *Hemopoiesis in Culture*, DHEW Publication No. (NIH) 74-205, p. 91. Robinson, W. A. (ed.), 1974.

14. Byron, J. W. 1973. Drug receptors and the haemopoietic stem cell. Nature (New Biol.), London *241*:152.

15. Calvo, W. 1968. The innervation of the bone marrow in laboratory animals. Amer. J. Anat. *123*:315.

16. Byron, J. W. Cholinergic mechanisms and the hemopoietic stem cell. In: *Erythropoiesis*. Univ. Tokyo Press, (Ed. Nakao, Fisher, and Takaku), 1975.

17. Ho, R. J., Russell, T. R., Asakawa, T., and Hucks, M. W. 1975. Inhibition of cyclic nucleotide phosphodiesterase activity by an endogenous factor. J. Cyclic Nucleotide Res. *1*:81.

Part III

CYCLIC NUCLEOTIDES AND THE REGULATION OF CELL PROLIFERATION

WHITFIELD et al.
The Positive Regulation of Cell Proliferation by a Calcium-Cyclic AMP Control Couplet

KIEFER and KANTOR
Cell Surfaces, Cyclic Nucleotides and Unrestrained Cell Division: A Natural Role Is Proposed for the Cell Surface Receptor of Vibrio cholerae *and* Escherichia coli *Enterotoxins*

RASKA
Cyclic AMP Metabolism in BHK21 Cells Infected or Transformed with Adenovirus Type 12

The Positive Regulation of Cell Proliferation by a Calcium-Cyclic AMP Control Couplet

J. F. Whitfield, A. L. Boynton,
J. P. Mac Manus, R. H. Rixon,
P. R. Walker, and U. Armato

The continuous cell proliferation needed to maintain the structure and function of some organs and tissues is restricted to certain regions such as the epidermal-dermal junction of skin and the epithelial-reticular framework of the thymic cortex. If, and when, they are forced out of these proliferative zones by population pressure, the proliferatively activated precursor cells sooner or later stop "cycling," switch off their proliferogenic cistrons (i.e., proliferatively deactivate), and initiate entirely different programs which convert them into functionally and morphologically distinctive entities such as erythrocytes, goblet cells, small lymphocytes, or squamous cells and keratinous squames. By contrast, the proliferogenic cistrons (which direct the synthesis, assembly and operation of the proliferative machinery) in normal cells of other organs do not function unless activated by cell loss (e.g., liver) or appropriate hormones (e.g., adrenal cortex, liver, uterus).

Proliferative development of every *normal* cell is started by *activators* and then guided by briefly accumulated *intracycle regulators* which "cue in" different phases of the proliferative program. The coordination of growth between the components of entire organ systems demands that cell proliferation also respond to endogenous regulatory molecules generated by the interaction of exogenous hormones with appropriate receptors on the cell surface (e.g., various polypeptide hormones) or in the cytosol (e.g., steroid hormones) (49,68,69). Thus, calcium and cyclic AMP, the mediators of such diverse processes as glandular secretions, many hormone actions, muscle contraction, and neurotransmitter release and other secretory processes (32,45,47,68,70), may also control cell proliferation.

The available evidence certainly indicates that calcium ions and cyclic AMP are proliferative regulators (17, 18,20,64,74,82,83,88,89). However, the widely held belief that cyclic AMP is an inhibitory regulator (64,74) is almost certainly incorrect, because it is founded on a strikingly nonphysiological type of experiment in which asynchronously proliferating heteroploid cells of established lines (e.g., the conditionally tumorigenic *BALB/3T3* fetal mouse endothelial cells) are exposed continuously to very high concentrations of dibutyryl cyclic AMP in media containing heterologous (e.g., fetal bovine) serum and supranormal (about 1.8 times the level in normal blood plasma) calcium activities. Unfortunately, a very important principle seems to have been overlooked by most workers in this field: any agent, a continuous exposure to which stops proliferation, might actually be a key *positive* regulator that under normal physiological conditions appears only briefly in the cell to trigger a specific phase of proliferative development without blocking other phases (88). Such misconceptions are best avoided by studying systems such as the nearly synchronous proliferation of hepatocytes in the incontestably physiological environment of the partially resected or hormonally stimulated liver.

Proliferative activation in the liver tissue remaining after partial resection triggers a progressive increase in the basal activity of the plasma membrane-bound adenylate cyclase which catalyzes the synthesis of cyclic AMP from ATP (Figure 1) (53,69). Activation must also somehow alter a parathyroid hormone/thyroxine-sensitive adenylate cyclase receptor, because there is a very striking transient increase in the enzyme's sensitivity to thyroxine and a transient, but less striking, increase in its responsiveness to low levels of parathyroid hormone, both of which peak around 4 h after surgery (Figure 1) (53).

Proliferative activation either by partial resection or hormone infusion (*TAGH* mixture: L-triiodothyronine, amino acids, glucagon, heparin (77), is also followed by two waves of cyclic AMP accumulation (Figure 2, see also Figure 11), the second of which crests immediately before the beginning of deoxyribonucleotide synthesis (Figures 2 and 3A) (33,55,56,78,85,88,89). It is tempting to attribute the first cyclic AMP surge to adenylate cyclase's briefly heightened sensitivity to ambient parathyroid and thyroid hormones. However, this is probably not the case, because undiminished cyclic AMP surges also occur in the partially resected livers of thyroparathyroidectomized rats (see Figure 16). The second cyclic AMP surge, on the other hand, is more likely due to the combination of an increased basal adenylate cyclase activity with a transient reduction of the cyclic nucleotide's degradation by cyclic AMP-phosphodiesterase (33).

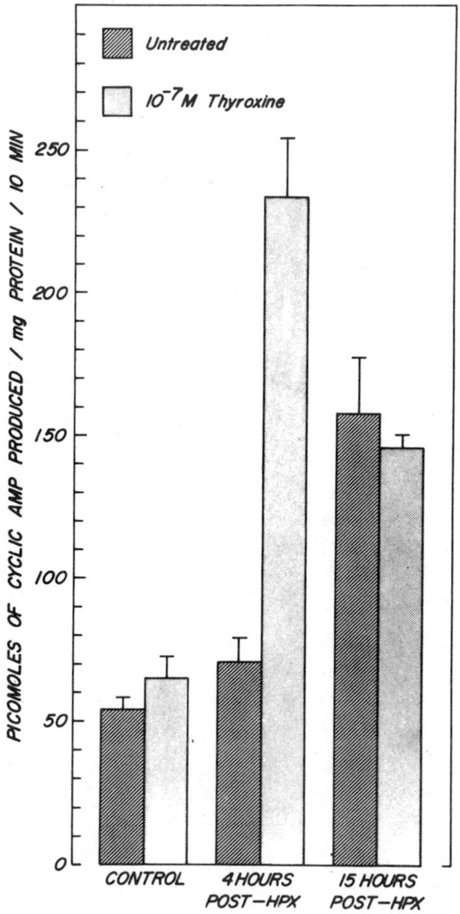

FIGURE 1. *The post-partial hepatectomy (post-HPX) increase in the basal activity of rat liver adenylate cyclase, and the transient, but yet striking, increase in the enzyme's responsiveness to thyroxine. Liver plasma membranes were isolated from homogenates of the liver tissue remaining after partial (70%) hepatectomy (HPX) of 200 g Wistar rats. The amounts of cyclic AMP formed when added ATP by the adenylate cyclase in such cell-free preparations was measured by the protein binding procedure of Gilman (38). The data (means ± SEM of values from three to four livers) are graphic syntheses of some of the entries in Table I of Leoni et al. (53).*

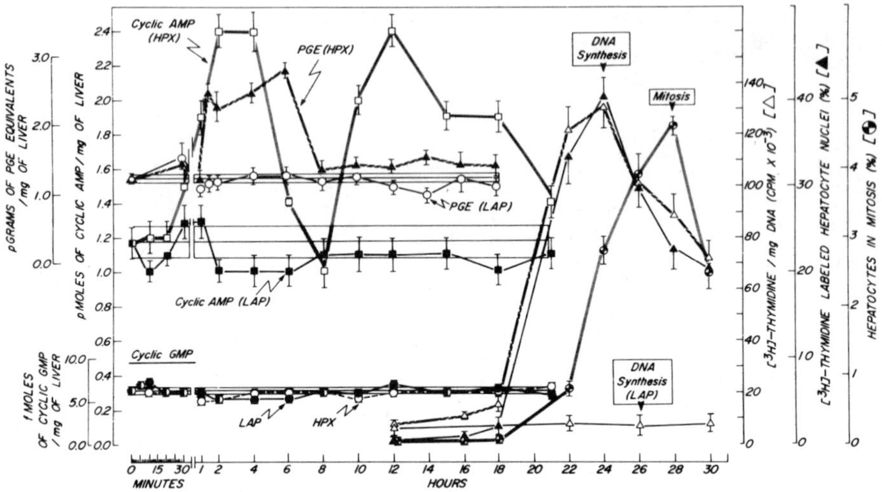

FIGURE 2. *The generation of two prereplicative waves of cyclic AMP accumulation, one prereplicative wave of accumulation of prostaglandin E (PGE) equivalents without any accompanying changes in cyclic GMP levels in (200 g Sprague-Dawley) rat hepatocytes proliferatively activated at time 0 by partial hepatectomy. The cyclic AMP was measured by protein-binding assay (38), and cyclic GMP by radioimmunoassay (87). Prostaglandin E equivalents were measured after silicic chromatography by radioimmunoassay (46) using antisera generously provided by Dr. H. Behrman (Merck, Sharpe and Dohme, Rahway, N.J.). DNA synthesis was measured autoradiographically according to Rixon and Whitfield (67) and MacManus et al. (55,56). Mitotic activity is expressed as the total percentage of cells in prophase, metaphase, anaphase and telophase. The shaded areas represent levels in normal animals. Abbreviations are: HPX, partial (70%) hepatectomy; LAP, laparotomy or sham hepatectomy. The points and shaded areas are means ± SEM of values from 8 to 24 livers.*

The activated hepatocyte's behavior also severely challenges the widely publicized belief that cyclic GMP is a universal proliferative activator. Thus, neither the *total* guanylate cyclase activity (48) nor the liver remnant's *total* cyclic GMP content (Figure 2) increases during the prereplicative period. However, according to Kimura and Murad (48) there is a decline in soluble guanylate cyclase activity which is offset by increased particulate (i.e., membrane-associated) activity during the first 12 h after partial resection. On the other hand, Goridis and Reutter (41) have observed an extremely large increase in membrane-associated guanylate cyclase activity, which appears between 16 and 24 h after surgery and persists for three days. If such an increase occurs under our conditions, it is certainly not associated with increased cyclic GMP levels in the whole liver remnant

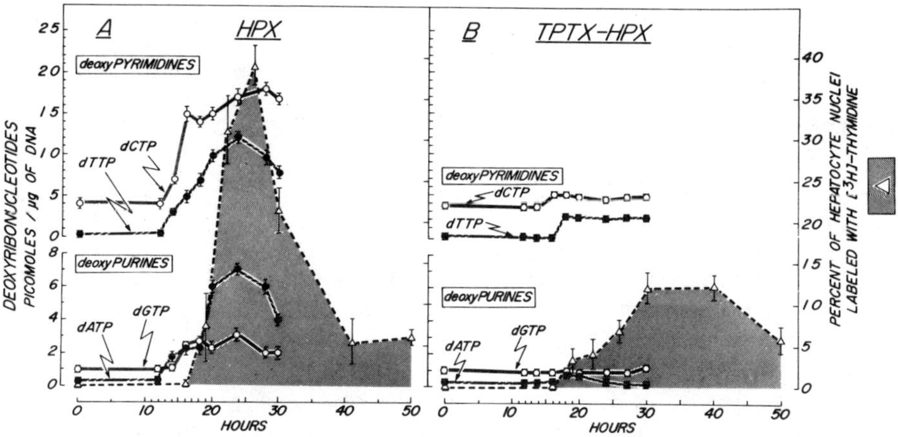

FIGURE 3. *The inability of activated hepatocytes in hypocalcemic, thyroparathyroidectomized rats to initiate syntheses of the four deoxyribonucleotide syntheses. A illustrates the initiation of deoxyribonucleotide syntheses in partially (70%) resected livers of normal rats. B illustrates the much reduced initiation of deoxyribonucleotide syntheses in rats whose thyroid-parathyroid complexes had been removed 72 h before HPX. The levels of deoxyribonucleotides were measured by the methods of Lindberg and Skoof (54) and Skoog (79) dCTP, deoxycytidine triphosphate; dGTP, deoxyguanosine triphosphate; dTTP, deoxythymidine triphosphate; TPTX-HPX, thyro-arathyroidectomized and partially hepatectomized 72 h later. The points are the means ± SEM of the values from 9 to 15 livers.*

(Figure 2). It is interesting to note that we will describe below another case in which the capacity of liver remnant homogenates to synthesize particular molecules (i.e., prostaglandin F equivalents) from exogenous precursor increases without a corresponding accumulation of these molecules in the intact remnant. However, interesting though it may be, such a delayed increase in the cyclic GMP-synthesizing potential is almost certainly too late to influence proliferative activation or significantly to affect the prereplicative reaction sequence.

One or both of the cyclic AMP surges may either participate in, or merely be side-effects of, prereplicative development. Elimination of the first surge with *dl*-propranolol and other adrenergic antagonists affects neither the second surge nor the initiation of DNA synthesis (55, 85). By contrast, delaying or reducing the second surge with *dl*-propranolol, phenoxybenzamine or phentolamine (which do not inhibit on-going DNA synthesis), delays or prevents the initiation of DNA under the conditions of MacManus *et al.* (55), but stopping this surge with pindolol, another adrenergic blocker, does *not* affect the initiation of DNA synthesis under the conditions of Thrower and Ord (85). However, we must put in a caveat that all of these antagonists probably have such a variety of effects that conclusions based on their gross actions in the whole animal must be treated with great caution. For example, pindolol's blockage of the second cyclic AMP surge without affecting DNA synthesis might be due to the antagonist *replacing* the inhibited cyclic AMP surge by itself stimulating transmembrane fluxes, or intracellular translocations of calcium ions, the "ultimate" triggers of DNA synthesis. Despite these many uncertainties, it seems that the first cyclic AMP surge is proliferatively dispensable, while the second surge may be involved in the initiation of DNA synthesis.

At first sight, the capability of cyclic AMP, but not cyclic GMP, serving as an initiator of DNA synthesis seems to be established by the demonstration of its ability to replace, and cyclic GMP's inability to replace, the (cyclic AMP-elevating) glucagon component of the *proliferatively activating TAGH* mixture (Figure 4) (78). However appealing this may be, it is probably a spurious indication of cyclic AMP's real function simply because endogenous cyclic AMP, unlike exogenous dibutyryl cyclic AMP, is not involved in proliferative activation.

The feasibility of cyclic AMP being an initiator of DNA synthesis (rather than a proliferative activator) is much more convincingly established by the ability of exogenous cyclic AMP (as well as its dibutyryl derivative) to stimulate DNA synthesis in cultures of *already proliferative activated* hepatocytes from chick embryos and neonatal rats (1-3,5,60). For example, exposure to a wide range of concentrations of exogenous cyclic AMP or dibutyryl cyclic AMP stimulates the flow of cultivated neonatal rat hepatocytes into the DNA-synthetic and mitotic phases of their growth-division cycles (Figures 5 and 6) after a mean delay of 12 to 14 h (Figure 7). By contrast, cyclic GMP by itself is much less stimulatory than cyclic AMP (Figures 5 and 6). However, cyclic AMP and cyclic GMP can act synergistically. Thus, extremely low concentrations of these nucleotides, which are individually ineffective, significantly stimulate hepatocyte proliferation when combined in equimolar proportions (Figure 8).

Cyclic AMP might induce the diamine and polyamine syntheses which accompany proliferative development of hepatocytes and many other types of cells (8,27,71,72,80) and are consequently candidate mediators of the cyclic nucleotide's proliferogenic action. Thus, both exogenous dibutyryl cyclic AMP and elevators of the endogenous cyclic AMP level, such as adenylate cyclase stimulators and cyclic AMP-phosphodiesterase inhibitors (e.g., adrenocorticotrophic hormone, glucagon, methylxanthines), stimulate synthesis of the short-lived (10 to 20 min half-life) enzyme, ornithine decarboxylase, which initiates the chain of polyamine syntheses by catalyzing putrescine formation (8,10,23,36,58,65,73,78). Moreover, two bursts of ornithine decarboxylase production, which seem to coincide with the two cyclic AMP surges, have been observed by some

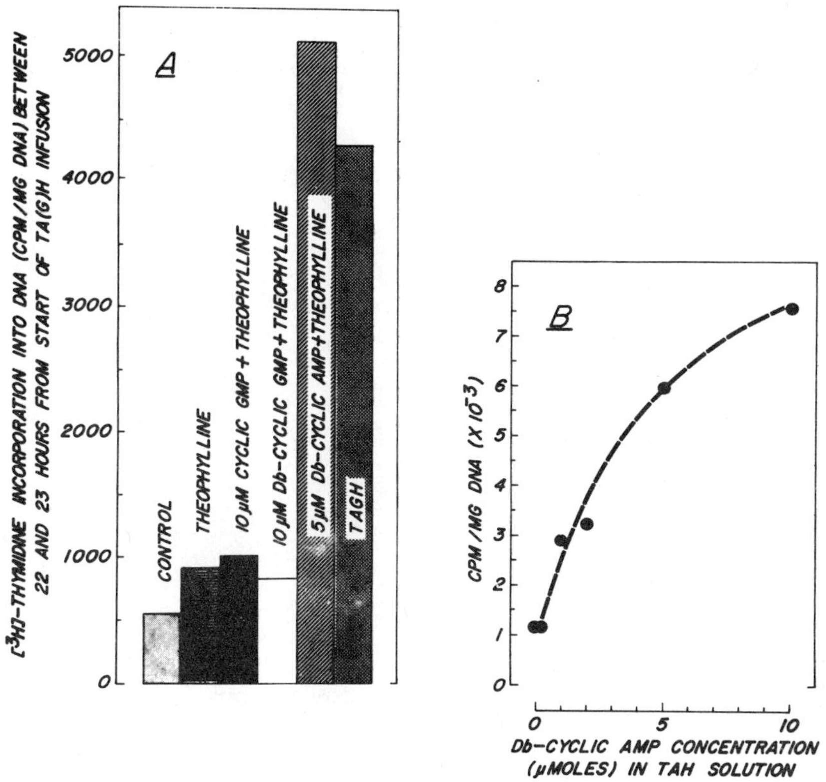

FIGURE 4. *The demonstration by Short et al. (78) of the ability of dibutyryl cyclic AMP, but not cyclic GMP, to replace the cyclic AMP-elevator, glucagon, in the TAGH (*L-*triodothyronine, amino acids, glucagon, heparin) mixture, infusion of which proliferatively activates part of the hepatocyte population in the* <u>*intact*</u> *liver of the female rat.* <u>*A*</u>. *The comparative effects of substitution of Db-cyclic AMP and cyclic GMP for glucagon. Ten ml of the appropriate mixture was infused for 3 h (at a rate of 3.3 ml/h) into the tail vein of female Fisher 344 rats weighing between 140 and 150 g.* <u>*B*</u>. *The effects of the substitution of various amounts of Db-cyclic AMP for the glucagon component of the TAGH mixture on the incorporation of [^{3}H]-thymidine into liver DNA between 22 and 23 h after beginning of mixture infusion. The incorporation of [^{3}H]-thymidine into DNA of the intact, mixture-treated livers was measured according to Short et al. (78). Db-cyclic AMP, N^6,$O^{2'}$-dibutyryl cyclic AMP; TAH, solution containing* L-*triiodothyronine, amino acids and heparin. These illustrations are graphic syntheses of tabular entries in Short et al. (78).*

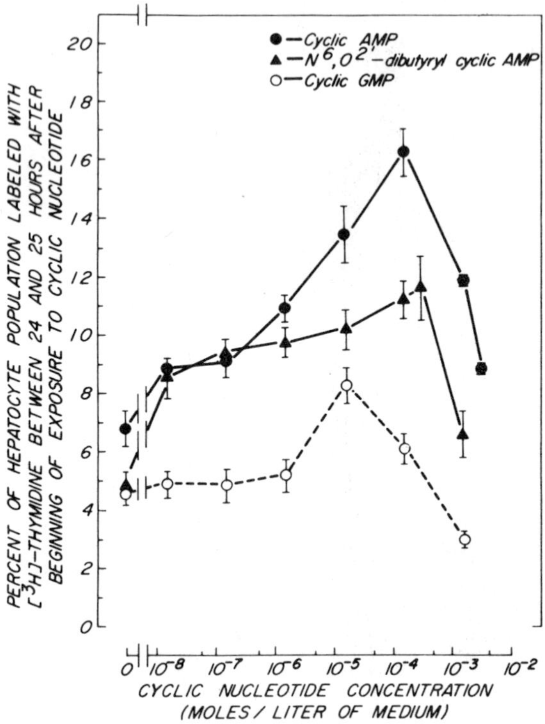

FIGURE 5. *The effects of various concentrations of cyclic AMP, dibutyryl cyclic AMP and cyclic GMP on the flow of cultivated hepatocytes into the DNA-synthetic phase of their growth-division cycle. Livers were removed from young Wistar rats and the hepatocytes cultivated according to Armato et al. The cells were cultivated for 4 days, at which time they were exposed to cyclic nucleotide. Twenty-four hours later the cells were exposed for 1 h to [^{3}H]-thymidine (0.5 μCi/ml of culture medium; specific activity 6.7 Ci/mmole), and fixed and prepared to autoradiography according to Armato et al. (3,5). The points are means ± SEM of values from 6 to 21 cultures.*

(36), but not by others (33). However, some doubt is cast upon a linkage between the cyclic AMP surges and ornithine decarboxylase synthesis by the observation of Thrower and Ord (85) that under their conditions this synthesis is unaffected by reduction of the cyclic AMP surges with adrenergic antagonists.

With all due respect to the dissenting voices of Ebina *et al.* (33) and Thrower and Ord (85), most of the available evidence still suggests that the proliferatively dispensable (55) first cyclic AMP surge is responsible for the large wave of putrescine accumulation in the liver remnant, which peaks between 8 and 10 h after surgery (Figure 9A). As expected, the remnant's spermidine con-

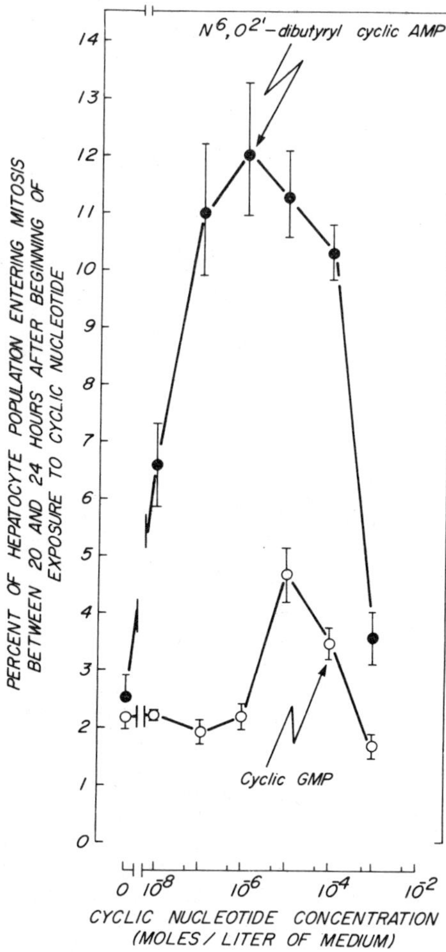

FIGURE 6. *The effects of various concentrations of cyclic GMP and dibutyryl cyclic AMP on the flow of cultivated hepatocytes into the mitotic phase of their growth-division cycle. Livers were removed from young Wistar rats and cultivated according to Armato et al. (3,5). After 4 days of cultivation, the cells were exposed to cyclic GMP or dibutyryl cyclic AMP. After 20 h of exposure to cyclic nucleotide, the medium was replaced with a cyclic nucleotide-free medium containing 0.1 mM colchicine to stop the progress of cells through mitosis at a point between prophase and metaphase (the so-called colchicine-"metaphase"). Four hours later, the cells were fixed in glacial acetic acid-ethanol (1:9, v/v), stained according to the Feulgen method for DNA, and the proportion of cells which had entered mitosis and collected at the colchicine-sensitive mitotic phase was determined. The points are the means ± SEM of the values from 6 to 12 cultures.*

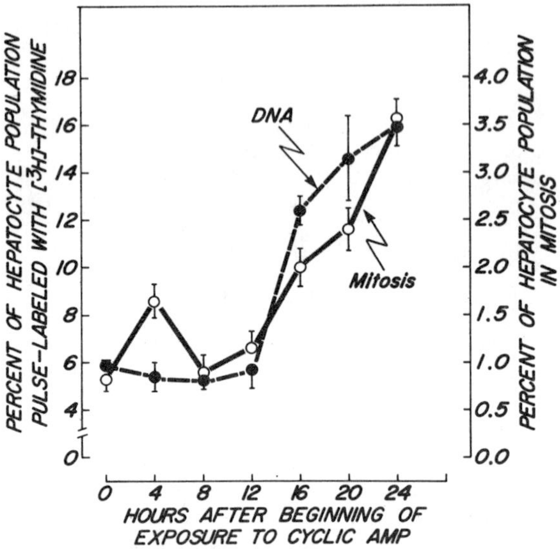

FIGURE 7. *The stimulation by exogenous cyclic AMP (1.5 × 10^{-4} M) of the flow of cultivated hepatocytes into the DNA-synthetic and mitotic phases of their growth-division cycle. Livers were removed from young Wistar rats and cultivated according to Armato et al. (3,5). After 4 days of cultivation, such primary cultures were exposed to cyclic AMP (not dibutyryl cyclic AMP) and the proportion of cells making DNA at various times thereafter was determined by exposing cultures to [^{3}H]-thymidine (0.5 μCi/ml of medium; specific activity, 6.7 Ci/mmole) for 1 h. The labeled cultures were then prepared for autoradiography according to Armato et al. (3). Mitotic activity is the sum of the proportions of cells which were in prophase, metaphase, anaphase or telophase at the time of exposure of the cultures to [^{3}H]- thymidine. The points are the means ± SEM of the values from four cultures.*

tent nearly triples during the first 8 h after surgery, and then remains at this elevated level for at least the next 22 h (Figure 9A). On the other hand, the remnant's spermine content remains normal for the first 22 h after surgery, and then declines as the hepatocytes start flowing into mitosis (Figure 9A).

It is probably premature to assess the candidacy of the polyamines for the role of mediators of cyclic AMP action, because it is not at all certain that they are specifically needed for prereplicative development of any cell type. Moreover, ornithine decarboxylase's presumptive inducer, the first cyclic AMP surge, can be eliminated without adversely affecting prereplicative development. On the other hand, ornithine decarboxylase should be expected to contribute to the hepatocyte's prereplicative development through its ability to stimulate RNA polymerase I (58). Furthermore, there is some evidence

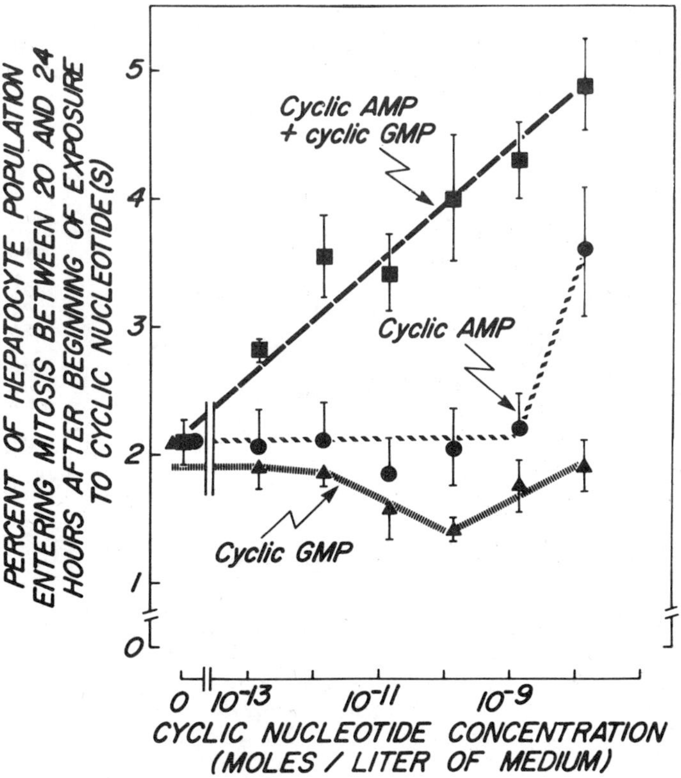

FIGURE 8. *The stimulation of the flow of cultivated hepatocytes into mitosis by equimolar combinations of very low cyclic AMP and cyclic GMP concentrations which are separately ineffective. Livers were removed from young Wistar rats and hepatocytes cultivated according to Armato et al. (3,5). After 4 days of cultivation, the cultures were exposed to cyclic AMP, cyclic GMP or an equimolar mixture of these nucleotides. After 20 h of exposure to the nucleotide(s), the medium was replaced with a cyclic nucleotide-free medium containing 0.1 mM colchicine to stop the progress of cells through mitosis at a point between prophase and metaphase (the so-called colchicine-"metaphase"). Four hours later, the cells were fixed in glacial acetic acid-ethanol (1:9, v/v), stained according to the Feulgen method for DNA, and the proportion of cells which had entered mitosis and collected at the colchicine-sensitive mitotic phase was determined. The points are the means ± SEM of the values from six cultures.*

that polyamines are needed for both the prereplicative development and DNA synthesis of cells such as *BALB/3T3* fetal mouse endothelial cells and WI-38 fetal human lung cells. Thus, exposure of these cells to *MGBG* (methylglyoxal-*bis*[guanyl hydrazone]), a multipotent spermidine analogue that inhibits polyamine synthesis, function and transmembrane transport (26,34,61.90), causes a blockage of the initiation of DNA synthesis which can be at least partly overcome by a brief exposure to exogenous polyamines (Figure 10) (19).

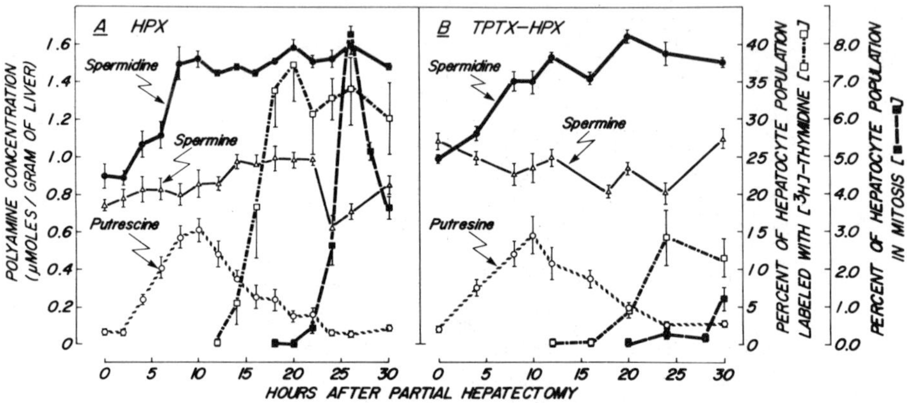

FIGURE 9. *The effects of HPX on the levels of the diamine putrescine and the polyamines spermidine and spermine in the remaining liver tissue in (A) normal and (B) hypocalcemic rats (with plasma calcium levels less than 7 mg/100 ml) which had been thyro-parathyroidectomized (TPTX) 72 h before the partial resection. The amine levels were determined according to the fluorometric dansyl-chloride (1-dimethylamino-naphthalene-5-sulfonyl chloride) method of Seiler and Weichmann (76). The proportions of hepatocytes making DNA were determined according to Rixon and Whitfield (66,67). Mitotic activity is expressed as the total percentage of cells in prophase, metaphase, anaphase and telophase. The points are the means ± SEM of the values from nine liver remnants.*

Prostaglandins are other candidates for the role of mediator of cyclic AMP's proliferative action(s). Partial hepatectomy prolongedly (for at least 12 h) increases the capacity of the remnant homogenate to synthesize prostaglandins E (PGE) and F (PGF) from arachidonate precursor. Despite this persistently elevated synthetic capacity of cell-free preparations, there is no accumulation of PGF equivalents, and only a transient accumulation of PGE equivalents in the whole liver remnant which peaks about the same time as the proliferatively dispensable first cyclic AMP surge and subsides between 6 and 8 h after surgery (Figure 2). An injection of indomethacin before HPX eliminates the PGE surge and blocks the initiation of DNA synthesis without affecting the cyclic AMP surges (Figures 11 and 12).

Indomethacin may stop the initiation of DNA synthesis either by itself, or indirectly via inhibition of prostaglandin synthesis. At first glance, the latter alterna-

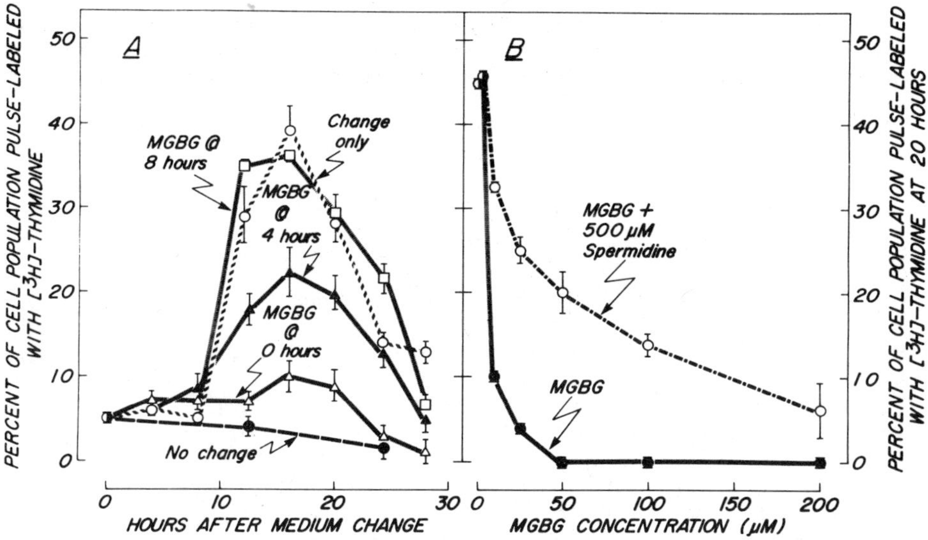

FIGURE 10. *Autoradiographic demonstrations of the inhibition of the initiation of DNA synthesis by the polyamine analogue MGBG (methylglyoxal-bis[guanyl hydrazone]), and the reversibility of this blockage by a brief exposure to exogenous spermidine.* A. *BALB/3T3 fetal mouse endothelial (11) cells were seeded (at a density of 0.4 × 10^4/cm^2) on coverslips in medium consisting of 90% Dulbecco's modified Eagle's medium plus 10% (v/v) bovine calf serum which was replenished every two days thereafter until the cultures had become confluent and the majority of the cells had stopped "cycling." Two days later, the medium was again replenished to stimulate the cells to initiate one growth-division cycle. Twenty μl of a 10 mM solution of MGBG (in phosphate-buffered saline) was added to give a final concentration of 100 μM at the indicated times after the stimulating medium change. The cells were labeled at the appropriate times by a 1 h exposure to [^{3}H]-thymidine (10 μCi/ml of medium; specific activity, 20 Ci/mmole).* B. *BALB/3T3 cells were seeded at a density of 1.0 × 10^4/cm^2 on coverslips in 2.0 ml of complete medium. The cultures were grown to confluence without feeding and then incubated an additional two days (also without feeding). The proliferatively quiescent cells were induced to start a single growth-division cycle by adding 0.5 ml of fresh bovine calf serum to the pre-existing 2.0 ml of complete medium. The various amounts of MGBG (dissolved in 20 μl of phosphate-buffered saline) were added at the time of stimulation (i.e., by addition of serum). Two hours later, the complete MGBG-containing medium was removed (and then stored at 37°C) and replaced by 2.0 ml of a solution of 500 μM spermidine in serum- and MGBG-free medium. Control cultures were exposed to the same serum-free medium without spermidine. One hour later, the spermidine-medium was discarded and replaced with the original MGBG-containing*

complete medium. This complicated procedure involving interrupted exposure to MGBG was needed because of the polyamine oxidases in bovine serum and the ability of MGBG competitively to inhibit the carrier-mediated transport of spermidine into the cell (26). The points in both A and B are the means ± SEM of the values from four cultures.

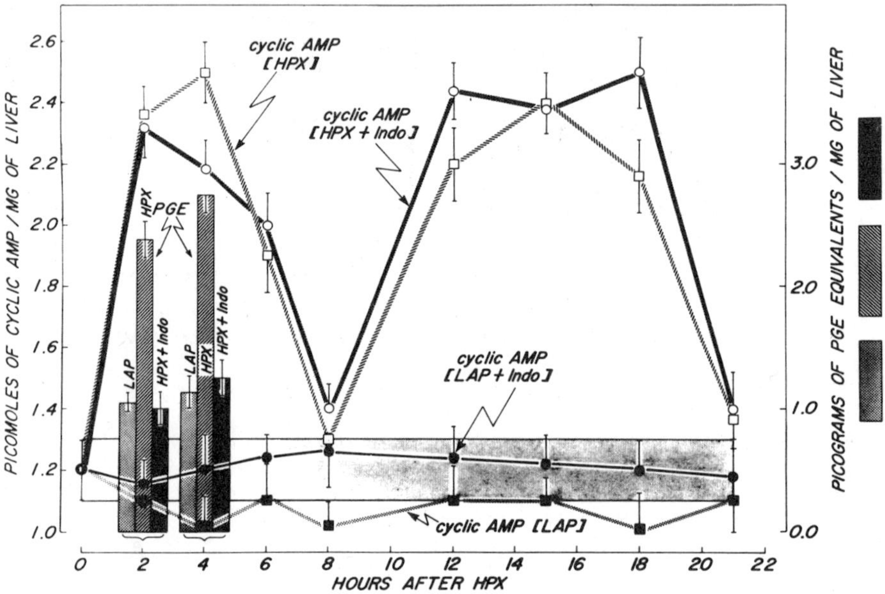

FIGURE 11. *The ability of indomethacin to stop the post-partial hepatectomy accumulation of PGE without affecting the two waves of cyclic AMP accumulation. Indomethacin was dissolved in propylene glycol and injected intraperitoneally (60 mg/100 g of rat) 4 h before HPX. The PGE levels (measured according to Jaffe and Behrman (46) at 2 and 4 h are shown as shaded bars. The cyclic AMP levels were measured by the protein-binding assay of Gilman (38). Indo, indomethacin; LAP, laparotomy or sham-HPX which also received propylene glycol; PGE, prostaglandin E. The points or bar heights are the means ± SEM of the values from 8 to 16 livers.*

tive, which implies a role for prostaglandins in the initiatory process, seems to be eliminated by the inability of exogenous prostaglandins (PGE_1, 15-methyl PGE_1, $PGF_{1\alpha}$) to overcome the indomethacin-imposed block (MacMan (MacManus and Braceland, unpublished observations). However, this choice may be incorrect, because indomethacin inhibits both the *synthesis* and *function* of prostaglandins (81,86). Regardless of these considerable uncertainties, indomethacin blockage may prove ultimately to be just another demonstration of calcium's key role in the initiation of DNA synthesis. Such a (perhaps unexpected) suggestion is based on the following facts: PGE can

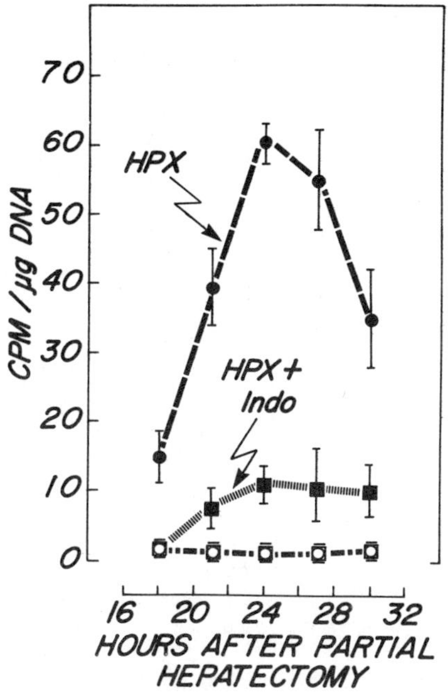

FIGURE 12. *The ability of indomethacin (which inhibits PGE accumulation without affecting the cyclic AMP surges) to block the initiation of DNA synthesis in partially (70%) resected rat liver. Indomethacin was injected (6 mg/100 g, intraperitoneally in propylene glycol) 4 h before surgery at 0-time. The level of DNA synthesis was estimated by measuring the incorporation of [^{3}H]-thymidine into DNA according to MacManus et al. (55,56). The points are the means ± SEM of the values from 6 to 18 livers.*

function as an ionophore which transfers calcium from intramitochondrial deposits to the cytosol (24,57); the single preoperative injection of indomethacin which blocks DNA synthesis also prolongedly lowers the plasma calcium level (MacManus and Braceland, unpublished observations); and indomethacin blocks the mobilization of bone calcium by parathyroid hormone (MacManus and Braceland, unpublished observations) which, like the initiation of DNA synthesis (see below), involves translocations of calcium ions between mitochondria and cytosol (24,51,52).

Calcium ions are initiators of hepatocyte DNA synthesis, and the second cyclic AMP surge introduces them into the prereplicative reaction sequence. Consequently, increasing preoperative (i.e., preactivation) exposure to hypocalcemia in the parathyroprivic rat progressively increases the proportion of hepatocytes unable to accumulate the four deoxyribonucleotide precursors of DNA (Figure 3B), make DNA (Figures 3B, 13 and 14) and proliferate (Figures 3B, 13 and 15) after HPX (67). This increasing proliferative incompetence is due more to calcium deprivation than

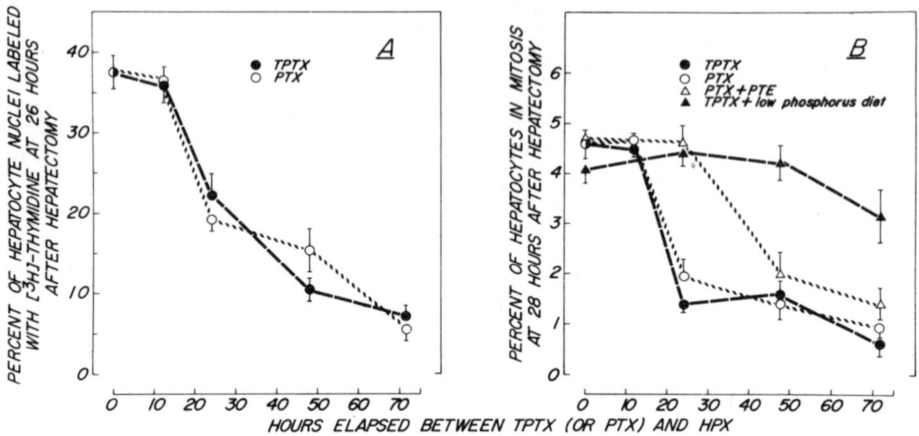

FIGURE 13. *The progressive, but reversible, decline of the proliferative competence of hepatocytes with increasing preactivation (i.e., partial hepatectomy) exposure to the hypocalcemic conditions in parathyroidectomized and thyro-parathyroidectomized rats.* A. *Either the parathyroid glands or the whole thyroid-parathyroid complexes of male, 200 g Sprague-Dawley rats were removed (by electrocautery and surgical excision, respectively). At an appropriate time thereafter, the livers of these animals were partially (70%) resected, and the proportion of the remnant hepatocytes making DNA 26 h later was determined autoradiographically according to Rixon and Whitfield (66, 67).* B. *Either the parathyroids, or the whole thyroid-parathyroid complexes were removed from rats. At an appropriate time thereafter, the animals were partially (70%) hepatectomized and 28 h later the total proportion of remnant hepatocytes in the four stages of mitosis was determined according to Rixon and Whitfield (66,67). Some animals received an injection of parathyroid extract (50 USP parathyroid hormone units/100 g of rat) immediately after HPX to prolongedly raise the plasma calcium level (67). The plasma calcium level in other animals was maintained at a normal level throughout the experimental period by feeding them a low phosphorus (36 mg P/100 g of feed in contrast with the normal 700 mg P/100 g of feed) diet beginning three days before TPTX and continuing until their sacrifice at 28 h after partial hepatectomy. The points are the means ± SEM of values from five to eight liver remnants. PTE, parathyroid extract; PTX, parathyroidectomy; TPTX, thyro-parathyroidectomy by surgical removal.*

parathyroid hormone deprivation, because the hepatocytes' proliferative capacity does not severely decline when hypocalcemia is prevented by feeding the parathyroidectomized rats a low phosphorus diet (Figure 13B).

Calcium seems exclusively to regulate the early stages of the DNA-synthetic process, because other responses to partial hepatectomy such as the two cyclic AMP

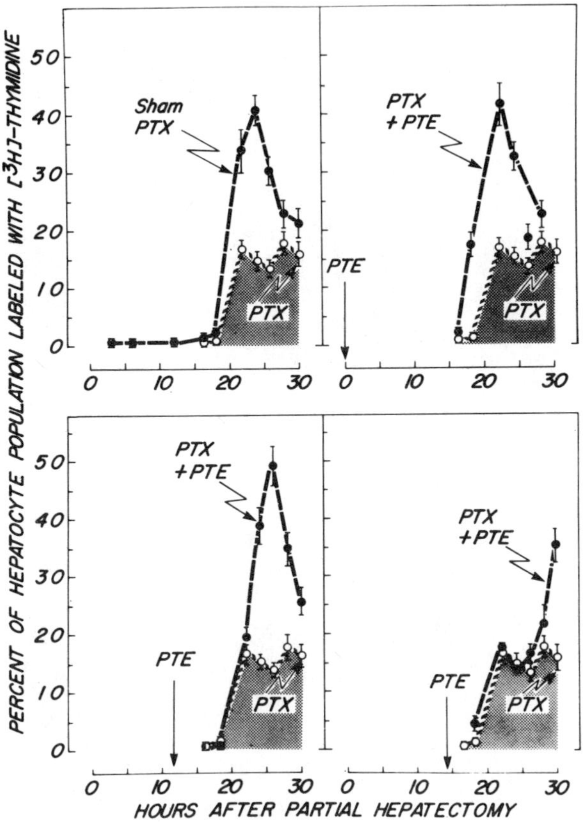

FIGURE 14. *The effect of an intraperitoneal injection of parathyroid extract on the ability of hepatocytes in parathyroidectomized rats to initiate DNA synthesis after partial hepatectomy. The partial hepatectomy was carried out 24 h after parathyroid removal, and PTE (50 USP parathyroid hormone units/100 g of rat) was injected at various times (as indicated by the arrows) after partial hepatectomy. Pooled control data from all experiments on the effects of partial hepatectomy in normal, sham-parathyroidectomized rats are shown separately and that in nonhormone treated parathyroidectomized rats (stippled area) are reduplicated in each graph for comparative purposes. The DNA synthetic activity is expressed as the percentage of the parenchymal cells in the right anterior lobe of the liver which incorporated [^3H] -thymidine for 1 h before sacrifice. The points are the means ± SEM of the values from 7 to 15 liver remnants. PTE, parathyroid extract (Eli Lilly and Company, Indianapolis).*

surges (Figures 16) and the accumulations of PGE equivalents, putrescine and spermidine (Figures 9B and 16) are totally unaffected even by prolonged preoperative exposure to hypocalcemia. Therefore, activated, but calcium-

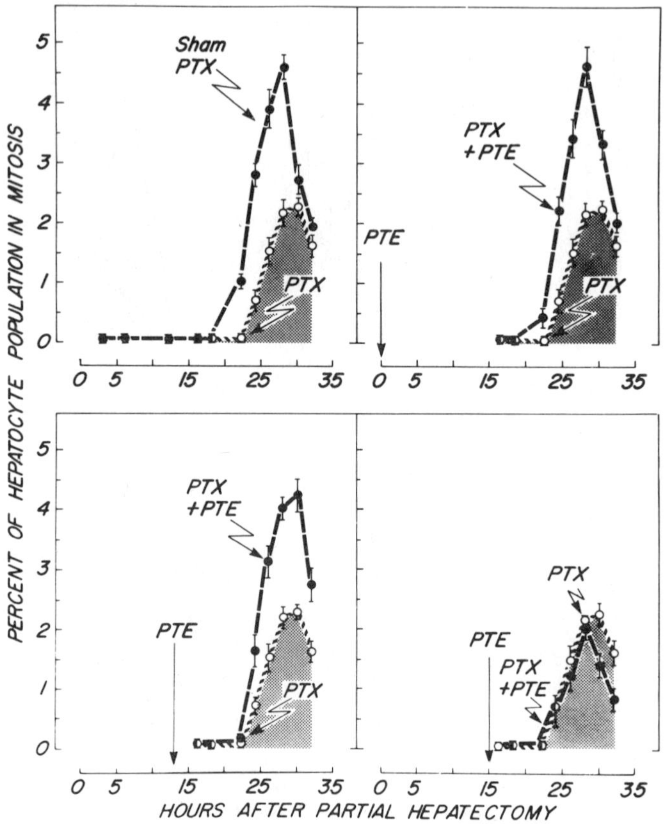

FIGURE 15. *The effect of an intraperitoneal injection of PTE on the ability of hepatocytes in PTX rats to flow into mitosis after partial hepatectomy. For details see the legend of Figure 14. The mitotic activity is expressed as the percentage of parenchymal cells in the liver remnant (i.e., the right anterior lobe) which were in mitosis at the times indicated. The points are the means ± SEM of the values from 7 to 15 liver remnants.*

deprived, hepatocytes reach a late stage of prereplicative development where they need only one injection of parathyroid hormone (Figures 13B, 14 and 15) or a single pulse of exogenous calcium ions (Figure 17) to complete their prereplicative preparations, initiate DNA synthesis and finally enter mitosis *at the normal time* (67). However, unless "rescued" by exogenous calcium ions or parathyroid hormone, the blocked, calcium-deprived hepatocytes lose their state of preparedness and revert to a nonstimulatable state (Figure 17).

For the calcium ion effectively to serve as a positive intracycle regulator, its intracellular concentration must rise briefly at the appropriate point in the prereplicative reaction sequence. The ionic calcium concentra-

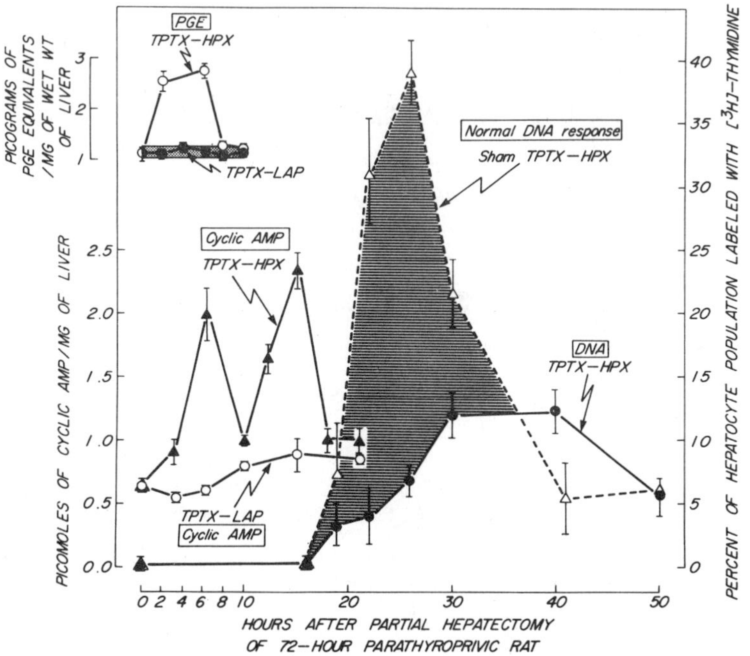

FIGURE 16. *The generation of normal waves of accumulation of cyclic AMP and type E prostaglandins by hepatocytes with severely reduced ability to initiate DNA synthesis in the partially (70%) resected livers of* hypocalcemic *rats from which the thyroid-parathyroid complex had been removed 72 h previously. The shaded area depicts the flow of hepatocytes into the DNA-synthetic phase of proliferative development in the partially (70%) resected livers of normocalcemic rats. The methods of measurement of cyclic AMP and prostaglandin levels, and the proportion of DNA-synthesizing hepatocytes are listed in the legend of Figure 2. PGE, prostaglandin E; TPTX-HPX, thyro-parathyroidectomized and then partially hepatectomized 72 h later; sham TPTX-HPX, sham neck surgery (without thyroid-parathyroid resection) and then partially hepatectomized 72 h later. The points are the means ± SEM of values from 4 to 10 livers.*

tion in the hepatocyte's cytosol is the net result of carrier-mediated transmembrane movements and a highly specific uptake of cytosol calcium ions by the mitochondria, which consequently *reversibly* sequester a large fraction of the cell's total calcium content (24,25,51,52). Since cyclic AMP releases calcium ions from their mitochondrial depots (12-14), each post-partial hepatectomy wave of cyclic AMP accumulation must transiently raise the level of calcium ions in the cytosol. The dispensability of the first cyclic AMP surge, and its attendant calcium ion pulse, is explained by the absence of calcium's target DNA-synthetic enzyme or enzyme complex in the early stages

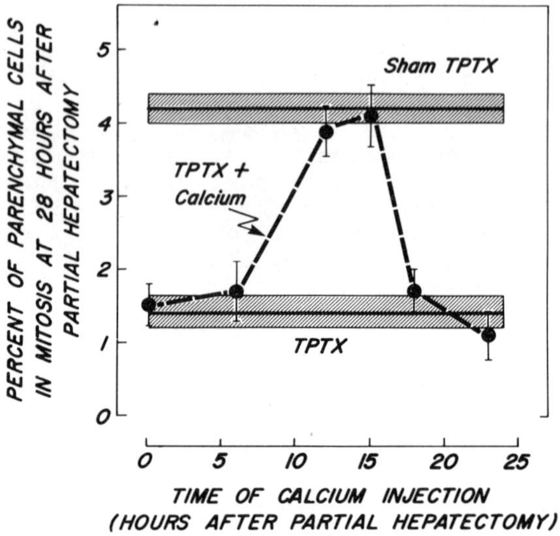

FIGURE 17. *The effect of calcium (0.25 mg/100 g of rat), injected intraperitoneally (as $CaCl_2$) at different intervals after HPX, on the ability of calcium-deprived hepatocytes to proliferate in TPTX rats. The partial liver resection was carried out 24 h after the TPTX operation. Hepatocyte proliferation is expressed as the total percentage of parenchymal cells in the remnant (right anterior lobe) which were in the four stages of mitosis 28 h after surgery. Each value is the mean ± SEM of six to eight rats. The hatched regions depict the mean ± SEM of 10 saline-treated TPTX rats or 12 sham-TPTX rats.*

of prereplicative development. However, the proliferatively important second cyclic AMP surge occurs after the appearance of such an enzyme or enzyme complex which needs the second calcium ion pulse to induce it to start making deoxyribonucleotides.

Besides lowering the extracellular calcium concentration, parathyroidectomy may lower the intracellular calcium activity by increasing the extracellular (and consequently the intracellular) phosphate concentration and it may reduce the supply of ionophoric calcium-binding proteins by interfering with vitamin D metabolism (12,25,63). These factors would reduce transmembrane calcium movements, lower the cytosol ionic calcium concentration and deplete the mitochondrial calcium stores, all of which would combine to prevent the second cyclic AMP surge from raising the calcium activity high enough to start deoxyribonucleotide synthesis. Provided the depletions of calcium and membrane ionophores have not progressed too far, the proliferative block can be overcome by a single pulse of exogenous calcium ions or by an injection of parathyroid hormone (67). On the other hand, blockage due to prolonged (more than 24 h) parathyroprivia and hypocalcemia can be overcome only by multiple injections of parathyroid hormone (67).

The control of hepatocyte proliferation by calcium is only one example of what may prove to be an ancient and wide-spread phenomenon. For example, calcium regulates the proliferation of the bacterium *Streptococcus lactis* which lacks such sophisticated calcium-sequestering organelles as mitochondria. In this example, transfer of proliferatively quiescent organisms to fresh medium is followed by an inward surge of extracellular calcium ions which trigger proliferation by somehow inactivating nisin, a polypeptide repressor (Figure 18) (42-44). Calcium is

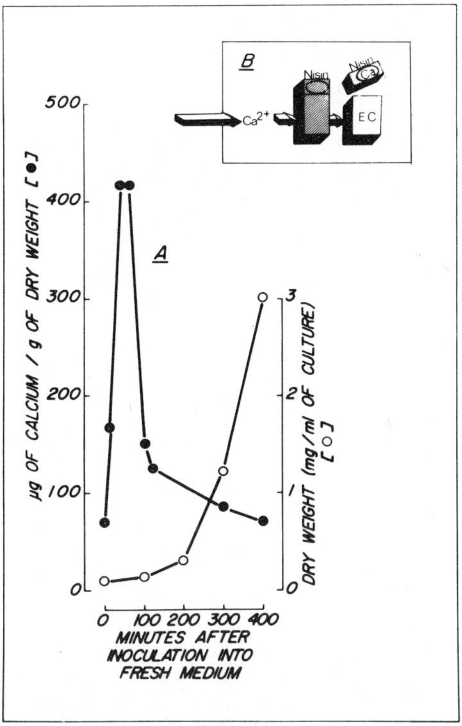

FIGURE 18. *The brief, proliferogenic accumulation of extracellular calcium by Streptococcus lactis, which we believe to be an early version of the more sophisticated cyclic AMP-induced intracellular calcium translocations that trigger DNA synthesis in more complex eukaryotic cells. During evolution, these calcium movements through the cell membrane are replaced by intracellular shifts of calcium ions between the cytosol and mitochondria.* *A. Proliferatively quiescent organisms were inoculated into tryptone-glucose broth at 0-time, and the subsequent changes in the total cell calcium content measured by flame spectrophotometry (44). These curves were derived (with permission of Dr. A. Hurst) from some of the data in Figure 1 of Hurst and Lazarus (44). B. A very sche-*

matic representation of a mechanism by which extracellular calcium enters the proliferatively quiescent organism where it turns on the proliferative machinery by somehow inactivating a protein repressor, nisin (43,44).

also needed for active growth and proliferation of the more highly evolved myxomycete, *Physarum polycephalum*, which does contain mitochondria (30,31). A role for mitochondrial calcium sequestration in regulating the proliferation of this more complex organism is suggested by the intramitochondrial calcium deposition associated with the inhibition of plasmodial growth and proliferation by illumination (30).

The possibility of the calcium-cyclic AMP control couplet being used by cells other than hepatocytes is indicated by the occurrence of a pre-DNA synthetic cyclic AMP surge in regenerating lens cells, *BALB/3T3* mouse endothelial (11) cells, *BHK* hamster kidney cells, Chinese hamster *CHO* and *V79* cells, *HeLa* human uterine cervical carcinoma cells, and *RPMI 8866* human lymploid cells (Figure 19) (28,29,59,62,73,75,84,92). At least one of these

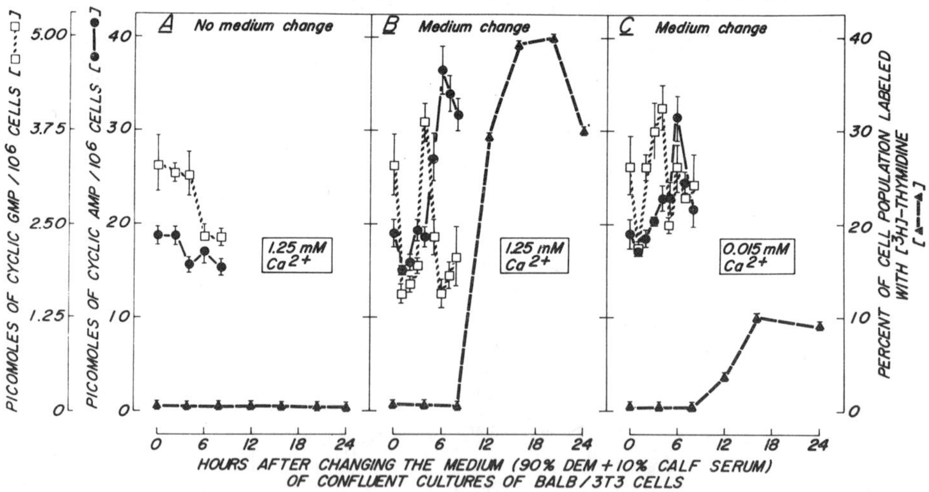

FIGURE 19. *The changes in the cyclic AMP and cyclic GMP contents of BALB/3T3 fetal mouse endothelial (11) cells after they have initiated a growth-division cycle in confluent cultures in normal and low-calcium medium. Cells were seeded in 60 mm petri dishes for cyclic nucleotide determinations, or on plastic coverslips in 35 mm petri dishes for autoradiography. They were plated in a complete medium consisting of 90% (v/v) Dulbecco's modified Eagle's medium and 10% (v/v) bovine calf serum, the total calcium concentration of which was 1.8 mM. The medium was changed every two days thereafter until the population became confluent and proliferatively quiescent. The confluent cultures were incubated for another two days, at which time the medium was discarded and replaced with*

fresh medium containing only 0.015 mM physiologically available (ionized or at least ionizable) calcium. Two days later, the dense, calcium-deprived cultures were either (A) left without a medium change; (B) exposed to fresh, complete medium containing 1.25 mM available calcium; or (C) exposed to fresh medium containing only 0.015 mM available calcium. The subsequent flow of cells into the DNA-synthetic phase of their growth-division cycle was measured autoradiographically following 1 h exposures at appropriate times to [^{3}H]-thymidine (10 µCi/ml of medium; specific activity, 20 Ci/mmole) according to Whitfield et al. (87). Cyclic nucleotide levels were determined by the radioimmunoassay procedures described by Whitfield et al. (87). The points are the means ± SEM of the values from four cultures.

cells, the preneoplastic (11) *BALB/3T3* mouse endothelial cell, also briefly accumulates cyclic GMP immediately before the pre-DNA-synthetic cyclic AMP surge (Figure 19) (75). In Chinese hamster *CHO* cells, the cyclic AMP surge coincides with increased translation of the message for a "type II" protein kinase and Costa *et al.* (28,29) suggest that the cyclic nucleotide promotes the subsequent proliferative processes by activating this kinase. More evidence for the wide-spread use of the couplet is the fact that calcium-deprived *BALB/3T3* cells *in vitro*, like calcium-deprived hepatocytes *in vivo*, progress as far as the generation of the pre-DNA-synthetic cyclic AMP surge, but they cannot make DNA (Figure 19) unless rescued by a timely elevation of the extracellular ionic calcium concentration (15,17,18,20).

All of these observations severely undermine the current belief in cyclic AMP as *the* inhibitory regulator of cell proliferation (39,40,50,64,74). As stated at the outset, this belief is precariously founded on the ability of a continuous exposure to dibutyryl cyclic AMP to inhibit proliferation of cells of several long-established lines, most of which are either fully neoplastic (e.g., HeLa human, and L strain mouse, cells) or nearly so (e.g., the *conditionally tumorigenic BALB/3T3* mouse cells). By contrast, continuous exposure to either cyclic AMP or dibutyryl cyclic AMP *stimulates* the proliferation of primary cultures of normal bone marrow cells, hepatocytes and thymic lymphoblasts (Figures 5 to 8) (1,2,21,22,88, 89) which have not been outside of the body long enough to have lost their responsiveness to normal proliferative controls. Moreover, even continuous exposure to such a nonphysiological cyclic AMP concentration as 1.5×10^{-4} *M* promotes the proliferation of freshly isolated human and rat adrenocortical (from the *zonae fasciculata* and *reticularis*) cells (4) (Figures 20 and 21) which are proliferatively activated by the cyclic AMP-elevating adrenocorticotrophic hormone in the animal (37). Perhaps the existence of the pre-DNA-synthetic cyclic AMP accumulation is the simplest and the best indication of the faultiness of the current dogma. Thus, we must discard the

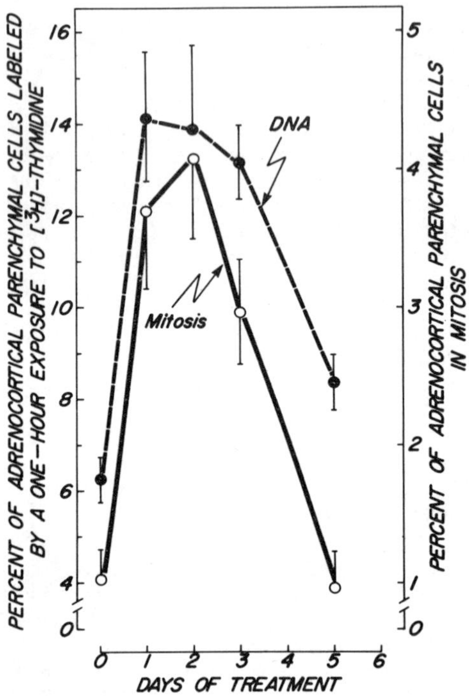

FIGURE 20. *The effects of exogenous cyclic AMP (1.5 × 10^{-4} M) on the flow of adult rat adrenocortical cells (from zonae reticularis and fasciculata) into the DNA-synthetic and mitotic phases of their growth-division cycles. Decapsulated rat adrenal cortices were cultured according to Armato and Nussdorfer (6). Beginning on the 16th day of cultivation, the cells were treated daily with cyclic AMP (dissolved in growth medium) for 1 to 5 days. The points are the means ± SEM of the values from five to eight cultures.*

old dogma and replace it with a new hypothesis according to which the tightly coupled series of prereplicative reactions in mammalian cells includes an elevation of the cyclic AMP level which is programmed to last just long enough to trigger a specific process, but at the same time to be sufficiently brief to avoid interfering with preceding or succeeding reactions.

Finally, a loss of the control of DNA synthesis by the calcium-cyclic AMP couplet may be part of the neoplastic transformation of at least some types of cells. For example, Rous sarcoma virus-infected adult chicken fibroblasts and tumorigenic cells from nickel sulfide-induced rat rhabdomyosarcomas proliferate rapidly and indefinitely in calcium-deficient (and homologous plasma-containing) media which do not sustain the proliferation of the corresponding normal cells (9,83). Moreover, the *conditionally tumorigenic BALB/3T3* mouse cells (which

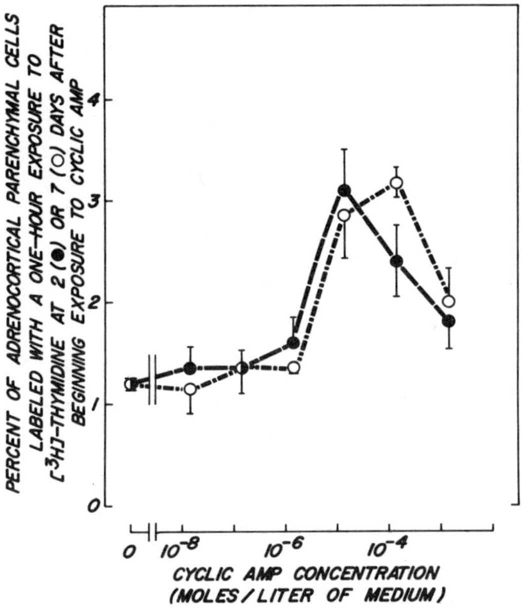

FIGURE 21. *The effects of various concentrations of exogenous cyclic AMP on the flow of adult human adrenocortical cells (from the zonae fasciculata and reticularis) into the DNA-synthetic phase of their growth-division cycle. Normal human adrenals obtained (from two consenting patients) during nephrectomy, were decapsulated and cultured according to Armato et al. (7). After 16 days of cultivation, the cells were exposed to various concentrations of cyclic AMP (dissolved in growth medium) for a further two or seven days, at which times they were exposed for 1 h to [^{3}H] -thymidine (0.5 μCi/ml; specific activity, 17 Ci/mmole) and then processed for autoradiography (3). Each day, control cultures received equivalent volumes of cyclic AMP-free growth medium. The points are the means ± SEM of the values from a total of 8 to 12 cultures.*

produce hemangioendotheliomas *only* when introduced into *BALB/c* mice after having been attached to glass beads) do not proliferate in an ionic calcium-free medium which indefinitely supports rapid proliferation of their chemically and virally transformed, *unconditionally tumorigenic* derivatives (16).

CONCLUSIONS

Calcium, cyclic AMP and cyclic GMP do not seem to be involved in the proliferative activation of a differentiated cell such as the rat hepatocyte. On the other hand, calcium and cyclic AMP form an "intracycle" regulatory couplet which controls a key phase of the prereplicative development of activated hepatocytes and several other types of actively cycling cell, and we propose the following model of the couplet's action (Figure 22).

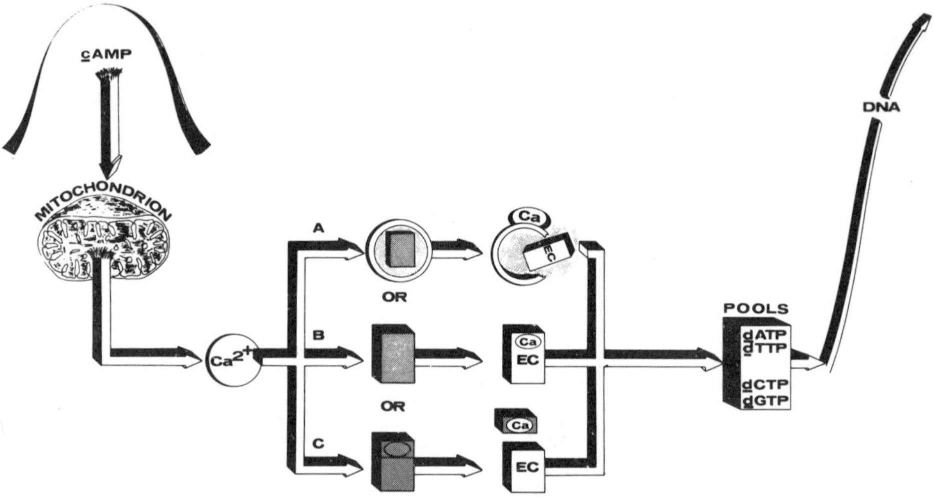

FIGURE 22. *A schematic depiction of some possible mechanisms by which the calcium-cyclic AMP intracycle regulatory couplet positively controls the initiation of DNA synthesis. Near the end of the prereplicative phase of proliferative development, the intracellular cyclic AMP level briefly rises and releases calcium ions from their sequestration in mitochondria. These liberated calcium ions might then trigger DNA synthesis in three possible ways, which are based on the ions' somewhat better understood roles in muscle contraction, secretion and initiation of streptococcal proliferation (32,45,70). Thus, the liberated calcium ions might (A), release a DNA-synthetic enzyme complex, EC, from functional or physical sequestration by membranous structures; or (B) induce an inactive EC to assume an active configuration; or (C) inactivate a repressor of EC. Ultimately, the released, or otherwise activated, EC starts making the four deoxyribonucleotide precursors of DNA, the excess cyclic AMP is destroyed, and the additional calcium ions are removed from the cytosol and redeposited in the mitochondria.*

Proliferative development begins when an activator(s) triggers the sequential derepression of the series of proliferogenic cistrons that direct, among other things, the synthesis and assembly of a calcium-activable complex of enzymes (such as ribonucleotide reductase and thymidylate synthetase) catalyzing the production of the four deoxyribonucleotide precursors of DNA (88,91). By analogy with other types of calcium-activated systems, the newly assembled complex is at first inactive because of the presence of a repressor (such as nisin in *Streptococcus lactis*)

(42-44); because of being in a specific allosteric configuration (such as the myosin: [troponin-tropomyosin-actin] system) (45); or because of being sequestered within, or on, membranous structures (such as those which sequester catecholamines and acetylcholine in adrenal medulla cells and at axon terminals (32,47,70) (Figure 22). When the activated cell nears the end of the G1 phase of its growth-division cycle, something happens (possibly a lowering of cyclic AMP-phosphodiesterase activity (33,59) which raises the cyclic AMP level. The additional cyclic AMP undoubtedly has many effects, one of which is to release calcium ions from mitochondrial depots. The liberated calcium ions then start deoxyribonucleotide synthesis by activating the target enzyme complex according to one of the three types of mechanism schematically presented in Figure 22. Having accomplished their tasks, the additional cyclic AMP is destroyed and the excess cytosol calcium ions are redeposited in the mitochondria to avoid affecting the succeeding reactions.

Finally, neoplastic transformation somehow removes the DNA-synthetic machinery from the influence of the calcium-cyclic AMP couplet. This loss of control may be due to reduced mitochondrial calcium sequestration which would "fix" the cytosol ionic calcium concentration at a higher, and permanently stimulatory, level. The elevated calcium content of the *unconditionally tumorigenic SV-BALB/3T3* cell (35) may be an example of such a change. On the other hand, the calcium-cyclic AMP couplet could be rendered functionless, and the DNA-synthetic system permanently activated, by failure of the cell to make a calcium-inactivable repressor or to sequester the newly synthesized DNA-synthetic enzymes in a calcium-mobilizable form.

ACKNOWLEDGEMENTS

We thank Dr. E. Draghi and P. G. Andreis of Padua for their excellent and dedicated assistance in carrying out a key portion of this work. We also thank D. J. Gillan of Ottawa for so successfully undertaking the arduous task of preparing the illustrations.

REFERENCES

1. Armato, U., Andreis, P. G., Belloni, A. S. and Draghi, E. 1974a. Long-term stimulatory effects of adenosine 3',5'-cyclic monophosphate on nuclear and cytoplasmic protein synthesis of rat hepatocytes in primary culture. Acta Anat. *88*:456-480.
2. Armato, U., Andreis, P. G. and Draghi, E. 1974b. Cyclic nucleotides influence the DNA synthesis of rat liver cells in primary culture. Digestion *10*:333-334.
3. Armato, U., Andreis, P. G. and Draghi, E. 1975a. Effect of adenosine 3',5'-cyclic monophosphate on the RNA and DNA synthesis and cell proliferation

of rat hepatocytes in primary culture: a radioautographic study. Chem.-Biol. Interact. *11*:67-90.

4. Armato, U., Andreis, P. G., Draghi, E. and Meneghelli, V. 1975b. Primary tissue culture of normal adult human decapsulated adrenal cortex. Radioautographic studies on the metabolic effects of ACTH (1-24). Hormone Res. *6*:105-115.
5. Armato, U., Draghi, E., Andreis, P. G. and Meneghelli, V. 1976. Stimulation by N , O -dibutyryl adenosine 3',5'-cyclic monophosphate of RNA and DNA synthesis and of cell proliferation of rat hepatocytes in primary tissue culture. J. Cell. Physiol. In press.
6. Armato, U. and Nussdorfer, G. G. 1972. Tissue culture of rat adult decapsulated adrenal glands. A methodological, ultrastructural and morphometric investigation. Z. Zellforsch. *135*:245-273.
7. Armato, U., Nussdorfer, G. G., Andreis, P. G., Mazzocchi, G. and Draghi, E. 1974c. Primary tissue culture of human adult adrenocortical cells: methodology and electron microscopic observations on ACTH-deprived and ACTH-treated cortical cells. Cell Tiss. Res. *155*:155-180.
8. Bachrach, U. *Function of Naturally Occurring Polyamines*. Academic Press, New York and London, 1973.
9. Balk, S. D., Whitfield, J. F., Youdale, T. and Braun, A. C. 1973. Roles of calcium, serum, plasma and folic acid in the control of proliferation of normal and Rous sarcoma virus-infected chicken fibroblasts. Proc. Nat. Acad. Sci. USA *70*:675-679.
10. Beck, U. T., Bellantone, R. A. and Canellakis, E. S. 1972. The *in vivo* stimulation of rat liver ornithine decarboxylase activity by dibutyryl cyclic adenosine 3',5'-monophosphate, theophylline and dexamethasone. Biochem. Biophys. Res. Commun. *48*:1649-1655.
11. Boone, C. W. 1975. Malignant hemangioendotheliomas produced by subcutaneous inoculation of BALB/3T3 cells attached to glass beads. Science *188*:68-70.
12. Borle, A. 1973. Calcium metabolism at the cellular level. Fed. Proc. *32*:1944-1950.
13. Borle, A. 1974. Cyclic AMP stimulation of calcium efflux from kidney, liver and heart mitochondria. J. Membrane Biol *16*:221-236.
14. Borle, A. 1975. Regulation of cellular calcium metabolism and calcium transport by calcitonin. J. Membrane Biol. *21*:125-146.
15. Boynton, A. L. and Whitfield, J. F. 1976a. The different actions of normal and supranormal calcium concentrations on the proliferation of BALB/c 3T3 mouse cells. In Vitro, in press.
16. Boynton, A. L. and Whitfield, J. F. 1976b. The different calcium requirements for the proliferation of conditionally and unconditionally tumorigenic mouse cells. Proc. Nat. Acad. Sci. USA, in

press.
17. Boynton, A. L., Whitfield, J. F. and Isaacs, R. J. 1976a. Calcium-dependent stimulation of BALB/c 3T3 mouse cell DNA synthesis by a tumor-promoting phorbol ester (PMA). J. Cell. Physiol. *87*:25-32.
18. Boynton, A. L., Whitfield, J. F. and Isaacs, R. J. 1976b. The different roles of serum and calcium in the control of proliferation of BALB/c 3T3 mouse cells. In Vitro *12*:120-123.
19. Boynton, A. L., Whitfield, J. F. and Isaacs, R. J. 1976c. A possible involvement of polyamines in the initiation of DNA synthesis by human WI-38 and mouse BALB/3T3 cells. J. Cell. Physiol.,in press.
20. Boynton, A. L., Whitfield, J. F., Isaacs, R. J. and Morton, H. J. 1974. Control of 3T3 cell proliferation by calcium. In Vitro *10*:12-17.
21 Byron, J. W. 1972. Evidence for a β-adrenergic receptor initiating DNA synthesis in hemopoietic stem cells. Exp. Cell Res. *71*:228-232.
22. Byron, J. W. 1973. Drug receptors and the hemopoietic stem cell. Nature New Biol. :152-154.
23. Byus, C. V. and Russell, D. H. 1974. Ornithine decarboxylase activity: control by cyclic nucleotides. Science *241*:650-652.
24. Carafoli, E. Mitochondrial uptake of calcium ions and the regulation of cell function. In: *Calcium and Cell Regulation*, Smellie, R. M. S. (ed.). The Biochemical Society, London, 1975, pp. 89-109.
25. Chausmer, A. B., Sherman, B. S. and Wallach, S. 1972. The effect of parathyroid hormone on hepatic cell transport of calcium. Endocrinology *90*: 663-672.
26. Clark, J. L. and Fuller, J. L. 1975. Regulation of ornithine decarboxylase in 3T3 cells by putrescine and spermidine: indirect evidence for translational control. Biochemistry *14*:4403-4409.
27. Cohen, S. S. *Introduction to the Polyamines*. Prentice Hall, Inc., New Jersey, 1971.
28. Costa, M., Gerner, E. W. and Russell, D. H. 1976a. Cell cycle-specific activity of type I and type II cyclic adenosine 3',5'-monophosphate-dependent protein kinases in Chinese hamster ovary cells. J. Biol. Chem., in press.
29. Costa, M., Gerner, E. W. and Russell, D. H. 1976b. G_1 specific increases in cyclic AMP levels and protein kinase activity in Chinese hamster ovary cells. Biochim. Biophys. Acta, in press.
30. Daniel, J. W. and Järlfors, U. 1972. Light-induced changes in the ultrastructure of a plasmodial myxomycete. Tissue and Cell *4*:405-426.
31. Daniel, J. W. and Rusch, H. P. 1961. The pure culture of *Physarum polycephalum* on a partially defined soluble medium. J. Gen. Microbiol. *25*:47-59.
32. Douglas, W. W. 1975. Involvement of calcium in exocytosis and exocytosis-vesiculation sequence. In: *Calcium and Cell Regulation*, Smellie, R. M. S.

(ed.). The Biochemical Society, London, pp. 1-28.
33. Ebina, Y., Iwai, H., Fukui, N., Ohtsuka, H. and Miura, Y. 1975. Prereplicative enzymic changes in regenerating rat liver. J. Biochem. *77*:641-645.
34. French, F. A., Freedlander, B. L., Hasking, A. and French, J. 1960. Carcinostatic activity of some dicarbonyl compounds and their bio-hydrazones. Acta Unio Internat. contra Cancrum *16*:614-624.
35. Fuller, G. M., Ellison, J. McGill, M. and Brinkley, B. R. 1975. The involvement of mitochondria and calcium in regulating microtubule assembly *in vitro*. J. Cell Biol. *67*:126a.
36. Gaza, D. J., Short, J. and Lieberman, I. 1973. On the possibility that the prereplicative increases in ornithine decarboxylase activity are related to DNA synthesis in liver. FEBS Lett. *251*:251-253.
37. Gill, G. N. 1972. Mechanism of ACTH action. Metabolism *21*:571-588.
38. Gilman, A. G. 1970. A protein binding assay for adenosine 3',5'-cyclic monophosphage. Proc. Nat. Acad. Sci. USA *67*:305-312.
39. Goldberg, N. D., Haddox, M. K., Estensen, R., White, J. G., Lopez, C. and Hadden, J. W. Evidence of a dualism between cyclic GMP and cyclic AMP in the regulation of cell proliferation and other cellular processes. In: *Cyclic AMP in Immune Response and Tumor Growth*, Braun, W., Lichenstein, L. M. and Parker, C. W. (eds.). Springer Verlag, New York, 1974, pp. 237-246.
40 Goldberg, N. D., O'Dea, R. F. and Haddox, M. Cyclic GMP. In: *Advances in Cyclic Nucleotide Research*, Vol. 3, Greengard, P. and Robison, G. A. (eds.). Raven Press, New York, 1973, pp. 155-223.
41. Goridis, C. and Reutter, W. 1975. Plasma membrane-associated increase in guanylate cyclase activity in regenerating rat liver. Nature *257*:698-700.
42. Hurst, A. 1968. Apparent destruction of nisin by the producer organism before initiation of growth in *Streptococcus lactis*. Nature *219*:403-404.
43. Hurst, A. and Kruse, H. 1970. The correlation between change in absorbancy, calcium uptake, and cell-bound nisin activity in *Streptococcus lactis*. Can. J. Microbiol. *16*:1205-1211.
44. Hurst, A. and Lazarus, W. 1968. Calcium uptake during growth of *Streptococcus lactis*. Nature *219*: 404-405.
45. Huxley, H. E., Weber, A. and Lehninger, A. L. Molecular basis of muscle contraction. In: *Functional Linkage in Biomolecular Systems*, Schmitt, F. O., Schneider, D. M. and Crothers, D. M. (eds.). Raven Press, New York, 1975, pp. 299-321.
46. Jaffe, B. M. and Behrman, H. R. Prostaglandins E, A and F. In: *Methods of Hormone Radioimmunoassay*, Jaffe, B. M. and Behrman, H. R. (eds.). Academic Press, New York, 1974, pp. 19-34.

47. Katz, B. *The Release of Neural Transmitter Substances.* Liverpool University Press, Liverpool, 1969.
48. Kimura, H. and Murad, F. 1975. Increased particulate and decreased soluble guanylate cyclase activity in regenerating liver, fetal liver, and hepatoma. Proc. Nat. Acad. Sci. USA *72*:1965-1969.
49. King, R. J. B. and Mainwaring, W. I. P. *Steroid - Cell Interactions.* University Park Press, Baltimore, 1974.
50. Kolata, G. B. 1973. Cyclic GMP: cellular regulatory agent? Science *182*:149-151.
51. Lehninger, A. L. 1970. Mitochondrial and calcium ion transport. Biochem. J. *119*:129-138.
52. Lehninger, A. L. Linked transport and binding functions in the mitochondrial membrane. In: *Functional Linkage in Biomolecular Systems*, Schmitt, F. O., Schneider, D. M. and Crothers, D. M. (eds). Raven Press, New York, 1975, pp. 165-179.
53. Leoni, S., Luly, P., Mangiantini, M. T., Spagnudo, S., Trentalance, A. and Verna, R. 1975. Hormone responsiveness of plasma membrane-bound enzymes in normal and regenerating rat liver. Biochim. Biophys. Acta *394*:317-322.
54. Lindberg, U. and Skoog, L. 1970. A method for the determination of dATP and dTTP in picomole amounts. Anal. Biochem. *34*:152-160.
55. MacManus, J. P., Braceland, B. M., Youdale, T. and Whitfield, J. F. 1973. Adrenergic antagonists and a possible link between the increase in cyclic adenosine 3',5'-monophosphate in DNA synthesis during liver regeneration. J. Cell. Physiol. *82*:157-164.
56. MacManus, J. P., Franks, D. J., Youdale, T. and Braceland, B. M. 1972. Increases in rat liver cyclic AMP concentrations prior to the initiation of DNA synthesis following partial hepatectomy or hormone infusion. Biochem. Biophys. Res. Commun. *49*:1201-1207.
57. Malmström, K. and Carafoli, E. 1975. Effects of prostaglandins on the interaction of Ca^{2+} with mitochondria. Arch. Biochem. Biophys. *171*:418-423.
58. Manen, C.-A. and Russell, D. H. 1976. Relationship of ornithine decarboxylase to RNA polymerase I activity. Life Sci., in press.
59. Matsumoto, T. and Uchida, T. 1975. Correlation between adenosine 3',5'-cyclic monophosphate levels, adenylate cyclase activity, and adenosine 3',5'-cyclic monophosphate phosphodiesterase activity in tissue culture cells stimulated by serum. J. Biochem. *78*:811-815.
60. Medoff, J. and Parker, N. 1971. Stimulation of growth by cyclic AMP in cultured embryonic tissue. J. Cell Biol. *55*:173a.

61. Mihich, E. 1963. Current studies with methylglyoxal-bis (guanylhydrazone). Cancer Res. *23*:1375-1389.
62. Millis, A. J., Forrest, G. A. and Pious, D. A. 1974. Cyclic AMP dependent regulation of mitosis in human lymphoid cells. Exp. Cell Res. *83*:335-343.
63. Omdahl, J. L. and DeLuca, H. F. 1973. Regulation of vitamin D metabolism and function. Physiol. Rev. *53*:327-372.
64. Pastan, I. H., Johnson, G. S. and Anderson, W. B. 1975. Role of cyclic nucleotides in growth control. Ann. Rev. Biochem. *44*:491-522.
65. Richman, R., Dobbins, G., Voina, S., Underwood, L., Mahaffee, D., Gitelman, H. J., Van Wyk, J. and Ney, R. L. 1973. Regulation of adrenal ornithine decarboxylase by adrenocorticotropic hormone and cyclic AMP. J. Clin. Invest. *52*:2007-2015.
66. Rixon, R. H. and Whitfield, J. F. 1974. Parathyroid hormone and liver regeneration. Proc. Soc. Exp. Biol. Med. *146*:926-930.
67. Rixon, R. H. and Whitfield, J. F. 1976. The control of liver regeneration by parathyroid hormone and calcium. J. Cell. Physiol. *87*:147-156.
68. Robison, G. A., Butcher, R. V. and Sutherland, E. W. *Cyclic AMP*. Academic Press, New York, 1971.
69. Rodbell, M., Walsh, D. A. and Bloom, F. E. Adenylate cyclase. In: *Functional Linkage in Biomolecular Systems*, Schmitt, F. O., Schneider, D. M. and Crothers, D. M. (eds.). Raven Press, New York, 1975, pp. 203-213.
70. Rubin, R. P. *Calcium and the Secretory Process*. Plenum Press, New York and London, 1974.
71. Russell, D. H. *Polyamines in Normal and Neoplastic Growth*. Raven Press, New York, 1973a.
72. Russell, D. H. 1973b. The roles of the polyamines, putrescine, spermidine and spermine in normal and malignant tissues. Life Sci. *13*:1635-1647.
73. Russell, D. H. and Stambrook, P. J. 1975. Cell cycle specific fluctuations in adenosine 3',5'-cyclic monophosphate and polyamines of Chinese hamster cells. Proc. Nat. Acad. Sci. USA *72*:1482-1486.
74. Ryan, W. L. and Heidrick, M. L. Role of cyclic nucleotides in cancer. In: *Advances in Cyclic Nucleotide Research*, Vol. 4, Greengard, P. and Robison, G. A. (eds.). Raven Press, New York, 1974, pp. 81-116.
75. Seifert, W. and Rudland, P. S. 1974. Cyclic nucleotides and growth control in cultured mouse cells: correlation of changes in intracellular 3',5' cGMP concentration with a specific phase of the cell cycle. Proc. Nat. Acad. Sci. USA *71*:4920-4924.
76. Seiler, N. and Weichmann, M. 1970. TLC analysis of amines as their DNAS-derivatives. Progress in Thin Layer Chromatography and Related Methods *1*:95-144.

77. Short, J., Brown, R. F., Husakova, A. G., Ibertson, J. R., Zemel, R. and Lieberman, I. 1971. Induction of deoxyribonucleic acid synthesis in the liver of the intact animal. J. Biol. Chem. *247*: 1757-1766.
78. Short, J., Tsukada, K., Rudert, W. A. and Lieberman, I. 1975. Cyclic adenosine 3',5'-monophosphate and the induction of DNA synthesis. J. Biol. Chem. *250*:3602-3606.
79. Skoog, L. 1970. An enzymatic method for the determination of dCTP and dGTP in picomole acmounts. Eur. J. Biochem. *17*:202-208.
80. Snyder, S. H., Krouz, D. S., Medina, V. J. and Russell, D. H. 1970. Polyamine synthesis and turnover in rapidly growing tissues. Ann. N. Y. Acad. Sci. *171*:749-777.
81. Sorrentino, L., Capasso, F. and DiRosa, M. 1972. Indomethacin and prostaglandins. Eur. J. Pharmacol. *17*:306-308.
82. Swierenga, S. H. H., MacManus, J. P. and Whitfield, J. F. 1976a. Regulation by calcium of the proliferation of heart cells from young adult rats. In Vitro *12*:31-36.
83. Swierenga, S. H. H., Whitfield, J. F. and Gillan, D. J. 1976b. Alteration by malignant transformation of the calcium requirements for cell proliferation *in vitro*. J. Natl. Cancer Inst., in press.
84. Thorpe, C. W., Bond, J. S. and Collins, J. M. 1974. Early events in lens regeneration: changes in cyclic AMP concentration during initiation of RNA and DNA synthesis. Biochim. Biophys. Acta *340*: 413-418.
85. Thrower, S. and Ord, M. 1974. Hormonal control of liver regeneration. Biochem. J. *144*:361-369.
86. Tolman, E. L. and Partridge, R. 1975. Multiple sites of interaction between prostaglandins and non-steroidal anti-inflammatory agents. Prostaglandins *9*:349-359.
87. Whitfield, J. F., MacManus, J. P., Boynton, A. L., Gillan, D. J. and Isaacs, R. J. 1974. Concanavalin A and the initiation of thymic lymphoblast DNA synthesis and proliferation by a calcium-dependent increase in cyclic GMP level. J. Cell. Physiol. *84*:445-454.
88. Whitfield, J. F., MacManus, J. P., Rixon, R. H., Boynton, A. L., Youdale, T. and Swierenga, S. 1976. The positive control of cell proliferation by the interplay of calcium ions and cyclic nucleotides. A review. In Vitro *12*:1-18.
89. Whitfield, J. F., Rixon, R. H., MacManus, J. P. and Balk, S. D. 1973. Calcium, cyclic adenosine 3', 5'-monophosphate and the control of cell proliferation. A review. In Vitro *8*:257-278.
90. Williams-Ashman, H. G., Coppoc, G. L., Schenone, A. and Weber, G. Aspects of polyamine biosynthesis

in normal and malignant eukaryotic cells. In: *Polyamines in Normal and Neoplastic Growth*, Russell, D. H. (ed.). Raven Press, New York, 1973, pp. 181-197.

91. Youdale, T. and MacManus, J. P. 1975. Failure of tritiated thymidine incorporation into DNA to reflect the autoradiographically demonstrable calcium-induced increase in thymic lymphoblast DNA synthesis. J. Cell. Physiol. *86*:495-502.

92. Zeilig, C. E., Johnson, R. A., Sutherland, E. W. and Friedman, D. L. 1974. Cyclic AMP levels in synchronized HeLa cells and a dual effect on mitosis. Fed. Proc. *33*:1391.

Cell Surfaces, Cyclic Nucleotides, and Unrestrained Cell Division: A Natural Role Is Proposed for the Cell Surface Receptor of Vibrio cholerae and Escherichia coli Enterotoxins

Helen Chilton Kiefer and Harvey S. Kantor

It is now clear that cyclic nucleotides are capable of exerting a profound effect *in vitro* on events of the mammalian cell cycle governing whether a cell proceeds to divide. Whether or not cyclic nucleotides have an *in vivo* regulatory role in cell division, is, of course, the central question considered in this volume and has been the subject of a number of recent reviews (1,2,15,21,37,44, 202). Much current evidence suggests that cyclic AMP may be involved *in vivo* as a negative signal, deterring cell division if it is elevated during certain critical portions of the cell cycle. If one considers from an evolutionary point of view the nature of the message conveyed by cyclic AMP as the "second messenger" of hormonal action (3), it becomes more evident why a similar signal might perhaps be used in cell growth regulation. A hormone produces a long distance signal to cells to give up a part of their autonomy and integrate their metabolic events with those of other cells for the survival of the whole organism. Perhaps when two primordial cells first decided to give up a portion of their autonomy in order to stop competing with each other over the available metabolites and divide the labors of metabolism for the good of their newly formed two-cell tissue, some coordinating message had to be given. Unrestrained division would obviously disrupt this coordination and return the tissue, which now has an obvious selective advantage, to rivalrous single cells. The first primitive coordinating message, with its concomitant division of metabolic labor and repression of cell division, could only be given over a short distance--either through direct cell-to-cell contact or through a closely confined chemotactic gradient, or more likely, both. A fitting signal would be a metabolically nonessential molecule, derived from a precursor always abundant in a viable cell.

This short philosophical discussion outlines a frame of reference for the hypothesis and supporting data to follow. It is suggested that hormonal second messenger use of cyclic AMP is "the chicken, and not the egg." That is, hormonal signals represent a later, highly specialized system which evolved from a more basic and primitive metabolic signalling system. Such specialization of primitive forms of signalling would be expected to occur as tissues evolved to more complex forms and the need for signalling over ever greater distances arose. Hence an evolutionary continuum is proposed for cyclic AMP: first, an intracellular metabolic regulator controlling the very basic integrating signals seen in bacteria in response to essential nutrient levels (4), and seen as an evolutionary carryover, perhaps, in the phosphorylase a--phosphorylase b system (5); second, the coordinator of metabolism between nearby cells, chemotactically as occurs in slime mold (*Dicostyleum discoideum*) (6,7), and such as almost certainly occurs, either through the sensing of a scarcity of metabolites or through direct cell-cell contact due to crowding in density dependent growth inhibition (8,9); third, a mediator of a longer range integration, coordinating the metabolic actions of whole groups of cells (tissues), i.e., the hormone system. Placed in this evolutionary continuum, cyclic AMP as an effector in cell division regulation takes on new meaning.

While several authors in this volume have concentrated on the first role of cyclic AMP, a wholly intracellular regulator of metabolism in prokaryotic organisms, we prefer to speculate at the next evolutionary rung of the ladder, possible cell-to-cell signals involving cyclic AMP. It is evident from much recent research that an internally mediated "bacteria-like" cyclic AMP signal probably occurs in mammalian cells facing limited basic nutrients from serum (1,10), perhaps restricting their rates of division. A signal of this form would be independent of the presence of adjacent cells except as they compete for, or prevent access to, these basic serum nutrients. Other results (11a,12) and some of the features of tumor cells themselves are more easily explained in terms of additional cell-to-cell contact signals complementing the basic cyclic AMP rise in response to limiting nutrients. It is our purpose for the balance of this discussion to characterize the essential behavior that a receptor should possess in order to serve as a cell surface signalling agent involved in cell division regulation. We describe a number of behaviors for a particular set of cell surface receptors that strongly suggest they govern a unique cyclic nucleotide regulatory system, perhaps evolutionarily separate and of different purpose than the well known hormonal system.

A review of the work of others suggests that a number

of changes mediated at the cell surface are frequently seen as features of tumor cells and often occur simultaneously when normal cells are transformed in tissue culture: altered cellular adhesiveness (13), defective "contact inhibition" of movement and growth (14), appearance of new surface antigens (15), and altered transport (16).

No proposal as yet has been forthcoming which directly relates specific changes in the cell surface to specific changes in cyclic nucleotide metabolism seen in unrestrained cell division. The work primarily of Hakomori and Murakami (17), Hakomori (18), and Brady and Fishman (19) indicates that a particular group of cell surface glycolipids, the gangliosides, become deranged as an early event in malignancy. The data from this laboratory, summarized herein, suggests that a particular ganglioside, G_{M1}[1] (see Figure 1) has unique behaviors which strongly suggest that it could be the direct link between cell surface aberrations and specific changes in cyclic nucleotide metabolism affecting *both cyclic AMP and cyclic GMP levels* when normal cells escape regulation of their growth. Furthermore, the data suggests that ganglioside G_{M1} governs some behaviors in normal cells which implicate it in the extent that cell-cell contact may be involved in growth regulation in normal tissue.

CELL SURFACE AND GANGLIOSIDE CHANGES BETWEEN NORMAL AND MALIGNANT CELLS

Cell surface changes have long been considered one of the earliest concomitants of malignancy (15,21). The apparent increase in fluidity of cancer cell membranes, first interpreted as the reason for lectin-agglutinability of malignant cells (22-24) has been substantiated by further studies on normal and transformed cell membranes. Edidin and coworkers have used fluorescent antibodies to show that the mobilities of particular surface antigens are far higher in transformed than normal cells (25). Barnett and coworkers have presented spin-label electron paramagnetic resonance studies indicating increased fluidity in the membrane lipids of transformed cells (26). Increased lectin agglutinability of normal cells after treatment with proteases (27,28) suggests that trypsin or other proteases remove normal cellular restraints on receptor mobility. There has been a suggestion that these restraints may be contributed in part by microfilaments and/or microtubules on the interior of the cell (29). Trypsin treatment frees normal cells for a round of division, suggesting that the positioning of receptors may be critical for growth regulation. Lateral mobility of both the proteins and the lipids occurs in normal cells with similar diffusion constants as has been demonstrated by paramagnetic resonance studies (30). Perhaps the major conclusion from all these data is that the enhanced fluidity of a cell membrane sub-

[1]The nomenclature of Svennerholm (Ref. 20) is used to designate the gangliosides.

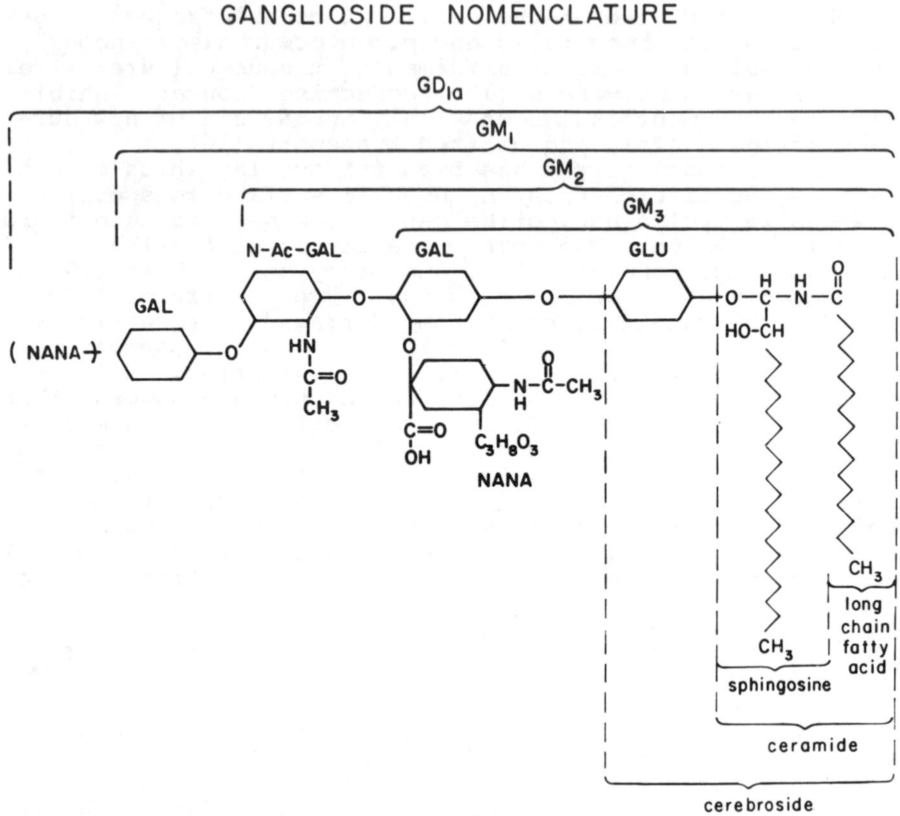

FIGURE 1. *Structure and nomenclature of gangliosides.*

sequent to a malignant event seems to allow ordered surface receptors, whether they are protein or lipid, to attain a state of disorder. This would suggest that the information normally conveyed through any ordered series of cell surface receptors would attain a similar state of disorder.

An important component of cell surface lipids is made up by the glycolipids. This class of lipids includes the gangliosides, which are located primarily in the plasma membrane (19). In addition to the apparent increases in general membrane fluidity, more specific membrane aberrations occur in malignant and transformed cells. Deranged biosynthesis of gangliosides is an early and marked event in transformation (17,19). Gangliosides are constructed by the sequential addition of sugar residues to a lipid core, a process which is mediated by specific transferases (31). The various gangliosides have different sequences of sugar residues, giving them their unique identity (see Figure 1). They possess a common lipid core, their presumed membrane attachment site. As unique cell surface entities, gangliosides have been invoked as recognition sites in immunology and cell-cell interactions.

Early studies of the significant changes of glycolipids in tumor cells resulted from their antigenicity. The ganglioside composed of galactosyl-glucosyl-ceramide was found to be a specific hapten for antisera directed against human tumor cell surface antigens (32,33). Subsequently, Hakomori and Murakami reported that the major ganglioside of baby hamster kidney cells (BHK), *N*-acetyl-neuraminyl-galactosyl-glucosyl ceramide G_{M3} (see Figure 1) decreased to one fourth its normal levels after transformation by polyoma virus (17). As a result of this finding, they suggested that the occurrence of incomplete carbohydrate chains might be a characteristic of malignant cells in general. Additional studies on SV40 and polyoma transformed BHK, 3T3 Swiss mouse cells, and human fibroblasts again verified that the more complex gangliosides, particularly the higher monosialogangliosides (G_{M1}) were greatly reduced in transformed cells as compared with normal cells (34,35). Immaturity of ganglioside biosynthesis does not seem to be an absolutely universal lesion of malignancy or transformation. Transformed chick embryo fibroblasts (19, 36) and certain hepatomas (37) do not seem to show aberrations in biosynthesis of the higher gangliosides. However, a number of recent reports, in newborn rat kidney (NRK) cells, human lung fibroblasts, HeLa human cervical carcinoma cells (19), human leukemic lymphocytes (38), and a series of rat embryo cells (39) all suggest that higher gangliosides, particularly ganglioside G_{M1} are deficient due to reduced biosynthesis when a cell is transformed.

Deficiencies in particular enzymes required for ganglioside biosynthesis are frequently observed in virally transformed cells, and thus probably account for altered patterns or immaturity of ganglioside biosynthesis (40,41). When a specific enzyme-mediated activity was determined to be the lesion occurring subsequent to chemical and x-ray transformation of BALB/c 3T3 mouse cells, the reduced enzyme activity was found to be kinetically similar to that of the normal enzyme. This finding implies that blockage of biosynthesis of the essential enzyme, rather than production of a less efficient mutant enzyme, has occurred (42). There was no indication of accelerated catabolism of gangliosides in SV40 transformed mouse cells (41). A detailed review of the behaviors of cell surface glycosyl transferases has recently been published (43).

The initial examination of ganglioside populations on transformed cells appears confusing, with certain ganglioside species appearing in greater amounts than normal, and other species appearing in lesser amounts. The picture becomes clearer when one realizes that sugar moieties are added stepwise by specific enzymes, and that "immature" gangliosides are the primary characteristic of transformed cells (17). Hence, blockage at one of the early synthetic steps prevents any sequential addition of sugars, resulting in accumulation of the less complete forms (G_{M3}, G_{M2}, and galactosyl-glucosyl ceramide). To these authors, it appears as if a feedback inhibition loop is not being completed, resulting in an excessive pile up of earlier ganglioside forms, and hyperactivity of the enzymes responsible for them, because the final product (monosialoganglioside G_{M1}, or higher disialogangliosides) do not appear on the scene in sufficient quantities to shut down the initial biosynthetic steps.

The relation of gangliosides to cell growth is currently unclear. As remarked in a recent review (44), "as yet no compelling functional role in transformation has been devised for the glycolipids and we will somewhat reluctantly pass them by." A series of studies in hamster cells indicated glycolipid populations which were reduced by viral transformation were also made in lesser amounts during active growth of normal cells (18,45). Rapidly dividing ileal epithelial cells show a gradient of differentiation from crypt to villus, crypt cells being more "primitive" and dividing more rapidly. At least two investigators indicate that more sophisticated ganglioside biosynthesis is occurring on the villus cells, and state that the mitotically active crypt cells have glycoproteins on their surfaces that are incomplete (46,47). However, when Brady and Fishman studied Swiss 3T3 mouse cells, no significant change in pattern and synthesis of gangliosides was seen over a 10-fold range of cell density, suggesting that rapidity of division did not affect ganglioside biosynthesis (19).

Exogenous gangliosides seem to be capable of becoming incorporated in cell membranes, and supply missing receptors in this way (38). Glycolipids from *Salmonella minnesota* R mutants bind to the plasma membrane of normal and transformed rat embryo cells, and inhibit growth, accompanied by a rise in cyclic AMP in the cells (48). In a more specific study, purified gangliosides and portions of gangliosides were studied for their efficacy in reducing division rates of normal and SV40 transformed 3T3 mouse cells. The monosialoganglioside G_{M1}, and to a lesser extent its precursor forms, were effective in reducing cell division rates when supplied exogenously. Ceramide alone, or the 5-sugar residue alone (see Figure 1) were ineffective, and higher (disialo-) gangliosides were less effective in reducing the division rates of cells in culture (49). Neuraminidase therapy (50) of tumors may work by some similar process, reducing higher disialogangliosides to the monosialoganglioside G_{M1}.

As mentioned before, there is currently no proposed role for glycolipids in cell division control. Contacts between cell surfaces seem critical in information transfer in a number of embryological events during development (as reviewed in Ref. 15). Extensive intercellular contact seems required for the complete expression of the morphological changes associated with passage through the cell cycle in Chinese hamster ovary cells (51,52). An early suggestion was made by Roseman (31,53) that cell-to-cell contact resulted in a transfer of specific sugar residues from one cell to the other, and this was a necessary event in recognition. Transglycosylation has been demonstrated experimentally between glycolipids on glass beads and hamster cells (54). Transglycosylation was not observed in the same cells subsequent to viral transformation. Specific changes in the transferases involved in ganglioside biosynthesis have been seen for normal cells at the early stage of cell contact (55). These changes were missing when the cells were virally transformed prior to contacting each other.

Cell surface changes are some of the earliest indications of the transition of a normal cell to a cell with malignant potential. These changes seem to occur in two major areas: (1) an increased general fluidity of the membrane, allowing disorder of possible surface information transmitters and receptors; and (2) deranged synthesis of a particular class of cell surface glycolipids, resulting in immature forms of the monosialoganglioside G_{M1} and omission of the higher disialogangliosides. Ganglioside G_{M1}, particularly, seems to function as a receptor when supplied exogenously to normal or transformed cells, and is apparently instrumental in reducing growth rates, accompanied by elevations in cyclic AMP. Some variations in ganglioside biosynthesis are seen as a function of the rapidity of division of normal cells. All in all, a firm case can be made for participation of ordered ganglioside receptors in the regulation of cell division.

ENTEROTOXINS AND CELL SURFACE GANGLIOSIDES

An important tie-in exists between the cell surface, particularly ganglioside G_{M1}, and intracellular concentrations of cyclic nucleotides. This connection was revealed inadvertently by the enterotoxins of *Vibrio cholerae* and *Escherichia coli*. The enterotoxins produce their biological effect in intact cells by first binding to a specific cell surface receptor, ganglioside G_{M1}, and subsequently activating adenylate cyclase. Excellent reviews are available that provide a comprehensive account of the actions of cholera and *E. coli* (heat labile) enterotoxins in man and experimental animals (93-100). Perhaps the most important observation for our understanding of the effects of these enterotoxins is the recently discovered association that their biological effects are cyclic AMP mediated subsequent to adenylate cyclase activation.

The key observation leading to this association was generated by the studies of Field *et al.* (101-103), using short circuited preparations of rabbit ileal mucosa, by the techniques of Ussing and Zerahn (104) in which the net ion fluxes across the membrane can be directly measured by means of isotope labeling (94). It was demonstrated that the addition of cyclic AMP or theophylline, an inhibitor of the destruction of cyclic AMP, rapidly caused a large increase in short circuit current accompanied by inhibition of sodium absorption and net secretion of chloride and bicarbonate. These changes were identical to those seen after a 30 to 45 min latent period following exposure to cholera toxin.

Other studies also showed an association of the action of cholera toxin with the action of cyclic AMP. Pierce *et al.* (105) demonstrated that cyclic AMP elevating agents such as prostaglandins and theophylline could cause fluid production in the small intestine of dogs when infused into the superior mesenteric artery. Similarly, lipolysis in fat cells (106), and glycogenolysis in platelets and liver homogenates (76), activities which are stimulated by cyclic AMP, were both induced by exposure to cholera toxin.

The first direct evidence of a relationship between intracellular cyclic AMP and cholera toxin was provided by the work of Schafer *et al.* (107) in 1970. Using dog small intestinal mucosal epithelial cells pretreated with crude cholera toxin *in vivo*, they found elevated levels of intracellular cyclic AMP after a lag of 1.5 h. Shortly thereafter Sharp and Hynie (108) used purified cholera toxin to demonstrate enhanced adenylate cyclase activity in a rabbit ileal loop model for cholera. These investigators also showed that the toxin had no effect on cyclic AMP phosphodiesterase. Reports that intestinal adenylate cyclase activity was elevated by cholera toxin in rat, guinea pig (109,110) and man (111) have confirmed the hypothesis and identified the molecular mechanism of cholera.

Based on the extensive information gathered from the study of cholera toxin, and a similar clinical syndrome of the small intestine produced by enterotoxigenic *E. coli* (112,113), it was subsequently determined that heat-labile *E. coli* enterotoxin shares a common molecular mechanism in that it also raises intracellular cyclic AMP levels by activation of intestinal epithelial cell adenylate cyclase (69,114). In a fashion similar to cholera toxin, *E. coli* enterotoxin activates intestinal adenylate cyclase after a lag period of approximately 2 h; this toxin also has no apparent effect on cyclic AMP phosphodiesterase (115). Activation of intestinal adenylate cyclase has also been shown in the dog (116) and human embryonic cells (64).

Competition studies have been performed for the combined addition of both cholera and *E. coli* enterotoxin to intestinal epithelial cells. Results of such studies suggest that both enterotoxins activate the same adenylate cyclase system, since activation by one enterotoxin precludes further activation by the other (117). Furthermore, such competition studies indicate that approximately the same adenylate cyclase velocity is achieved by the enzyme in a variety of cell types when activated maximally by either toxin (99).

In clinical syndromes, the enterotoxins are not ordinarily absorbed from the intestine into the general circulation in sufficient quantities to produce systemic manifestations (95). However, concentrations of the toxin as low as those which are effective in the intestine are capable of stimulating adenylate cyclase activity and of initiating biochemical responses characteristically mediated by cyclic AMP, in a variety of tissues; in fact, the enterotoxins appear to be activators of adenylate cyclase for each mammalian cell system in which they have been tested (97) (Table 1). These observations suggest that the toxins can selectively interact with ubiquitous components of all cells, and that the fundamental mechanism by which toxin-cell interactions modify the activity of membrane-bound adenylate cyclase may be very similar in these cells.

Exposure of cells to toxins for 30-60 s (followed by removal of free toxin from the cell surface by thorough washing) results in increased adenylate cyclase activity which may persist for hours (64). This very rapid fixation of toxin, the persistent nature of the effect after such a brief exposure, and the ability of the toxin to selectively activate an enzyme in the plasma membrane suggests very tight and specific binding to receptor structures located in the cell membrane.

A number of observations from several different laboratories have now indicated that G_{M1} ganglioside functions as a tissue receptor for cholera toxin. It has been observed that gangliosides can interact specifically with cholera toxin so as to block the binding of the toxin to the cell membrane (118-121). In particular, G_{M1} is a highly specific and potent inhibitor of cholera toxin binding; G_{D1a} and G_{M2} are at least 50-fold, and G_{M3} at least 250-fold less effective than G_{M1} (118,122) (see Figure 1). There is a biologically inactive subunit of cholera toxin, choleragenoid, which has cell membrane binding properties almost identical to the intact toxin (123,124) and which can prevent intestinal cells from responding to the toxin (121,125). Choleragenoid is fixed by G_{M1} with the same efficiency as cholera toxin (119,126). Furthermore, evidence for the receptor function of G_{M1} is indicated by the finding that incubation of fat cells (118) and rabbit small intestine (127) with exogenous G_{M1} increases the toxin binding ability of the cells and also the respective lipolytic and diarrheogenic response which suggests that exogenous ganglioside was incorporated as a functional receptor into the cell membrane. It has been shown that the highly specific toxin-binding determinants of the gangliosides appear to reside primarily or exclusively in the carbohydrate end of the ganglioside molecule (119,128). Cuatrecasas has used this observation to construct columns containing ganglioside-agarose derivatives which can selectively extract cholera toxin at concentrations as low as 10^{-11} *M* (129).

Haksar *et al.* (130) found a 50 percent increase in cyclic AMP formation in isolated adrenal cells exposed to neuraminidase prior to cholera toxin treatment. Neuraminidase presumably converts neuraminidase-sensitive di- and tri-sialosylgangliosides to the resistant monosialoganglioside (G_{M1}) (20,127) and in effect, makes available additional membrane receptor sites for toxin.

In an attempt to determine whether the binding sites of cell membranes are identical for both cholera and *E.*

TABLE 1. *Extra-intestinal effects of the enterotoxins of Vibrio cholerae and Escherichia coli**

Cell Type	Enterotoxins *V. cholerae*	Enterotoxins *E. coli*	Effect	Reference
Ascites cancer cells	+		Adenylate cyclase ↑	56
Adrenal cells	+	+	Morphologic changes	57,58
	+	+	Steroidogenesis ↑	
	+	+	Cyclic AMP ↑	59,60
Brain	+		Dopamine-like effect	61
			hyperthermia	62
Epithelioid carcinoma	+	+	Adenylate cyclase ↑	63,64
Erythrocytes				
toad	+		Adenylate cyclase ↑	65
rat	+		Adenylate cyclase ↑	65
pigeon	+		Adenylate cyclase ↑	66
turkey	+		Adenylate cyclase ↑	67
Fat cells	+	+	Adenylate cyclase ↑	68,69
	+	+	lipolysis	70,69
Fibroblasts	+		DNA synthesis ↓	71
Gall bladder	+		Cyclic AMP ↑	72
			fluid absorption ↓	
HeLa cells	+		Spreading ↓	73
Kidney	+		Adenylate cyclase ↑	74

TABLE 1 - cont'd

Cell Type	Enterotoxins *V. cholerae*	*E. coli*	Effect	Reference
Laryngeal carcinoma cells	+	+	Adenylate cyclase ↑	63,64
Leukocytes	+		Adenylate cyclase ↑	75
			Immune-mediated histamine release ↓	
Liver	+	+	Glycogenolysis ↑	76,77
	+		Adenylate cyclase ↑	78
		+	Cyclic AMP ↑	77
	+		Alkaline phosphatase ↑	79
Lymphocyte	+		DNA, RNA, protein synthesis ↓	80
	+		Humoral and cell-mediated immune response to alloantigens ↓	81
	+		Membrane ganglioside redistribution	38
Mast cells	+		Degranulation	82
Melanoma cell	+		Growth inhibition Differentiation tyrosine activity ↑	83
Myocardium		+	Adenylate cyclase ↑	84
Neuroblastoma	+		Adenylate cyclase ↑	84a
Ovary (Chinese hamster)	+	+	Cyclic AMP ↑	85

TABLE 1 - cont'd

Cell Type	Enterotoxins *V. cholerae*	*E. coli*	Effect	Reference
Pancreas	+		Cyclic AMP ↑	86
			Adenylate cyclase ↑	87
			Exocrine secretion ↑	86,87
Platelets	+		Glycogenolysis ↑ phosphorylase a ↑	76
Skin	+	+	Permeability ↑ (erythema, induration)	88,89
Thymocytes	+	+	Adenylate cyclase ↑	91
			Cyclic AMP ↑	90
	+		amino acid transport ↑	90
Thyroid	+	+	Adenylate cyclase ↑ glucose oxidation ↑	92

**The extra-intestinal activities of cholera enterotoxin have previously been reviewed up to 1973 (97). This table summarizes the literature since that time and includes the activities of E. coli enterotoxin (heat labile).*

coli enterotoxins, choleragenoid has been used to block the cellular receptor sites. Choleragenoid prevented the effect of cholera toxin but not that of *E. coli* toxin suggesting that different binding sites might be involved (121,125). In further attempts to delineate the binding properties of *E. coli* enterotoxin, affinity for gangliosides has been studied. The same ganglioside, G_{M1} binds best to both *E. coli* and cholera toxin (125). In addition, preincubation of G_{M1} with *E. coli* toxin completely prevented the biological effect of the toxin when examined with thymocytes (91), Chinese hamster ovary cells (131), and adrenal tumor cells (132). These results suggest that both toxins compete with each other for the same binding site, G_{M1}. Resolution of the discrepancies that exist with respect to the nature of the *E. coli* toxin receptor site will require examination with radioactively-labeled purified preparations of toxin. Such experiments have not yet been reported.

In conclusion, much data suggests that specific cell surface glycolipids, particularly ganglioside G_{M1}, interact with enterotoxins. By binding to the specific ganglioside receptor, enterotoxins are able to activate adenylate cyclase in essentially every mammalian cell system. It is suggested that enterotoxins may have inadvertently served as an experimental probe revealing the presence and the actions of a critical cell surface receptor.

BIDIRECTIONAL CYCLIC NUCLEOTIDE CONTROL

In 1972, Goldberg and his associates (133) proposed the Yin Yang hypothesis of biological control: that a reciprocal dualism existed between the opposing action of cyclic AMP and cyclic GMP, so that the ratio of the two cyclic nucleotides within cells might thus regulate many critical cellular functions. More recently, the dualism between cyclic nucleotides has been extended to the regulation of cell division (15,134-136) as cyclic AMP and cyclic GMP have been implicated in the control of growth of various types of cells in tissue culture.

It has been shown that intracellular levels of cyclic AMP, under a variety of conditions, are inversely related to growth rate (10,137-139 and see Ref. 1). Rapidly growing cells have low cyclic AMP levels and slowly growing cells have high levels (137). The levels of cyclic AMP become particularly high as cell growth becomes inhibited at high densities. On the other hand, cyclic AMP levels fail to rise in a variety of transformed cells even when they grow to very high densities (137).

Cyclic GMP has been postulated as one of the intracellular mediators triggering cell growth (135). Agents that induce proliferation of cultured cells, such as serum and mitogenic stimulation have been found to also stimulate increases in intracellular cyclic GMP (135,140, 146). Other findings have been that exogenous cyclic GMP stimulates DNA synthesis in lymphocytes (141) and in spleen cells (142). Increases in guanylate cyclase activity have been found in regenerating liver following partial hepatectomy (143,144). The increase in activity was specific for enzyme in the particulate fraction, with no change (143) or a decrease in activity (144) found for the soluble enzyme. A similar activation of membrane-associated guanylate cyclase was found in fibroblasts stimulated by fibroblast growth factor (145).

The bidirectionality of cyclic nucleotide influence on biological regulation thus far appears characteristic of more basic cellular functions (e.g., chemotaxis and cell division regulation), while responses to the polypeptide hormones are, for the most part, unidirectionally regulated. Cyclic nucleotide dualism would, therefore, be a desirable feature in any proposed system of cell division regulation. Recent findings in this laboratory suggest that the enterotoxin receptor, ganglioside G_{M1}, is capable of mediating some features of the dual reciprocal changes in cyclic AMP and cyclic GMP levels which constitute the Yin Yang phenomenon (149).

Five to six hours subsequent to cholera toxin exposure, at a time when adenylate cyclase activation is maximal, a concomitant reduction in guanylate cyclase is seen, as shown in Figure 2. Hence the production of cyclic AMP will increase and the production of cyclic GMP will decrease within the cell. A dose-dependent reduction in cyclic GMP-phosphodiesterase activity is also observed after the same time interval, as shown in Figure 3. It is not yet clear whether this reduction is due to interaction with cholera toxin or is simply a response to the reduction in cyclic GMP levels occurring in the cells. A similar response is known to occur with cyclic AMP phosphodiesterase when intracellular cyclic AMP levels fall (147, 148). A study determining the relative rates at which guanylate cyclase and cyclic GMP phosphodiesterase fall would elucidate the temporal and possible causal relationship between the two behaviors seen in Figures 2 and 3. While the three enzymes assayed change their behavior subsequent to exposure to cholera toxin, additional information is necessary to infer causality. For this reason, theoretical curves have been constructed to fit the observed dose-response behaviors using the general binding equation:

$$v = \frac{V\ max}{1 + Kd/(T)^*} \quad [1]$$

(* or (T)/Kd, depending on the "sense" of the curve).

In this equation, V max is the maximal limiting enzyme velocity achievable, (T) is the ligand concentration, in this case cholera toxin, and Kd is the dissociation constant describing the interaction between ligand and binding site. Linearization of equation [1] occurs by taking the reciprocal of each side, and data plotted according to this formulation is shown in the inset in Figure 3. A precise calculation of the kinetically determined Kd has

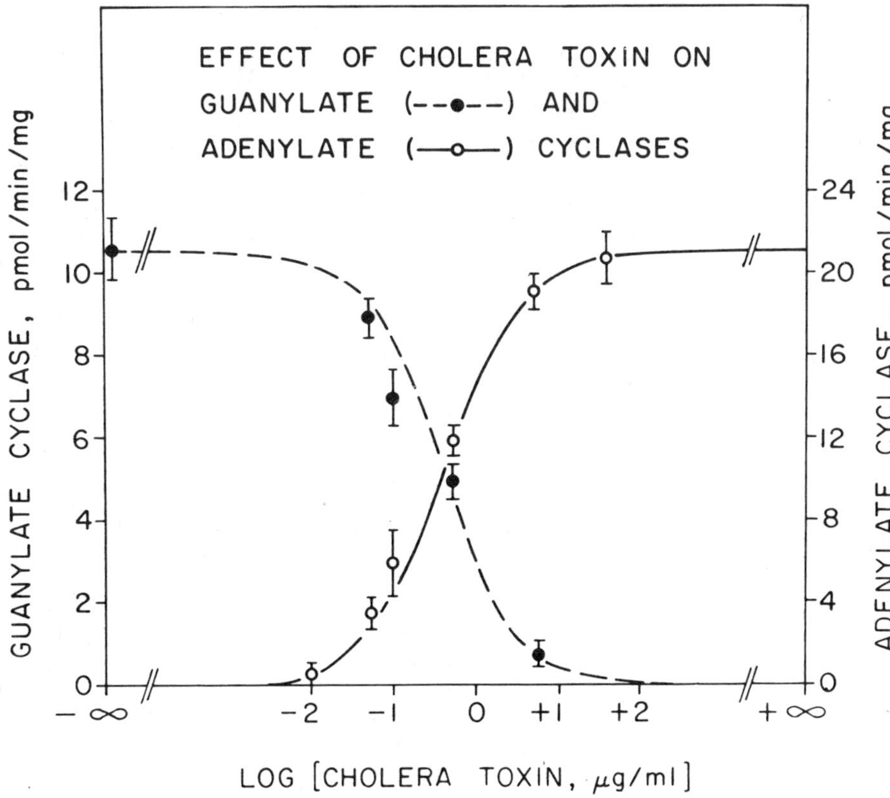

FIGURE 2. *Changes in guanylate (●) and adenylate cyclase activities in rabbit intestinal epithelial cells subsequent to 5 h incubation with varying concentrations of cholera toxin. Experimental points are contrasted with theoretical binding curves: guanylate cyclase (---); adenylate cyclase (——). Both assays were performed using whole cell homogenates with substrate at 10-fold excess over* K_M. *Identical results were obtained for incubation times of 4 and 6 h. For details of experimental conditions see (149). From Kiefer, H. C. et al. (1975), Biochem. Biophys. Res. Commun. 66:1017-1023.*

been obtained for each of the three enzymic responses from such theoretical curve fitting, and the results are shown in Table 2 and in the theoretical fits of the data in Figures 2 and 3. These data imply that binding of a ligand to a *single class* of *equal affinity ligand binding sites* regulates all three enzymic phenomena. The experimentally determined binding constants agree well with each other--and with the known dissociation constant for cholera toxin determined by equilibrium binding studies (128).

As reviewed above, activation of adenylate cyclase and

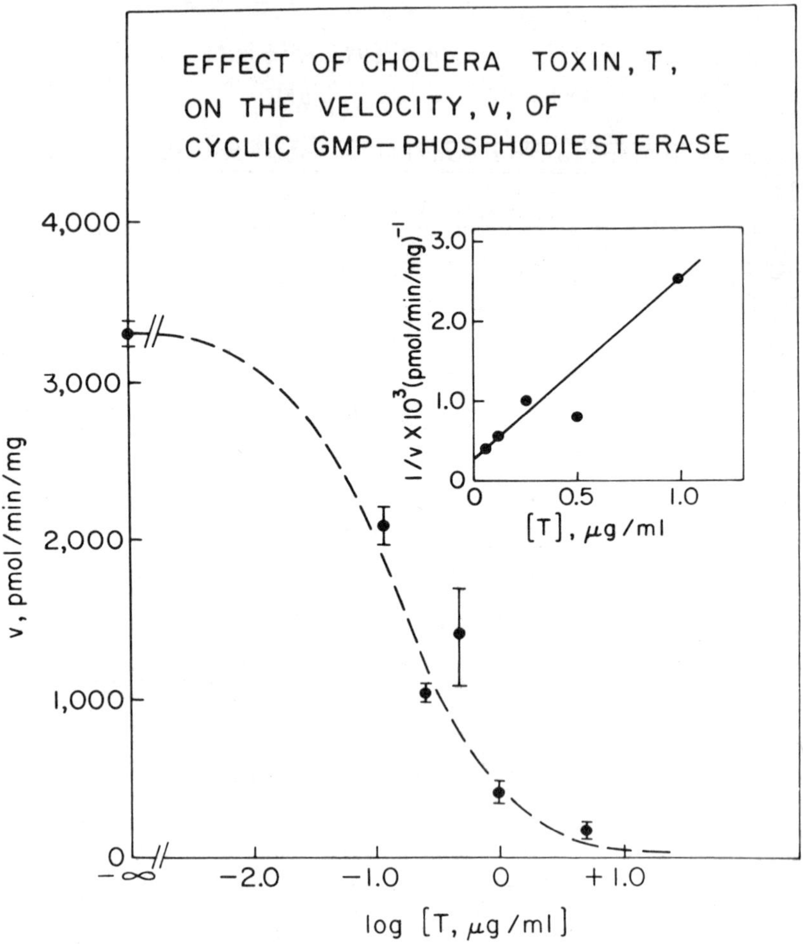

FIGURE 3. *Changes in cyclic GMP phosphodiesterase activity levels (●) in rabbit intestinal epithelial cells subsequent to 5 h incubation with varying concentrations of cholera toxin. The inset shows a linear double-reciprocal binding determination. Experimental points are contrasted with a theoretical binding curve (---); see text. Assays were performed on whole cell homogenates. Identical results were obtained for incubation times of 4 and 6 h. For details of experimental conditions see (149). From Kiefer, H. C. et al. (1975), Biochem. Biophys. Res. Commun. 66:1017-1023.*

deactivation of guanylate cyclase are anticipated signals for cell division arrest. Ganglioside G_{M1} seems capable of mediating these events. It is not known if G_{M1} can mediate the other half of the Yin Yang phenomenon: the reduction of adenylate cyclase activity and increase of guanylate cyclase activity. Simple release of the ligand-

TABLE 2. *Correlation of apparent affinity of cholera enterotoxin for biological activity and equilibrium binding activity*

Apparent Dissociation Constants (K_d) for Cholera Toxin		
Tissue	Biological Activity	K_d
Intestinal epithelial cells	Adenylate cyclase activation	5.0×10^{-9} *M*
	Guanylate cyclase inhibition	4.4×10^{-9} *M*
	cGMP phosphodiesterase	2.1×10^{-9} *M*
	(Average Affinity)	$3.8 \pm 1.3 \times 10^{-9}$ *M*
Fat cell - [^{125}I] - Cholera Toxin*		4.6×10^{-10} *M*
Liver membrane - [^{125}I] - Cholera Toxin*		1.1×10^{-9} *M*

**From Cuatrecasas, P. 1973. Biochemistry 12:3547 (128).*

receptor complex would be expected to return the levels to normal. However, data of several investigators suggests that the stimulus to a cell to divide may involve additional features--elevation of intracellular levels of cyclic GMP above normal (135) and a rapid, apparently active ejection of cyclic AMP from the cell into the surroundings (209). It remains to be seen whether G_{M1} can mediate any cyclic nucleotide behaviors which accompany a stimulus to division, or whether it regulates only those behaviors which occur when cell division becomes limited.

SERUM: EFFECTS ON CELL DIVISION

Considerable effort has been devoted in the last few years to a search for a messenger that could relay information to the cell membrane or the interior of the cell in order to regulate cell division. Serum is an indispensible component in a culture medium for the survival and growth of a majority of mammalian cells; it was, therefore, an obvious place to look for some growth-promoting factor. In 1965, Todaro *et al.* (150) observed that a change in culture medium induced cell division, and that induction was dependent on the concentration of serum used. They suggested that some factor in serum might act directly at the cell surface to modify the interaction between cells in contact; or, that it could act intracellularly on a process leading to DNA synthesis. It probably also acted at the level of RNA synthesis, since RNA was observed to increase very rapidly after serum addition.

One of the characteristics of transformed cells is that they have little or no serum requirement for growth. Burk (151) found evidence for a growth inhibitor which was apparently produced by normal cells and whose effect was counteracted by serum; transformed cells might lack this inhibitor. On the other hand, since growth of normal cells in the presence of fresh serum can go well beyond confluency, and since the saturation density of cells is proportional to the concentration of serum added, Holley and Kiernan (152) have suggested that a factor (or factors) in serum, probably protein in nature, must be the limiting factor for growth of normal, but not transformed cells. This result could be interpreted to mean either that some serum component becomes limiting for a crowded layer of cells and transformed cells have escaped the need for this component; or, that a substance in serum competes with the adjacent cells for an inhibitory surface attachment site. Transformed cells, having few inhibitory surface attachment sites, would show little serum effect.

Castor (153,154) developed a system for cultivating cells with perfused medium and following the progress of the cells with simultaneous cinematographic observation. In this way, depletion of the medium was avoided and inhibitors which might have been produced by the cells would be rapidly removed. Nontransformed cells were again found to cease dividing unless the serum was elevated, while transformed cells were not as sensitive to crowding, suggesting that the production of inhibitors does not limit cell growth, and that normal cells come to equilibrium with a specific serum concentration. Wester-Mark (155) studied growth of glia-like cells under steady-state medium conditions, and he came to the same conclusion as Castor about inhibition of growth due to cell contacts with subsequent release of the inhibition by serum. The importance of cell contact in growth limitation is reinforced by the work of Todaro *et al.* (150). These experimentors "wounded" (i.e., removed a center strip of cells from) confluent growth inhibited cultures, and found that cells adjacent to the "wound" began dividing and filled the area despite the fact that there was *no overall change in serum concentration* from that of the growth-inhibited state. Tritium labeling of newly synthesized metabolites was used to indicate which cells divided. Humphreys (156, 157) also considered cell-cell contacts of prime importance for regulation of growth control. He believed that serum added to confluent cultures separated cells from each other, so that they were able to move again, to divide and pile up; the serum independence of transformed cells could then be simply explained by their lack of adhesiveness.

Serum independence need not be an exclusive property of transformed cells, for "flat" variants have been obtained from transformed cells which have lost most of their transformed phenotype: "flat" variants are morphologically similar to nontransformed cells, poorly agglutinable, are contact inhibited and grow to a low saturation density; also, they are nontumorigenic. These variant cells, however, can grow in medium depleted of serum (158; see also 159-162), i.e., some of the properties generally associated with the transformed phenotype can be dissociated from each other.

It is generally agreed that serum promotes cell division, yet it is totally unknown by what means. Serum activation is certainly a complex process since a number of serum growth factors have been isolated, but none have been fully characterized. Investigations by Holley and others (152,163-165) have resulted in the isolation of a number of serum factors, some of which must act concurrently, since each factor affects "survival" (165), growth, or migration only (164). The fact that a single component of serum might act on migration, i.e., breaking contact only, suggests that increased mobility is not alone sufficient to induce cell division (167). Although proteases have been shown to induce cell growth, most serum factors tested so far have no common protease or esterase activity (166,167). Specific action on particular protein substrates, such as surface glycoproteins has yet to be ruled out.

Serum may express some of its growth regulatory phenomena via the cyclic nucleotides. Serum starvation causes a rise in cyclic AMP (168), even in cells that are not touching each other (10). As previously discussed, serum increases production of cyclic GMP (140), a possible stimulant to cell division. One of the earliest events seen following the addition of serum to quiescent cells is a rapid fall in cyclic AMP levels. This occurs prior to the biosynthesis of DNA in preparation for division (169, 170). Treatment of cells with certain proteases may also lower cyclic AMP, as it induces cell division (171). If the stimulus to division provided by either serum or protease treatment is followed by artificial elevation of intracellular cyclic AMP, the cells do not proceed to divide (139,172-174). Apparently a decrease in cyclic AMP forms an essential part of the signal provided by either agent.

Insulin is frequently a component of serum, and a suggestion has been made that serum effects on growth occur via insulin. Insulin was first reported in 1924 to have a growth-stimulating effect on cells in culture (175). Although, in contrast to serum, insulin is a well-defined entity, its effect on growth is not yet better understood. Temin (176) showed that addition of insulin enabled chicken embryo fibroblasts to grow in medium depleted of serum factors; however, a growth-stimulating factor which he isolated from serum was different from insulin since it could not be inactivated with anti-insulin antiserum (177). On the other hand, Griffiths (178) suggested that stimulation of growth by insulin was not a primary effect, but secondary to the general increase in transport and metabolic functions produced by this hormone.

Insulin can replace some of the factors in serum required for growth of fibroblasts (176). Such treatment results in rapidly decreased levels of cyclic AMP and increased division rates (138,179,180). The insulin effect is apparently due to inhibition of adenylate cyclase. Physiological concentrations of insulin inhibit adenylate cyclase activity in liver and fat cell crude membrane preparations (181-183) as well as the adenylate cyclase activity of fibroblasts (145,179). Insulin has also been shown to stimulate membrane-bound cyclic AMP phosphodiesterase in liver (184) and fat cells (185).

The literature summarized points to the diversity and complexity of the effects of serum. The salient question is still whether serum acts as a nutrient, either directly as a limiting substance, or indirectly, for instance by permitting other molecules to enter; or, whether serum acts at the cell surface, so to speak, coming between cells. Serum may trigger growth through disturbing the microenvironment of a cell, or by attaching to the surface membrane and setting in motion cyclic nucleotide mediated or other unknown processes.

Acknowledging all the complex factors in serum and the multiple ways in which they could and do interact with cells, we would like to focus our attention on the specific ways in which serum factors might interact with cell surface glycolipids. Cuatrecasas indicates that pretreatment of cholera toxin with serum or specific gangliosides prevents the binding to cell membranes and resultant biological effects (128). The presence of gangliosides in serum has been suggested (186), or some other component of serum may explain this observation. A similar inhibition phenomenon has been seen in this laboratory for the interaction of serum with *E. coli* enterotoxin (64). When different concentrations of serum are added to human embryonic ileal epithelial cell cultures, parallel log-dose response curves for the activation of adenylate cyclase, as seen in Figure 4, occur with *E. coli* enterotoxin treatment. The shift in the parallel log dose response curves with increasing serum concentrations indicates competitive inhibition, further suggesting that a component in serum resembles the natural binding site, ganglioside G_{M1}.

The above findings leads us to set up two competing equilibria to explain the inhibition, shown schematically in Figure 5. As seen from this figure, the concentration of adenylate cyclase activated at any concentration of enterotoxin will simply be a function of the concentration of serum factor, F, and the two equilibrium constants, K_T and K_{serum}. Hence, any level of unreacted G_{M1} receptor can be explained by a mathematical combination of the two competing equilibria. If the substance designated by T is considered to be "tissue" instead of "toxin," and K_T represents a specific binding constant describing the interaction of the tissue receptor and ganglioside G_{M1} on an adjacent cell, the intracellular cyclic AMP concentra-

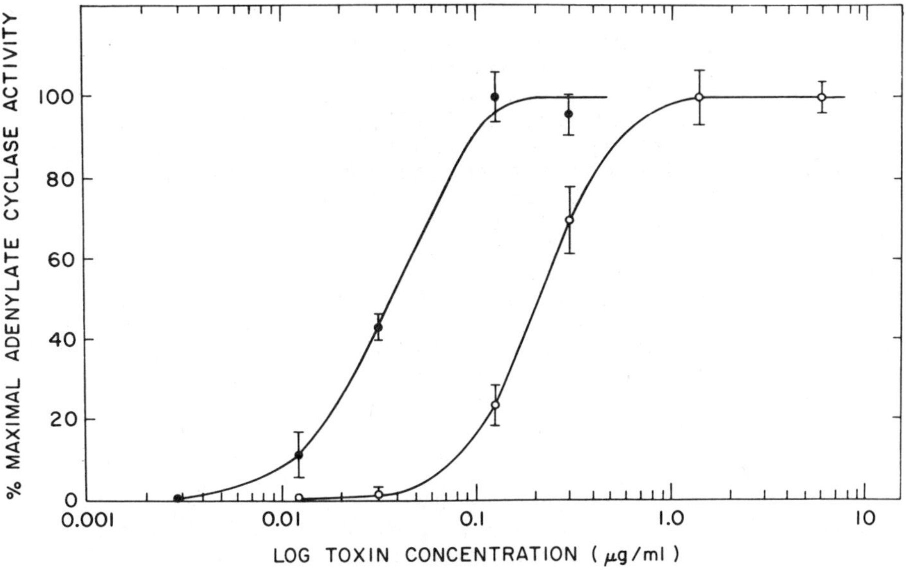

FIGURE 4. *Effect of fetal calf serum on the activation of adenylate cyclase by E. coli enterotoxin (heat-labile). Confluent monolayers of intestinal cells were incubated for 18 h with varying concentrations of enterotoxin added to Hank's basal minimal essential medium either in the absence (●), or in the presence (O) of 15% serum. Both experiments were carried out simultaneously. Each point represents the mean ± SEM of three different experiments. For details of experimental conditions see (64). From Kantor, H. S. et al. 1974. Infect. Immun. 9:1003-1010.*

tion will become a direct result of the serum concentration and the unique strength of two cell's interaction, measured by K_T.

PARALLELS OF CYCLIC AMP METABOLISM IN GROWTH-INHIBITED AND ENTEROTOXIN-STIMULATED CELLS

The elevation of cyclic AMP in response to hormones in cells and tissues is frequently transient. Much of the newly synthesized cyclic AMP is rapidly degraded by cyclic AMP phosphodiesterase to 5'-AMP. Experiments have also suggested that the activation of adenylate cyclase is itself a transient phenomenon. Ho and Sutherland (187) found that rat adipocytes exposed to adrenaline developed a refractory period during which the cells failed to respond to further adrenaline stimulation with elevations of cyclic AMP. The ability of diploid fibroblasts to maintain their refractoriness to adrenergic agents (188) while responding with cyclic AMP elevations to Prostaglandin E_1 verified that some form of "fatigue" of an adenylate cyclase receptor, and not high intracellular levels of phosphodiesterase, was the source of the

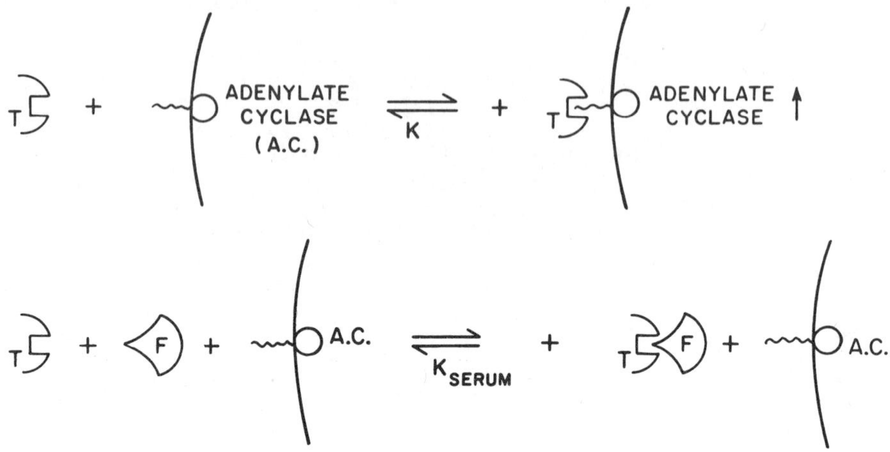

FIGURE 5. *Schematic representation of competing equilibria between serum factor (F) and ganglioside G_{M1} (~~~~~) binding of enterotoxin (T). Equilibrium constants for each interaction are designated by K. Adenylate cyclase activation occurs in response to enterotoxin binding of the ganglioside receptor.*

refractory period. This type of refractory period has been suggested to be a form of tachyphylaxis (188).

The cyclic AMP response to hormones described above seems to differ markedly from the cellular response in two other situations: that due to treatment with enterotoxins, and that found in normal cells approaching high density in culture. In a number of intact cell types exposed to enterotoxins it has been observed that cyclic AMP rises after a variable lag period (0.5-4 h) and continues at high levels for extended periods of time. This persistence is in part due to sustained elevation of adenylate cyclase without any evident refractory period, followed for up to 18 h in ileal cells (63) and embryonic intestinal cells (64), and in melanoma cells for a number of days (83). The persistent elevations in cyclic AMP seem to arise in part also from a cooperative event by cyclic AMP phosphodiesterase. A small number of reports of experiments with single determinations (generally 4 h after exposure to either *E. coli* or cholera enterotoxin) have shown the activity of cyclic AMP phosphodiesterase to be unchanged from basal levels, despite doubled levels of intracellular cyclic AMP (108,115). The only cell line known to these authors that does *not* show sustained high levels of cyclic AMP subsequent to enterotoxin exposure is the isolated thymocyte (91,191). In isolated thymocytes, rapid rises in phosphodiesterase activity occur in response to increases in cyclic AMP elicited by enterotoxins. In other mammalian cells, however, sustained activation of adenylate cyclase and a probable sustained repression of phosphodiesterase, seem to account for the prolonged elevations of cyclic AMP seen subsequent to enterotoxin exposure.

Normal cells approaching high density in culture show a characteristic cyclic AMP response (as reviewed in Ref. 1). Cyclic AMP rises along with the density of the cells, generally due to an increase in adenylate cyclase activity in each cell. Adenylate cyclase remains elevated for days, with no apparent refractoriness. Phosphodiesterase activity seems to be arrested in its pursuit of the rising levels of cyclic AMP, as repeatedly demonstrated in the work of Pastan *et al.* (1). Hence, cyclic AMP produced in response to increasing cell density persists, and seems to be immune to causing a feedback-like response which would destroy it (as occurs with hormonal stimulation). Pastan pointed out recently that the failure of contact-inhibited cells "to increase their phosphodiesterase levels in the face of rising cyclic AMP levels needs to be clarified" (1).

While only a limited number of studies have been performed, cells not subject to density-dependent inhibition of growth seem to grow through confluency in culture without elevations in cyclic AMP, whether they are embryonic cells, cells transformed by viruses or carcinogens, or malignant cells. This behavior can result from changes in adenylate cyclase, changes in phosphodiesterase, or changes in both (1). While there are multiple lesions possible in the abnormal cells, there seems to be one clearcut finding characterizing normal contact-inhibited cells. They do not have rises in phosphodiesterase which accompany the cyclic AMP elevations occurring along with growth cessation.

In reviewing the literature, it appeared to us that the majority of normal mammalian cells (and a few mammalian cancer cells, as well) were repeating this same pattern after exposure to enterotoxin: elevations of cyclic AMP did not seem to "trigger" synthesis of phosphodiesterase. We set out to verify this phenomenon and examine its cause in several tissues, with the expectation, by analogy, or providing a hypothetical explanation of the behavior seen in normal contact-inhibited cells. The tissues examined were rabbit ileum, canine thyroid, mouse thymocytes and Ehrlich Ascites tumor cells. A simple experimental design (Figure 6) allowed us to investigate this type of phosphodiesterase behavior (189).

Three reasonable explanations anticipated for the failure of phosphodiesterase to rise in response to toxin-induced levels of cyclic AMP were: [1] interaction of the hormone itself, with some component of its receptor complex is necessary for production of cyclic AMP phosphodiesterase activity; [2] toxin treatment produces an inhibitor which prevents activation of phosphodiesterase; or [3] cyclic AMP generated subsequent to exposure to enterotoxins is "compartmentalized" in the cell, preventing it from inducing phosphodiesterase activity (147). The experiments presented below were designed to distinguish between these three alternatives.

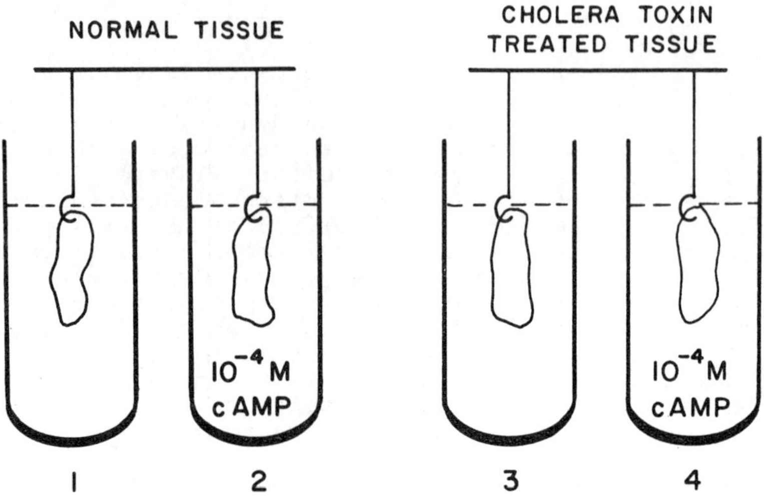

(a) 1 = 3 = 4, no PDE ; 2, PDE : Inhibitor

(b) 1 = 3, no PDE ; 2 = 4, PDE : Compartmentalization

FIGURE 6. *Schematic of experimental tissue incubation procedure. Isolated tissue baths containing oxygenated Krebs-Ringer phosphate medium, pH 7.4, are maintained at 37°C by a Haake circulating water thermostatic pump. Ileal tissues, either pretreated in vivo with cholera enterotoxin (2 μg/ml for 5 h) or control buffers are suspended on tissue hooks in the baths for easy access facilitating removal of portions of tissues at varying intervals. Test substances (e.g., 10^{-4} M cyclic AMP), Prostaglandin E_1 and Vasoactive Intestinal Polypeptide can be added to the incubating media (see text).*

Tissue preincubated *in vivo* with purified cholera enterotoxin (192) for 4-6 h, along with sham-incubated control tissues, were placed in Krebs-Ringer phosphate solution at 37°C and gassed with 95% O_2:5% CO_2. Various additives (hormones, exogenous cyclic AMP, prostaglandins) were made to alternate tissue baths. Phosphodiesterase assays (193) were performed on cell fractions removed from tissues at intervals ranging from 0-180 min. As indicated in Table 3, cholera toxin treated material maintains sustained high intracellular cyclic AMP levels over the entire test period. Figure 7 shows the identical response of normal and enterotoxin treated material when exposed to exogenous cyclic AMP, verifying the presence of an inducible phosphodiesterase, and suggesting that reason [1] above, that the presence of the hormone is necessary for phosphodiesterase induction, cannot account for the data. The identical response of normal and enterotoxin treated material exposed to exogenous cyclic AMP clearly indicates that no inhibitory or blocking process (number

TABLE 3. *Intracellular accumulation of cyclic AMP in response to various stimuli in rabbit ileal epithelial cells*

	Intracellular Cyclic AMP, picomoles/mg protein								
	Time, in min								
Tissue Treatment	0	10	15	20	30	45	60	80	120
Control	27.1	--	--	--	25.2	--	--	26.2	27.5
Prostaglandin E_1 (PGE_1) (10^{-5} *M*)	27.3	30.1	63.8	33.1	--	55.3	--	--	--
Vasoactive Intestinal Polypeptide (VIP) (50 μg/ml)	26.7	32.3	46.1	34.8	34.2	--	--	--	--
Cholera Enterotoxin Pretreatment*	64.5	--	--	--	--	63.4	--	66.2	62.0
Exogenous Cyclic AMP (10^{-4} *M*)	26.2	--	--	27.5	--	29.8	32.8	63.8	37.5

Each value is the average of duplicate determinations

**Cells were pretreated in vivo with 2 μg/ml of cholera enterotoxin for 5 h prior to immersion in tissue incubation baths (Figure 6).*

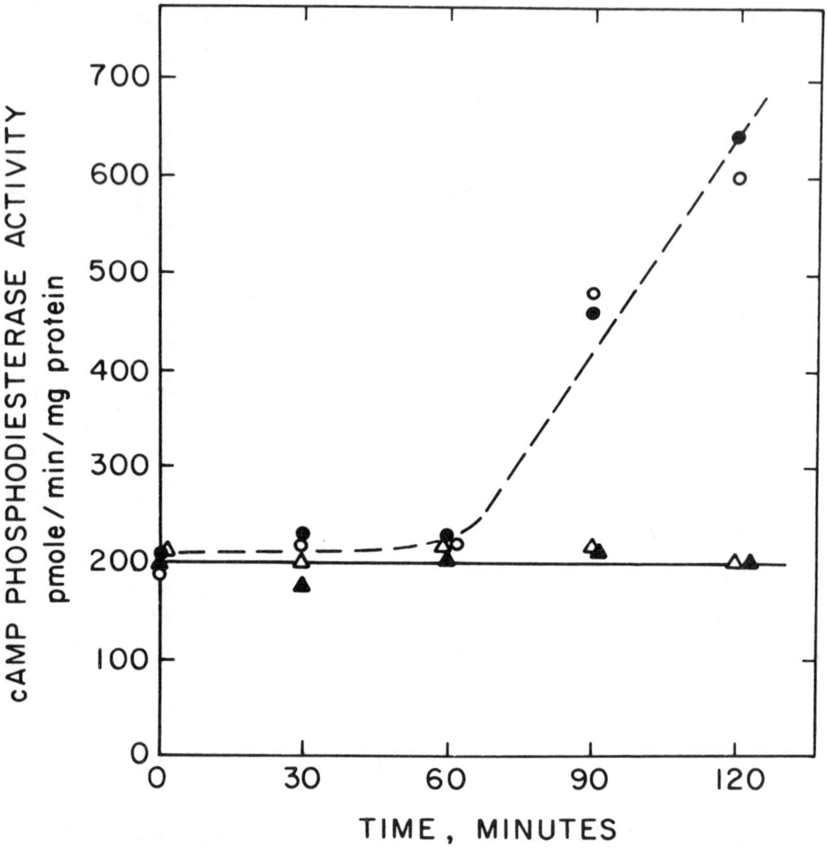

FIGURE 7. *Elevation of phosphodiesterase activity in response to incubation with 10^{-4} M cyclic AMP. Phosphodiesterase activity (193) in picomoles cyclic AMP hydrolyzed/min/mg protein is shown for cholera toxin pretreated (Δ) and control (▲) tissues incubated in Krebs-Ringer phosphate medium with no additives. Similar determinations are shown for cholera toxin pretreated (o) and control (●) tissues incubated in medium containing 10^{-4} M cyclic AMP. Each point is the average of triplicate determinations, and the entire experiment has been repeated four times with essentially identical findings. For details of experimental design see (189) and legend to Figure 6.*

[2] above) occurs in toxin treated tissue. A similar experiment, with the exception that a specific hormone, Vasoactive Intestinal Polypeptide (VIP) or prostaglandin E_1 is used to cause endogenous cyclic AMP accumulation, is shown in Figure 8. Intracellular cyclic AMP levels are summarized in Table 3 for all experiments with ileal epithelial cells. The rather rapid alternating rise and fall of cyclic AMP, correlating with an alternating rise and

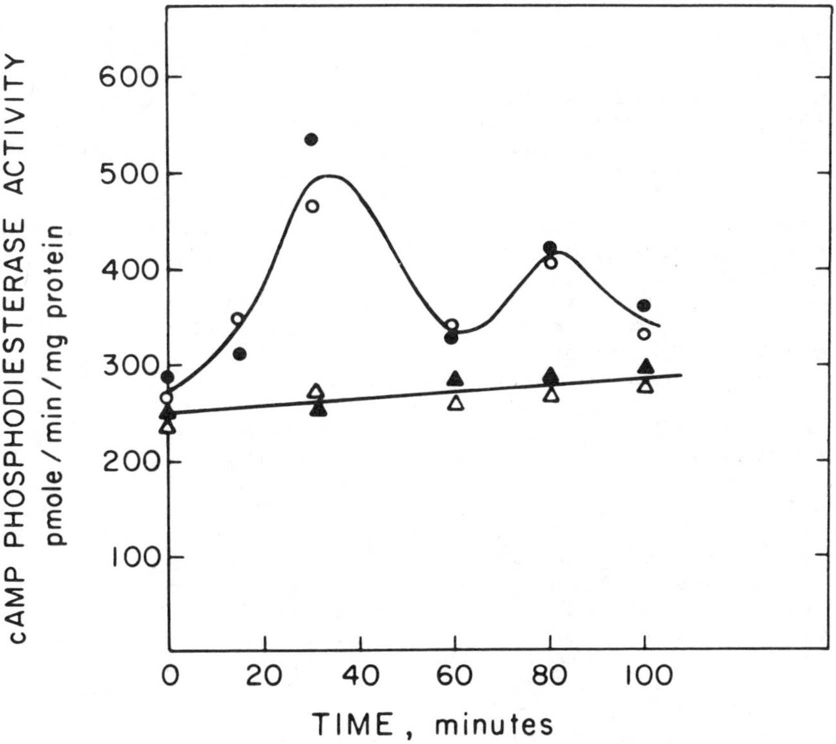

FIGURE 8. *Variations in phosphodiesterase activity in response to incubation with 10^{-5} M Prostaglandin E_1. Phosphodiesterase activity (193) in picomoles cyclic AMP hydrolyzed/min/mg protein is shown for cholera toxin pretreated (Δ) and control (▲) tissues incubated in Krebs-Ringer phosphate medium with no additives. Similar determinations are shown for cholera toxin pretreated (o) and control (●) tissues incubated in medium containing 10^{-5} M Prostaglandin E_1. Each point is the average of triplicate determinations, and the entire experiment has been repeated two times with essentially identical findings. For details of method see (189) and legend to Figure 6.*

fall in phosphodiesterase activity, may be a characteristic of ileal tissues, in which both VIP and Prostaglandin E_1 cause a cyclic AMP-mediated net fluid production. The data summarized in Table 3 clearly indicate that hormone-produced intracellular cyclic AMP levels of the same magnitude or less than toxin-produced cyclic AMP levels elicit a concomitant rise in phosphodiesterase.

The discrepancy between the time course of phosphodiesterase induction with exogenous cyclic AMP and hormones was noted (compare Figures 7 and 8) and examined further. Because of the apparent differential against diffusion of cyclic AMP into the cells, indicated by a comparison of extracellular (10^{-4} *M*) and intracellular (ca. 10^{-6} to 10^{-7} *M*) cyclic AMP, the lag in Figure 7 was interpreted as due to the increased time required for diffusion of cyclic AMP into the cells. To test this hypothesis, intracellular cyclic AMP levels were plotted vs elicited phosphodiesterase activity for both exogenous cyclic AMP and hormones, as shown in Figure 9. While not a complete correlation, this figure suggests that phosphodiesterase is activated in a dose-dependent fashion by the level of intracellular cyclic AMP generated, whether caused by exogenous cyclic AMP diffusing in or by endogenous production upon hormonal stimulation.

Careful fractionation of rabbit ileal epithelial cells has shown enterotoxin and Prostaglandin E_1 binding and adenylate cyclase activation to reside in the same cells (194) and to cause the identical cyclic AMP-mediated process. For this reason, results presented herein must be considered to indicate that cyclic AMP generated by enterotoxin stimulation following binding to the G_{M1} ganglioside receptor is sequestered in a protective way to avoid eliciting phosphodiesterase activity with subsequent hydrolysis; hormonally generated cyclic AMP, in the same cell, enjoys no such immunity.

The notion that cyclic AMP may exist inside the cell, either in metabolic pools, or actually "walled off" in compartments, is not a new proposal. Kuo and deRenzo (195) were among the earlier proponents of the compartmentalization of cyclic AMP in cells. While analyzing the cyclic AMP response to a number of agents in adipocytes, they found results "consistent with the hypothesis that cyclic AMP exists in different intracellular compartments, only one of which, comprising a small fraction of total cellular cyclic AMP, is involved in activation of the hormone-sensitive lipase." Since that time, a number of investigators have pointed out that apparent disparity between the 10^{-6} and 10^{-7} *M* cyclic AMP levels existing as a steady-state phenomenon in cells, and that picomole levels of cyclic AMP contributed to this when hormonal stimulation occurs. This disparity is pointed out for ACTH stimulation of steroidogenesis, in which the accumulation of cyclic AMP required 20- to 40-fold as much hormone as the stimulation of the auspiciously mediated process (196). Such "nonlinearity" of physiological response to cyclic AMP may most easily be resolved by some notion of compartmentalization.

The existence of a pool of cyclic AMP in HeLa cells which appears protected from phosphodiesterase degradation has suggested to Hilz *et al.* that basal cyclic AMP in these cells is somehow protected, and does not rapidly turn over (197). An additional suggestion that basal cyclic AMP is different from hormonally stimulated cyclic AMP comes from the work of Exton *et al.* in studies of perfused liver (198). Their findings indicated that much of the cyclic AMP generated intracellularly in livers perfused with glucagon or epinephrine passed relatively freely across the liver cell membrane, whereas much of the cyclic AMP in untreated livers was retained in the cells.

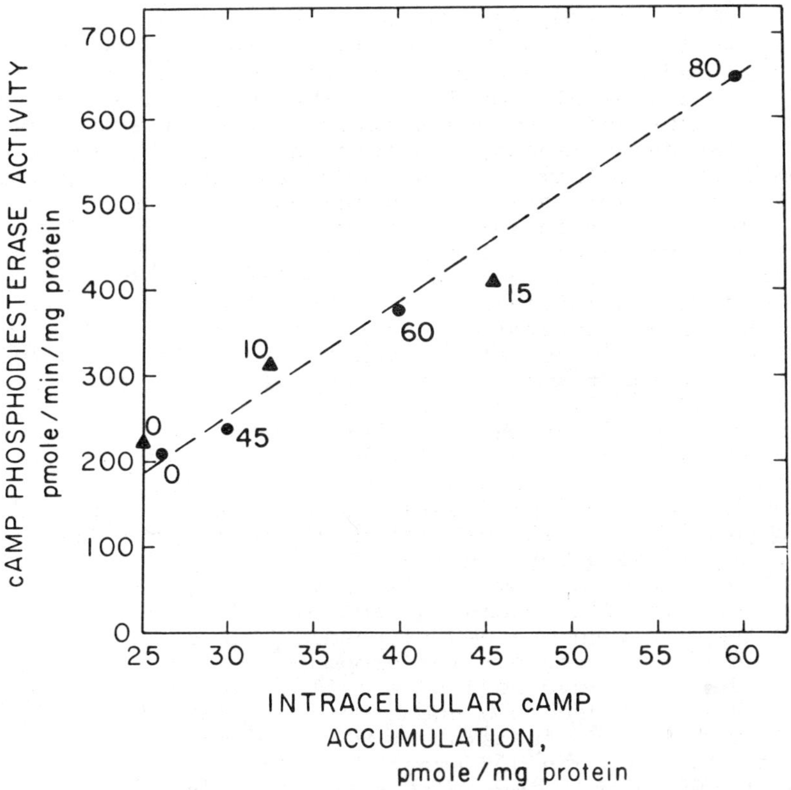

FIGURE 9. *Induction of phosphodiesterase activity (193) as a function of intracellular levels of cyclic AMP (204) achieved in the presence of 10^{-4} M cyclic AMP (●) or 50 μg/ml of the Vasoactive Intestinal Polypeptide (▲) in the incubating medium. The numbers adjacent to each point indicate the incubation period, in min, required to achieve the indicated intracellular concentration of cyclic AMP for the particular stimulating substance present in the medium. Results indicate the average of duplicate determinations for each parameter. For details of method see legend to Figure 6.*

These results would be consistent with a restrained pool of basal cyclic AMP and a mobile pool of hormonally stimulated cyclic AMP.

We have also presented cellular behavior most readily interpreted as compartmentalization of cyclic AMP. Although data presented here are for ileal epithelial cells, preliminary experiments indicate that a similar phenomenon occurs in other tissue such as thyroid (191). Since the literature previously cited presents a number of instances in which cyclic AMP response to enterotoxins is highly protracted, some meaningful explanation must be made for each case. The "compartmentalization" proposed herein may be the most logical explanation. We would like to make the added suggestion that compartmentalization reveals two totally different cellular functions of cyclic AMP: one that is later in evolution, is specialized, and regulates hormonal function; and, one that is evolutionarily more primitive, represents a more protected function, is implicated in cell-cell social organization, and, by analogy with the work of others, is related to basal cyclic AMP.

In summary, generation of cyclic AMP in contact-inhibited normal cells shares two identical features with ganglioside G_{M1}-mediated cyclic AMP production: [1] the lack of a refractory behavior of adenylate cyclase upon extended activation; and [2] failure of phosphodiesterase to rise in the face of intracellular accumulation of cyclic AMP. A parallel can be made between the stimulus to cyclic AMP accumulation in dense cultures and the stimulus to cyclic AMP accumulation subsequent to enterotoxin exposure. According to this analogy, the integrity of some compartmentalizing phenomenon, either a specific binding protein or a membraneous division, becomes a possible lesion in the multiple-lesion complex that is malignancy. Preliminary results by this group indicate that at least one malignant cell, the Ehrlich Ascites tumor cell, appears to be missing the protective compartment, allowing cholera toxin-mediated cyclic AMP rises to induce markedly high levels of phosphodiesterase activity. Elevated cyclic AMP levels in neuroblastoma cells, reported by Prasad (199,200) may follow a similar pattern. In these cells, phosphodiesterase hyperactivity seems to mediate the lack of differentiation and malignant potential, and phosphodiesterase inhibition is more significant than adenylate cyclase activation in encouraging differentiation and reduction of growth rate (201).

DISCUSSION

The foregoing sections have attempted to present selected reviews of aberrations seen when normal cells develop features of unrestrained growth. A number of these aberrations occur in cell surface properties and in cyclic nucleotide metabolism. However, no hypothesis for a direct relationship between the cell surface and cyclic nucleotide metabolism, in either normal or malignant cells has been proposed to date. While glycoproteins have been implicated as critical cell surface receptors in malignancy, they currently have no known interaction with cyclic nucleotide metabolism and are frequently found in interior membranes, as well as in the plasma membrane (202). Glycolipids, on the other hand, occur almost exclusively in the plasma membrane (19), have a high degree of recognition specificity by virtue of their unique sugar sequences, and have been proposed as cell surface receptors in immunological and cell-cell interactions (202). That specific ganglioside glycolipids might regulate adenylate cyclase in cell growth regulation has been previously proposed (99,203). The early derangement in ganglioside biosynthesis when viral or carcinogen-induced transformation occurs is strongly supportive of the thesis that gangliosides may be critically involved in regulating normal growth patterns.

A number of features of a particular cell surface glycolipid, ganglioside G_{M1}, have been outlined. This glycolipid has the apparently unique property of serving as a binding site for *V. cholerae* and probably *E. coli* enterotoxin, through which these toxins activate adenylate cyclase in ileal epithelial cells *in vivo* to cause their pathological sequelae. When other mammalian cells are exposed *in vitro* to toxins, the cells also respond with adenylate cyclase elevation, via toxin binding to ganglioside G_{M1} (207,208). Since it is highly unlikely that mammals would have developed a specific receptor simply to suffer the misfortunes of cholera, one must make the assumption that enterotoxins are inadvertently revealing a normal cell surface receptor with a specific function.

For the past four years the intent of research in this laboratory has been to amass sufficient information concerning the behavior of this receptor to define its specific function. A summary of the known properties is enlightening:

1. The apparent omnipresence of this receptor on mammalian cells (see Table 1) compared with the narrow tissue specificity of hormonal receptors, is suggestive of a more basic, or *more primitive* function for this receptor than for the adenylate cyclase system involved in hormonal function.

2. As previously mentioned, ganglioside G_{M1} has a deranged synthetic pattern following transformation of a number of cell types suggesting that it could not adequately convey cellular information once a malignant event has occurred in these cells. Furthermore, when added to cell cultures in log phase, it can be incorporated into and become a functional structure of the cell membrane with subsequent reduction of division rates.

3. The activation of adenylate cyclase via ganglioside G_{M1} can be competively blocked by some factor in serum. Such an event would lower functioning intracellular cyclic AMP levels in direct proportion to the amount of serum present. (A direct inference from Figures 4 and 5, summarized in Figure 10.)

4. Ganglioside G_{M1} apparently governs several events of cyclic nucleotide metabolism which have been demonstrated in ileal epithelial cells: [1] adenylate cyclase activation with concomitant guanylate cyclase reduction (the Yin Yang phenomenon) (134) and either a direct or resultant lowering of cyclic GMP phosphodiesterase activity in the cell; [2] an ability to mediate the production of cyclic AMP into a protected environment in which it cannot elicit the rises in cyclic AMP phosphodiesterase which would destroy it. While these latter results have been characterized to date in one cell type, the work of

others (Table 1) and preliminary experiments from this laboratory suggest they will be seen in other tissues.

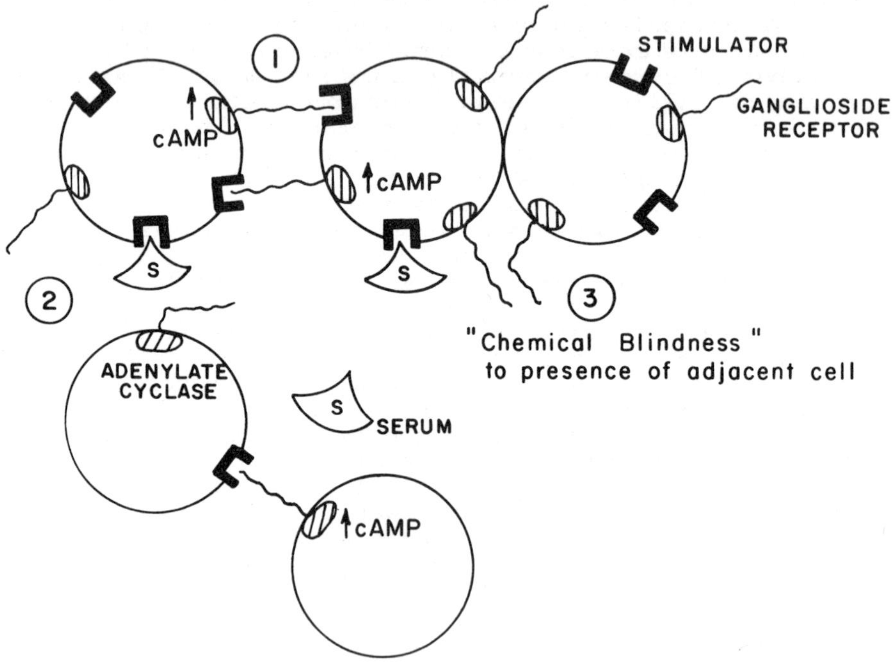

FIGURE 10. *Schematic representation of the manner in which the G_{M1} ganglioside receptor could be involved in cell-cell interactions relating to cell division regulation. The cell surface structure ([) noted as "stimulator" resembles enterotoxin in its ability to bind to ganglioside (〰) and activate adenylate cyclase (⬭). A possible growth stimulatory role for a serum factor (▽) is indicated schematically.*

These features of ganglioside G_{M1} present a striking similarity to many behaviors seen during cell division regulation, and suggest a number of possible lesions which may contribute to unrestrained growth in malignancy. Some of these features and lesions are summarized in a schematic (Figure 10). Obviously some sort of cell-cell contact is envisioned in this schematic. While cell-cell contact is currently a moot point in theories of cell division regulation, it almost certainly is at least partially involved in a number of cell systems (11,53). Perhaps a role for ganglioside G_{M1} as a noncontact receptor could be envisioned, but direct cell-cell contact allows the full value of this receptor for explaining certain developmental and tumor phenomena to be more clearly evident.

Advantages of "Unidirectional Negative Turnoff Signal"

As has previously been mentioned, it is too simplistic a notion to assume that any phenomenon as important as the integration of relative rates of division of the varying cells composing an organism would be relegated to a single, fragile receptor system. However, regardless of the number of systems actually involved, a unidirectional negative signalling device has certain basic topological features which make it uniquely attractive in accounting for certain aspects of growth and development. This section is an attempt to deal with the logical implications, or dialectic properties of such a signal.

Figure 10 depicts in schematic form a number of the properties of a unidirectional negative signal: [1] a "partner" cell has a structure ([) that tells an adjacent cell possessing a receptor (~~~) to elevate its cyclic AMP, and hence, by current dogma, reduce its division rate; an important corolary is that the first cell is not at all affected by having given a negative growth signal to the adjacent cell (i.e., the message is *unidirectional*); [2] geographical alignment of the "stimulator" structure and receptor is essential to the transfer of the message--misalignment, or fluidity of membranes allowing randomness of receptor and structure, would produce cells "chemically blind" to each other's presence; [3] a substance competing for the structure ([) can prevent the negative growth signal from being given to an adjacent cell; a substance competing for the receptor (~~~) may accidentally trigger the negative signal.

The rigorous implication of the above criteria are particularly useful in constructing hypotheses for two puzzling situations: directional tissue growth during development of an embryo, and benign vs malignant tissue overgrowth. In development of tissues, cells frequently overgrow each other in layers, oblivious to the presence of underlying cells only to stop abruptly when fusing at a predetermined media line (e.g., in the embryologic formation of the structures of the upper lip). Predetermined cell surface receptor patterns providing for "chemical blindness" to underlying cells, with matching negative receptors for fusion at the median line could provide such a pattern of growth. Surface-surface recognition among embryo cells (205,206) is a well known phenomenon, suggesting some form of coded recognition. While the set of receptors described is obviously too simplistic to explain all developmental patterns, it does offer a simplified hypothesis for *directional* growth control in development.

Any theory of growth regulation should be capable of explaining benign vs malignant tissue overgrowths. Such a distinction comes automatically with a unidirectional negative receptor. Single point mutation or absence of single proteins is the simplest known lesion in cell metabolism. Therefore, one can assume, as the simplest event, that a point mutation occurs contributing to a defective ganglioside receptor or to a defective "stimulator" structure (see Figure 10). Ample evidence of deficiencies in single specific transferases which contribute to disordered ganglioside biosynthesis in malignancy has been presented. If the ganglioside receptor is defective, the cell cannot have its cyclic AMP elevated via this receptor, and hence, by current dogma, may continue to divide. It will be invasive, as it cannot be turned off; if it breaks free and migrates to another area, it cannot be turned off by normal cells, since it is "chemically blind" to their presence (metastasis). On the other hand, a defect in the "stimulator" structure occurring while a cell has normal ganglioside receptors would cause a tissue overgrowth with benign characteristics. Such cells, when adjacent, would not be capable of giving signals repressing growth to each other, and hence would continue dividing; however, when adjacent to normal cells, they would be easily turned off by the normal cells (encapsulated, nonmetastasizing).

CONCLUSION

In conclusion, we have characterized the behavior of a particular glycolipid cell surface receptor, ganglioside G_{M1}. Its plasma membrane location, its widespread capability in mammalian cells of affecting cyclic nucleotide metabolism in unique fashions, its interaction with serum factors, its ability to limit growth when added exogenously to tissue cultures, and its early aberrations in malignancy all point to a possible role in cell division regulation. The advantage of such a receptor system in suggesting the nature of certain events in developing systems and in benign vs malignant overgrowths has been presented. Further, it is suggested that this receptor complex defines an evolutionarily unique cyclic AMP system having a position intermediate between prokaryotic cells and the long range tissue-to-tissue signals seen in hormone action.

ACKNOWLEDGEMENT

We would like to express our appreciation for the time and effort generously given by Diane Hanner and her staff in the preparation of this manuscript.

REFERENCES

1. Pastan, I. H., Johnson, G. S., and Anderson, W. B. 1975. Ann. Rev. Biochem. *44*:491-522.
2. Clarkson, B. and Baserga, R. (eds.) *Control of Proliferation in Animal Cells*. Cold Spring Harbor Conf. Cell Proliferation, Vol. 1, Cold Spring Harbor Lab. (1974).
3. Sutherland, E. W. 1972. Science *177*:401-408.
4. Pastan, I. and Perlman, R. 1970. Science *169*:339-344.

5. Walsh, D. A., Perkins, J. P., and Krebs, E. G. 1968. J. Biol. Chem. *243*:3763-3765
6. Bonner, J. T. 1970. Proc. Nat. Acad. Sci. USA *65*: 110-113.
7. Robertson, A., Drage, D. J., and Cohen, M. H. 1972. Science *175*:333-335.
8. Abercrombie, M. and Heaysman, J. E. M. 1954. Exp. Cell Res. *6*:293-306.
9. Stoker, M. G. P. and Rubin, H. 1967. Nature *215*: 171-172.
10. Seifert, W. and Paul, D. 1972. Nature New Biol. *240*:281-283.
11. Dulbecco, R. and Elkington, J. 1973. Nature *246*: 197-200.
11a. Mannino, R. J. and Burger, M. M. 1975. Nature *256*:19-22.
12. Levine, E. M., Jeng, D., and Chang, Y. 1974. J. Cell Physiol. *84*:359-363.
13. Abercrombie, M. and Ambrose, E. J. 1962. Cancer Res. *22*:525-548.
14. Stoker, M. 1967. Current Topics in Developmental Biology *2*:107-128.
15. Burger, M. M. 1973. Current Topics Cell Regulation *3*:135-193.
16. Pardee, A. B. 1964. Nat. Cancer Inst. Monogr. *14*: 7-20.
17. Hakomori, S. and Murakami, W. T. 1968. Proc. Nat. Acad. Sci. USA *59*:254-261.
18. Hakomori, S. 1970. Proc. Nat. Acad. Sci. USA *67*: 1741-1747.
19. Brady, R. O. and Fishman, P. H. 1974. Biochim. Biophys. Acta *355*:121-148.
20. Svennerholm, L. 1963. Neurochem. J. *10*:613-623.
21. Pardee, A. B. 1975. Biochim. Biophys. Acta *417*: 153-172.
22. Pollack, R. E. and Burger, M. M. 1969. Proc. Nat. Acad. Sci. USA *62*:1074-1076.
23. Inbar, M. and Sachs, L. 1969. Proc. Nat. Acad. Sci. USA *63*:1418-1425.
24. Nicolson, G. L. 1971. Nature New Biol. *233*:244-246.
25. Edidin, M. and Weiss, A. 1972. Proc. Nat. Acad Sci. USA *69*:2456-2459.
26. Barnett, R. E., Furcht, L. T., and Scott, R. C. 1974. Proc. Nat. Acad. Sci. USA *71*:1992-1994.
27. Rosenblith, J. Z., Ukena, T. E., Yin, H. H., Berlin, R. O., and Karnovsky, M. J. 1973. Proc. Nat. Acad. Sci. USA *70*:1625-1629.
28. Burger, M. M. and Goldberg, A. R. 1967. Proc. Nat. Acad. Sci. USA *57*:359-366.
29. Edelman, G. M., Yahana, I., and Wang, J. L. 1973. Proc. Acad. Sci. USA *70*:1442-1446.
30. Edidin, M. 1974. Rev. Biophys. Bioeng. *3*:179-201.
31. Roseman, S. 1970. Chem. Phys. Lipids *5*:270-297.
32. Rapport, M. M., Graf, L., Skipski, V. P., and Alonzo, N. F. 1959. Cancer *12*:438-445.
33. Tal, C. 1965. Proc.Nat.Acad.Sci.USA *54*:1318-1321.

34. Hakomori, S., Teather, C., and Andrews, H. 1968. Biochem. Biophys. Res. Commun. *33*:563.
35. Mora, P. T., Brady, R. O., Bradley, R. M., and McFarland, V. W. 1969. Proc. Nat. Acad. Sci. USA *63*:1290.
36. Warren, L., Critchley, D., and MacPherson, I. 1972. Nature *235*:275-277.
37. Emmelot, P. 1973. Eur. J. Cancer *9*:319-333.
38. Revesz, T. and Greaves, M. 1975. Nature *257*:103-106.
39. Langenbach, R. 1975. Biochim. Biophys. Acta *388*: 231-242.
40. Brady, R. O., Fishman, P. H., and Mora, P. T. 1973. Fed. Proc. *32*:102-108.
41. Cumar, F. A., Brady, R. O., Kolodny, E. H., McFarland, V. W., and Mora, P. T. 1970. Proc. Nat. Acad. Sci. USA *67*:757-764.
42. Fishman, P. H., Brady, R. O., and Todaro, G. J. 1975. J. Biol. Chem. *250*:55-60.
43. Shur, B. D. and Roth, S. 1975. Biochim. Biophys. Acta *415*:473-512.
44. Pollack, R. E. and Hough, P. V. C. 1974. Ann. Rev. Med. *25*:431-446.
45. Sakiyama, H., Gross, S. K., and Robbins, P. W. 1972. Proc. Nat. Acad. Sci. USA *69*:872.
46. Weiser, M. W. 1973. J. Biol. Chem. *248*:2536-2541.
47. Isselbacher, K. J. 1974. Ann. Int. Med. *81*:681-686.
48. Brailovsky, C., Trudel, M., Lallier, R., and Nigam, V. N. 1973. J. Cell Biol. *57*:124-132.
49. Keenam, T. W., Schmid, E., Franke, W. W., and Wiegandt, H. 1975. Exp. Cell Res. *92*:259-270.
50. Rosato, F. E., Brown, A. S., Miller, E. E., Rosato, E. F., Mullis, W. F., Johnson, J., and Moskovitz, A. 1974. Surg. Gyn. Ob. *139*:675-682.
51. Porter, K. R., Prescott, D., and Frye, J. 1973. J. Cell Biol. *57*:815-836.
52. Rubin, R. W. and Everhart, L. P. 1973. J. Cell Biol. *57*:837-844.
53. Roseman, S. 1973. In: *The Neurosciences: III*, Study Program (Schmitt, F. O., ed.), MIT Press, Cambridge, Mass., pp. 795-804.
54. Yogeeswaran, G., Laine, R. A., and Hakomori, S. 1974. Biochem. Biophys. Res. Commun. *59*:591-599.
55. Yogeeswaran, G. and Hakomori, S. 1975. Biochemistry *14*:2151-2156.
56. Bitensky, M. W., Wheeler, M. A., Mehta, H., and Miki, N. 1975. Proc. Nat. Acad. Sci. USA *72*: 3844-3848.
57. Donta, S. T., King, M., and Sloper, K. 1973. Nature New Biol. *243*:246-247.
58. Donta, S. T., Moon, H. W., and Whipp, S. C. 1974. Science *183*:334-336.
59. Kwan, C. N. and Wishnow, R. M. 1974. Exp. Cell Res. *87*:436-438.
60. Kwan, C. N. and Wishnow, R. M. 1974. Infect. Immun. *10*:146-151.

61. Miller, R. J. and Kelly, P. H. 1975. Nature *255*: 163-166.
62. Clark, W. G., Cumby, H. R., and Davis, H. E., IV. 1974. J. Physiol. *240*:493-504.
63. Kantor, H. A. Unpublished observations.
64. Kantor, H. S., Tao, P., and Wisdon, C. 1974. Infect. Immun. *9*:1003-1010.
65. Bennett, V. and Cuatracasas, P. 1975. J. Membr. Biol. *22*:1-28.
66. Gill, D. M. 1975. Proc. Nat. Acad. Sci. USA *72*: 2064-2068.
67. Field, M. 1974. Proc. Nat. Acad. Sci. USA *71*:3299-3303.
68. Sahyoun, N. and Cuatrecasas, P. 1975. Proc. Nat. Acad. Sci. USA *72*:3438-3442.
69. Evans, D. J., Chen, L. C., Curlin, G. T., and Evans, D. G. 1972. Nature New Biol. *236*:137-138.
70. Vaughn, M., Pierce, N. G., and Greenough, W. B., III. 1970. Nature *225*:658-659.
71. Hollenberg, M. D. and Cuatrecasas, P. 1973. Proc. Nat. Acad. Sci. USA *70*:2964-2968.
72. Mertens, R. B., Wheeler, H. O., and Mayer, S. E. 1974. Gastroenterology *67*:898-906.
73. Davies, J., Tyrrell, A. J., Ramsden, D. B., Louis, L. N., and Milne, R. G. 1973. Exp. Molec. Pathol. *18*:1-9.
74. Kurokawaki, K., Friedler, R. M., and Massry, S. G. 1975. Kidney Internat. *7*:137-144.
75. Lichtenstein, L. M., Henney, C. S., Bourne, H. R., and Greenough, W. B., III. 1973. J. Clin. Invest. *52*:691-697.
76. Zieve, P. D., Pierce, N. F., and Greenough, W. B.,III. Johns Hopkins Med. J. *129*:299-303.
77. Hynie, S., Raskova, H., Sechser, T., Vanecek, J., Matejovska, D., Matejovska, V., Trea, M., and Polak, L. 1974. Toxicon *12*:173-179.
78. Gorman, R. E. and Bitensky, M. W. 1972. Nature *235*:439-440.
79. Pierce, N. F., Graybill, J. R., Kaplan, M. M., and Bouwman, D. L. 1972. J. Lab. Clin. Med. *79*:145-156.
80. Sultzer, B. M. and Craig, J. P. 1973. Nature New Biol. *244*:178-180.
81. Henney, C. S., Lichtenstein, L. M., Gillespie, E. *et al.* 1973. J. Clin. Invest. *52*:2853-2857.
82. Tsuchiya, M., Kamisaka, Y., Oda, M., Asakura, H., Ohashi, M., and Shimada, T. 1972. Biochim. Biol. Sperim. *10*:107-113.
83. O'Keefe, E. and Cuatrecasas, P. 1974. Proc. Nat. Acad. Sci. USA *71*:2500-2504.
84. Dorner, F. and Mayer, P. 1975. Infect. Immun. *11*: 429-435.
84a. Levinson, S. L. and Blume, A. J. Personal communication.

85. Guerrant, R. L., Brunton, L. L., Schnaitman, T. C., Rebhun, L. I., and Gilman, A. G. 1974. Infect. Immun. *10*:320-327.
86. Smith, P. A. and Case, R. M. 1975. Biochim. Biophys. Acta *399*:277-290.
87. Kempen, H. J. M., dePont, J. J. H. H. M., and Bonting, S. C. 1974. Biochim. Biophys. Acta *392*:276-287.
88. Craig, J. P. 1965. Nature *207*:614-616.
89. Evans, D. J., Jr., Evands, D. G., and Gorbach, S.L. 1973. Infect. Immun. *8*:725-730.
90. Boyle, J. M. and Gardner, J. D. 1974. J. Clin. Invest. *53*:1149-1158.
91. Zenser, T. V. and Metzger, J. F. 1974. Infect. Immun. *10*:503-509.
92. Mashiter, K., Mashiter, G. D., Hauger, R. L., and Field, J. B. 1973. Endocrinology *92*:541-549.
93. Pierce, N. F., Greenough, W. B., III, and Carpenter, C. C. J. 1971. Bacteriol. Reviews *35*:1-13.
94. Field, M. 1971. New Eng. J. Med. *284*:1137-1144.
95. Carpenter, C. C. J. 1971. Amer. J. Med. *50*:1-7.
96. Sharp, G. W. G. 1973. Ann. Rev. Med. *24*:19-28.
97. Finkelstein, R. A. 1973. CRC Crit. Rev. Microbiol. *3*:553-623.
98. Banwell, J. G. and Sherr, H. 1973. Gastroenterology *65*:467-497.
99. Kantor, H. S. 1975. J. Infect. Dis. *131*:S22-S32.
100. Sack, R. B. 1975. Ann. Rev. Microbiol. *29*:333-353.
101. Field, M., Plotkin, G. R., and Silen, W. 1968. Nature *217*:469-470.
102. Field, M., Fromm, D., Al-awquati, Q., Greenough, W. B., III. 1972. J. Clin. Invest. *51*:796-804.
103. Field, M., Plotkin, G., and Silen, W. 1968. Gastroenterology *54*:1233-1241.
104. Ussing, H. H. and Zerahn, K. 1951. Acta Physiol. Scand. *23*:110-127.
105. Pierce, N. F., Carpenter, C. C. J., Elliot, H. L., Greenough, W. B., III. 1971. Gastroenterology *60*:22-32.
106. Vaughan, M., Pierce, N. F., and Greenough, W. B., III. 1970. J. Infect. Dis. *121*:S111-S113.
107. Schafer, D. E., Lust, W. D., Sircar, B., and Goldberg, N. D. 1970. Proc. Nat. Acad. Sci. USA *67*:851-856.
108. Sharp, G. W.G. and Hynie, S. 1971. Nature *229*:266-269.
109. Kimberg, D. V., Field, M., Johnson, J., Henderson, A., and Gershon, E. 1971. J. Clin. Invest. *50*:1218-1230.
110. Chen, L. C., Rohde, J. E., and Sharp, G. W. G. 1971. Lancet *1*:939-941.
111. Chen, L. C., Rohde, J. E., and Sharp, G. W. G. 1972. J. Clin. Invest. *51*:731-740.
112. Gorbach, S. L., Banwell, J. G., Chaterjee, B. D., Jacobs, B., and Sack, R. B. 1971. J. Clin. Invest. *50*:881-889.

113. Banwell, J. G., Gorbach, S. L., Pierce, N. F., Mitra, R., and Mondal, A. 1971. J. Clin. Invest. *50*:890-900.
114. Gorbach, S. L. and Kantor, H. S. 1973. Proc. 8th Joint Conference, U.S.-Japan Cooperative Medical Science Program, Symposium on Cholera 1972. pp. 39-45.
115. Kantor, H. S., Tao, P., and Gorbach, S. L. 1974. J. Infect. Dis. *129*:1-9.
116. Guerrant, R. L., Ganguly, U., Casper, A. G. T., Moore, E. J., Pierce, N. F., and Carpenter, C. C. J. 1973. J. Clin. Invest. *52*:1707-1714.
117. Kantor, H. S. 1974. Clin. Res. *22*:614A.
118. Cuatrecasas, P. 1973. Biochemistry *12*:3558-3566.
119. Holmgren, J., Lonroth, I., and Svennerholm, L. 1973. Infect. Immun. *8*:208-214.
120. King, C. A. and van Heyningen, W. E. 1973. J. Infect. Dis. *127*:639-647.
121. Pierce, N. F. 1973. J. Exp. Med. *137*:1009-1023.
122. Holmgren, J., Lonroth, I., and Svennerholm, L. 1973. Scand. J. Infect. Dis. *5*:77-78.
123. Cuatrecasas, P. 1973. Biochemistry *12*:3577-3581.
124. Holmgren, J., Lindholm, L., and Lonroth, I. 1974. J. Exp. Med. *139*:801-819.
125. Holmgren, J. 1973. Infect. Immun *8*:851-859.
126. Cuatrecasas, P., Parikh, I., and Hollenberg, M. D. 1973. Biochemistry *12*:4253-4263.
127. Holmgren, J., Lonroth, I., Mansson, J. E., and Svennerholm, L. 1975. Proc. Nat. Acad. Sci. USA *72*:2520-2524.
128. Cuatrecasas, P. 1973. Biochemistry *12*:3547-3557.
129. Cuatrecasas, P. 1973. Biochemistry *12*:4253-4264.
130. Haksar, A., Maudsley, D. C., and Peron, F. G. 1974. Nature *251*:514-515.
131. Guerrant, R. L., Brunton, L. L., and Gilman, A. G. 1974. Clin. Res. *22*:443A.
132. Donta, S. T. and Viner J. P. 1975. Infect. Immun. *11*:982-985.
133. Goldberg, N. D., Haddox, M. K., Hartle, D. K., and Hadden, J. W. 1973. Fifth Int. Congr. Pharmacol., Karger, Basel. pp. 146-169.
134. Goldberg, N. D., Haddox, M. K., Dunham, E., Lopez, C., and Hadden, J. W. In: Clarkson, B. and Beserga, R. (eds.). *The Cold Spring Harbor Symposium on Regulation of Proliferation in Animal Cells*. Academic Press, Inc., New York. pp. 609-625.
135. Hadden, J. W., Hadden, L. M., Haddox, M. K., and Goldberg, N. D. 1972. Proc. Nat. Acad. Sci. USA. *69*:3024-3027.
136. Goldberg, N. D., Haddox, M. K., Nicol, S. E., Glass, D. B., Sanford, C. H., Kuehl, F. A., and Estensen, R. 1975. Adv. Cyclic Nucleotide Res. *5*:307-330.
137. Otten, J.,Johnson, G. S., and Pastan, I. 1971. Biochem. Biophys. Res. Commun. *44*:1192-1198.

138. Otten, J., Johnson, G. S., and Pastan, I. 1972. J. Biol. Chem. *247*:7082-7087.
139. Burger, M. M., Bombik, B. M., Breckenridge, B. M., and Sheppard, J. R. 1972. Nature New Biol. *239*:161-163.
140. Seifert, W. E. and Rudland, P. S. 1974. Nature *248*:138-140.
141. Weinstein, Y., Chambers, D. A., Bourne, H. R., and Melmon, K. L. 1974. Nature *251*:352-353
142. Diamantstein, T. and Ulmer, A. 1975. Exp. Cell Res. *93*:309-314.
143. Goridis, C. and Reutter, W. 1975. Nature *257*:698-700.
144. Kumura, H. and Murad, F. 1975. Proc. Nat. Acad. Sci. USA *72*:1965-1969.
145. Rudland, P. S., Gospodarowicz, D., and Seifert,W. 1974. Nature *250*:741-742.
146. Rudland, P. S., Seeley, M., and Seifert, W. 1974. Nature *251*:417-419.
147. Bourne, H. R., Tomkins, G. M., and Dion, S. 1973. Science *181*:952-954.
148. Paulson, L. G., Lovell-Smith, C. J., Manganiello, V. C., and Vaughan, M. 1974. Proc. Nat. Acad. Sci. USA *71*:1639-1642.
149. Kiefer, H. C., Atlas, R., Moldan, D., and Kantor, H. S. 1975. Biochem. Biophys. Res. Commun. *66*:1017-1023.
150. Todaro, G. J., Lazar, G. K., and Green, H. 1965. J. Cell Comp. Physiol. *66*:325-334.
151. Burk, R. R. 1966. Nature *212*:1261-1262.
152. Holley, R. W. and Kiernan, J. A. 1968. Proc. Nat. Acad. Sci. USA *60*:300-304.
153. Castor, L. N. 1969. J. Cell Physiol. *72*:161-162.
154. Castor, L. N. 1971. Exp. Cell Res. *68*:17-24.
155. Westermark, B. 1971. Exp. Cell Res. *69*:259-264.
156. Humphreys, T. 1972. In: *Cell Interactions*, Silvestri, L. G. (ed.), 3rd Lepetit Colloq, North Holland Publ., Amsterdam, pp. 264-276.
157. Baker, J. B., and Humphreys, T. 1971. Proc. Nat. Acad. Sci. USA *68*:2161-2164.
158. Dulbecco, R. 1970. Nature *227*:802-806.
159. Inbar, M., Rabinowitz, Z., and Sachs, L. 1969. Int. J. Cancer *4*:690-696.
160. Pollack, R. E. and Burger, M. M. 1969. Proc. Nat. Acad. Sci. USA *62*:1074-1076.
161. Pollack, R. E., Green, H., and Todaro, G. J. 1968. Proc. Nat. Acad. Sci. USA *60*:126-133.
162. Smith, H., Scher, C. D., and Todaro, G. J. 1971. Virology *44*:359-370.
163. Holley, R. W. and Kiernan, J. A. 1971. *Growth Control Cell Cult*, Ciba Foundation Symp., 1970, pp. 3-15.
164. Lipton, A., Klinger, I., Paul, D., and Holley, R. W. 1971. Proc. Nat. Acad. Sci. USA *68*:2799-2801.
165. Paul, D., Lipton, A., and Klinger, I. 1971. Proc. Nat. Acad. Sci. USA *68*:645-648.

166. Frank, W., Ristow, H. J., and Zabel, S. 1970. Eur. J. Biochem. *14*:392-398.
167. Frank, W., Ristow, H. J., and Schwalb, S. 1972. Exp. Cell Res. *70*:390-396.
168. Oey, J., Vogel, A., and Pollack, R. 1974. Proc. Nat. Acad. Sci. USA *71*:694-698.
169. Schiffman, S., Rapaport, S. I., Ware, A. G., and Mehl, J. W. 1960. Proc. Soc. Exp. Biol. Med. *105*:453-455.
170. Ratnoff, O. D. and Davie, E. W. 1962. Biochemistry *1*:677-685.
171. Levitzki, A., Stallcup, W. B., and Koshland,D.E.,Jr. 1971. Biochemistry *10*:3371-3378.
172. Frank, W. 1972. Exp. Cell Res. *71*:238-241.
173. Froehlich, J. E. and Rachmeler, M. 1974. J. Cell Biol. *60*:249-257.
174. Willingham, M. C., Johnson, G. S., and Pastan, I. 1972. Biochem. Biophys. Res. Commun. *48*:743-748.
175. Gey, G. O. and Thalhimer, W. 1924. J. Amer. Med. Assoc. *82*:1609.
176. Temin, H. M. 1967. J. Cell Physiol. *69*:377-384.
177. Pierson, R. W. and Temin, H. M. 1972. J. Cell Physiol. *79*:319-330.
178. Griffiths, J. B. 1970. J. Cell Sci. *7*:575-585.
179. Jimenez de Asua, L., Surian, E. S., Flawia, M. M., and Torres, H. N. 1973. Proc. Nat. Acad. Sci. USA *70*:1388-1392.
180. Sheppard, J. R. 1972. Nature New Biol. *236*:14-16.
181. Hepp, K. D. and Renner, R. 1972. FEBS Lett. *20*:191-194.
182. Hepp, K. D. 1972. Eur. J. Biochem. *51*:266-276.
183. Illiano, G. and Cuatrecasas, P. 1972. Science *175*:906-908.
184. House, P. D. R., Poulis, P., and Weidemann, M. J. 1972. Eur. J. Biochem. *24*:429-437.
185. Manganiello, V. and Vaughan, M. 1973. J. Biol. Chem. *248*:7164-7170.
186. Yu, R. K. and Ledeen, R. W. 1972. J. Lipid Res. *13*:680-686.
187. Ho, R. J. and Sutherland, E. W., Jr. 1971. J. Biol. Chem. *246*:6822-6827.
188. Franklin, T. J. and Foster, S. J. 1973. Nature New Biol. *246*:146-148.
189. Kiefer, H. C., Katz, M. R., Shepard, M. R., and Kantor, H. S. 1975. Clin. Res. *23*:479A.
190. Kiefer, H. C., Katz, M. R., Shepard, M. R., and Kantor, H. S. Manuscript in preparation.
191. Kiefer, H. C. and Katz, M. R. Unpublished data.
192. Finkelstein, R. A. and Lo Spalluto, J. J. 1970. J. Infect. Dis. *122*:S63-S72.
193. Thompson, W. J. and Appleman, M. M. 1971. Biochemistry *10*:311-316.
194. Schwartz, C. J., Kimberg, D. V., and Ware, P. 1975. Gastroenterology *68*:94-104.
195. Kuo, J. F. and DeRenzo, E. C. 1969. J. Biol. Chem. *244*:2252-2260.

196. Lang, U., Karlaganis, G., Vogel, R., and Schwyzer, R. 1974. Biochemistry *13*:2626-2633.
197. Hilz, H., Kankel, E., Wiegers, U., and Fuhrmann, U. 1975. Biochem. Biophys. Res. Commun. *64*:519-527.
198. Exton, J. H., Lewis, S. B., Ho, R. J., Robinson, G. A., and Park, C. R. 1971. Ann. N. Y. Acad. Sci. *185*:85-100.
199. Prasad, K. N. and Hsie, A. W. 1971. Nature New Biol. *231*:141-142.
200. Prasad, K. N. and Kumar, S. *Control of Proliferation in Animal Cells*, Clarkson, B. and Baserga, R. (eds.). Cold Spring Harbor Conf. Cell Proliferation, Vol. 1, Cold Spring Harbor Lab, 1974, pp. 581-594.
201. Prasad, K. N. 1974. Differentiation *2*:367-369.
202. Rapin, A. M. C. and Burger, M. M. 1974. Adv. Cancer Res. *20*:1-91.
203. Hollenberg, M. D., Fishman, P. H., Bennett, V., and Cuatrecasas, P. 1974. Proc. Nat. Acad. Sci. USA *71*:4224-4228.
204. Gilman, A. G. 1970. Proc. Nat. Acad. Sci. USA *67*:305-312.
205. Gottlieb, D. I., Merrell, R., and Glaser, L. 1974. Proc. Nat. Acad. Sci. USA *71*:1800-1802.
206. Moscona, A. A. 1968. Develop. Biol. *18*:250-277.
207. Peterson, J. W. 1974. Infect. Immun. *10*:157-166.
208. Gascoyne, N. and Van Heyningen, W. E. 1975. Infect. Immun. *12*:466-469.
209. D'Armiento, M., Johnson, G. S., and Pastan, I. 1973. Nature New Biol. *242*:78-80.

Cyclic AMP Metabolism in BHK21 Cells Infected or Transformed with Adenovirus Type 12

Karel Raska, Jr.

Low frequency of oncogenic transformation of hamster cells with adenovirus type 12 (Ad12) makes the study of the biochemical mechanism of events leading to establishment of the transformed state difficult. The use of abortively infected cells, however, is useful because this abortive infection can ultimately lead to the establishment of stable transformation. It is probable that virus genes responsible for transformation are expressed in infected cells prior to occurrence of the actual event of transformation.

Infection of baby hamster kidney (BHK21) cells, arrested in the G1 phase of the cell cycle by growth in medium containing a low concentration of serum, with an oncogenic Ad12 results in such an abortive infection (28). In the infected cells early virus mRNA is transcribed (10, 19), the infected cells synthesize T antigen and a round of DNA synthesis is induced (26). No viral DNA (5,18), late adenovirus mRNA or viral capsid proteins can be detected after infection (18). During this process, the integration of viral DNA into the host genome occurs (5). The cells enter mitosis, but the chromosomes become fragmented prior to cytokinesis and most of the cells die (27). Among the cells that survive the infection, the frequency of transformation is quite high (30).

With reports that cAMP metabolism is affected in transformed cells (4), and that addition of cAMP conspicuously affected the growth of transformed cells (23,25) we added cAMP analog to Ad12 infected BHK21 cells and to BHK21 cells transformed with Ad12 *in vitro*.

SYNTHESIS OF VIRUS-SPECIFIC mRNA AND INDUCTION OF CELLULAR DNA SYNTHESIS IN Ad12-INFECTED BHK21 CELLS

The infection of BHK21 cell monolayers arrested in G1 phase by 50 h growth in medium containing 0.5 percent total calf serum results in a marked increase of RNA synthesis (19). The rates of synthesis of cellular ribosomal RNA and tRNA are increased in infected cells. From 6 h.p.i. the synthesis of adenovirus mRNA can be detected in infected cells, it reaches a maximum at 10 h.p.i. The relative proportion of newly synthesized RNA which is virus-specific is highest between 10 and 14 h.p.i. and at 18 h it already has decreased (Table 1). Hybridization-competition analysis (18), and hybridization of this RNA to restriction endonuclease fragments of Ad12-DNA (24) indicate that it is early in nature. Results summarized in Figure 1 illustrate that RNA isolated from abortively infected BHK21 cells failed to compete with hybridization of late Ad12 mRNA.

TABLE 1. *Hybridization with Ad12 DNA of ^{3}H RNA from G1 arrested BHK21 cells at varying times after Ad12 infection*[a]

Time After Infection (h)	Specific Activity of RNA (cpm/1 μg)	Total Radioactivity Added (cpm)	Radioactivity in Hybrid (cpm/filter)
6	2,678	110,126	12
10	2,339	87,980	161
12	4,598	155,440	351
14	2,345	120,080	268
18	14,473	665,800	239

[a]*RNA was labeled in BHK21 cells by addition of uridine-5-^{3}H to the incubation medium (5 μCi/ml) at 5, 9, 11, 13, and 17 h, respectively. After 60 min, the cells were harvested and RNA was isolated. ^{3}H-labeled RNA was then hybridized with 4 μg of Ad12 DNA immobilized on nitrocellulose filters.*

The synthesis of DNA is induced by adenovirus infection, beginning at 12 h.p.i. and reaching a maximum between 16 and 20 h.p.i. (18). This induction of DNA synthesis is accompanied by increase of deoxyribonucleotide synthesizing enzyme machinery in the infected cells (32). The hybridization analysis indicates that this DNA is only of cellular origin (5,18). The only virus-specific protein so far identified in the infected BHK21 cells is the T-antigen (26).

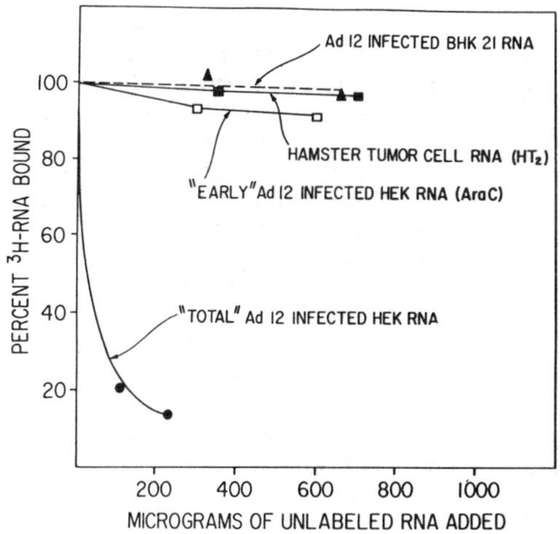

FIGURE 1. *Hybridization-competition of different unlabeled RNAs and labeled "late" (pulse-labeled) Ad12 mRNA with Ad12 DNA. The filters contained 0.1 μg of Ad12 DNA; 98,000 cpm (78 μg) of ^{3}H-labeled "late" (pulse-labeled) mRNA from HEK cells was added to each reaction.* ■ *, unlabeled RNA from hamster tumor cells;* ▲*, unlabeled RNA from Ad12-infected BHK21 cells (21 h after infection)* □ *, unlabeled RNA from HEK cells infected with Ad12 in the presence of 20 μg of AraC/ml (21 h after infection);* ●*, unlabeled RNA from HEK cells infected with Ad12 (21 h after infection).*

EFFECTS OF DIBUTYRYL cAMP ON INDUCTION OF CELLULAR DNA BY Ad12

When dibutyryl-cAMP (at a final concentration of 2 m*M*) was added to G1-arrested Ad12-infected BHK21 cells at 2 h.p.i., the incorporation of ^{3}H-thymidine into DNA was reduced by 95 percent as determined at the time of maximal synthesis, 18 h.p.i. (Figure 2). AMP at the same concentration had a stimulatory effect (33). When dibutyryl-cAMP in the medium was not replenished, there was a gradual release from the inhibitory effect so that 32 h.p.i. ^{3}H-thymidine incorporation had increased in the ditbutyryl-cAMP treated infected cells to 50 percent of that in the untreated infected cells at maximum incorporation (Figure 2). When cultures were refed with serum-free medium containing 2 m*M* db-cAMP at 22 h, release from inhibition was reversed and incorporation dropped back to the strongly inhibited levels. The inhibition of DNA synthesis was real and not a result of an artifact of interference with thymidine uptake (8), and was confirmed in cells labeled with other precursors of DNA synthesis and by determinations of

DNA contents in treated cultures by chemical means.

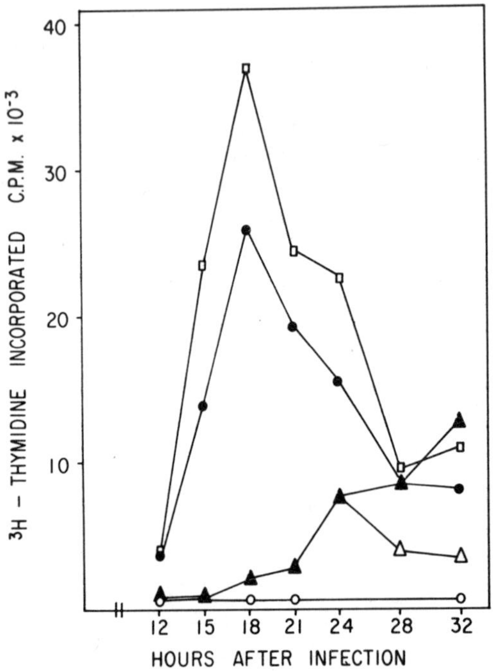

FIGURE 2. *Inhibition of adenovirus type 12 DNA synthesis induced by dibutyryl-cAMP. BHK21 wells were arrested in G1 by 50 h growth in HT medium supplemented with 0.5% fetal calf serum and were infected with Ad12 as previously described, at a MOI of 200 p.f.u./cell. T-antigen was detected in more than 99% of control infected cells. Dibutyryl-cAMP or AMP, both at a concentration of 2 mM, was added 2 h after infection. Control uninfected and Ad12-infected cultures as well as cultures containing nucleotides were pulse-labeled with ^{3}H-thymidine for 60 min and then harvested at the indicated times. The cells were solubilized with SDS (final concentration 0.5%) and precipitated with ice cold TCA. Acid-precipitable material was collected on Millipore filters, dried, and radioactivity on filters was determined in a Packard-Tricarb Liquid Scintillation Spectrometer. ●, Ad12-infected cells; □ , Ad12-infected cells with 2 mM AMP added at 2 h PI; ▲, Ad12-infected cells with 2 mM dibutyryl-cAMP added to the medium 2 h PI: Δ, Ad12-infected cells with 2 mM dibutyryl-cAMP added at 2 h PI and refed with serum-free medium containing 2 mM cAMP at 22-23 h PI: O, G1-arrested uninfected cells.*

DNA contents in treated cultures by chemical means.

To obtain some information about the possible relation of the effects of db-cAMP to virus-induced events in the infected cells, the drug was added at varying times after infection (Table 2). The inhibitory effects were

TABLE 2. *Effects of dibutyryl-cAMP added at varying times after infection upon induction of DNA synthesis*

Time of Addition of Dibutyryl-cAMP (h after infection)	Incorporation of ^{3}H-Thymidine/ Culture (c.p.m.)	Inhibition (%)
No cAMP	14,061	
4.5	2,711	89
6.0	2,268	92
9.5	7,622	53
12.0	7,784	49
14.0	13,391	6
Uninfected control, No cAMP	1,250	

BHK21 cells were arrested in low serum and infected at a multiplicity of infection 200 p.f.u./cell. T-antigen was detected in 98% of untreated infected cells. Dibutyryl-cAMP was added at times indicated and cells were pulse-labeled with ^{3}H-thymidine from 16 to 17 h PI.

maximal when the cyclic nucleotide was added up to 6.0 h.p.i. Later the effect decreased. When added at 14 h. p.i., no significant inhibitory effects were noted.

EFFECTS OF DIBUTYRYL-cAMP ON FREQUENCY OF T-ANTIGEN

Adenovirus type 12 induces T antigen in abortively infected BHK21 cells (26). This protein, that binds strongly to double-stranded DNA (21), is synthesized only in cells infected with functional virus as can be inferred from the fact that ultraviolet-irradiated Ad12 fails to induce T antigen in this system (19). The treatment of cells with db-cAMP significantly reduces the frequency of T antigen in infected cells (Table 3). While 84-88 percent of untreated cells infected at a given multiplicity of infection were positive, only 1-7 percent of cAMP treated cells were positive. Failure to replenish the drug led to a delayed appearance of T antigen (33).

EFFECTS OF DB-cAMP ON KILLING OF BHK21 CELLS BY AD12

Although the virus does not replicate in this system, most of the cells are killed (27). Treatment with db-cAMP which inhibits the induction of DNA synthesis delays the killing of infected cells by 24-36 h, but does not prevent it.

EFFECTS OF DB-cAMP ON THE SYNTHESIS OF VIRUS-SPECIFIC RNA

The results in Table 1 indicate that Ad12-specific RNA can be detected first at 6 h.p.i. under the conditions used. It may be significant that the inhibition of

TABLE 3. *Effects of cAMP on the frequency of cells with positive T-antigen immunofluorescence staining (%)*

Infected, Untreated	Infected, + 2 m*M* Dibutyryl-cAMP
86, 84, 88	7, 1, 3

G1-arrested BHK21 cells were infected with Ad12 at a multiplicity of infection of 100 p.f.u./cell. At 24 h PI, the coverslips on which the cells were grown were hydrated with PBS, fixed with acetone (-70°C) for 10 min, and stained. The medium and nucleotides were changed in all experiments at 21-22 h. Results in the table represent mean values in three independent experiments. A minimum of 200 cells was counted for each culture.

DNA synthesis is maximal if db-cAMP is added up to 6 h.p.i., that is, prior to onset of Ad12 mRNA synthesis. Later the effect decreases, and failure of cAMP to inhibit coincides roughly with the peak of synthesis of virus-specific RNA and onset of cellular DNA synthesis. The primary role of db-cAMP could be the inhibition of adenovirus mRNA synthesis, because the replication of the virus itself in permissive fibroblastic cells is inhibited by the drug (2).

The quantitative analysis of adenovirus-specific RNA synthesis in db-cAMP treated Ad12-infected BHK21 cells indicates that no significant differences can be found between the control and db-cAMP treated A12-infected BHK21 cells (Zimmerman and Raska, unpublished). The hybridization-competition analysis has shown no significant differences in sequences transcribed. It was not determined however, that in db-cAMP treated cells the virus-specific RNA is transported to the cytoplasm and is associated with polysomes.

EFFECTS OF Ad12-INFECTION ON THE INTRACELLULAR LEVELS OF cAMP IN G1-ARRESTED BHK21 CELLS

The results summarized above suggest that Ad12 infection may lead to a reduction of cAMP levels in the infected cells. A drop in cAMP levels was reported as a requirement for initiation of S-phase in G1-arrested cells (3). The addition of external analog would bypass the virus-induced change in cAMP concentration and the cells so treated would remain in the G1-arrested state.

When the concentration of cAMP was determined in infected cells (Figure 3), it was found that a first significant decrease is detected at 8 h.p.i. Further decreases occur and a minimum level is reached at 14 h.p.i. This level remained in the infected cells through at least 22 h.p.i. (14). Similar observations were made in G1-arrested cells infected with SV40 virus (20); in their experiments, however, the drop of cAMP level was followed by a later increase. This difference is probably a result of virus-induced cell killing occurring in our experimental

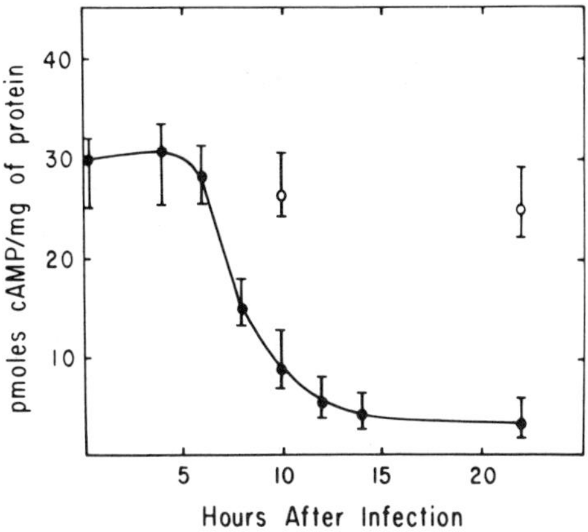

FIGURE 3. *Levels of cyclic AMP in G1-arrested BHK21 cells infected with Ad12. The points represent averages of 3-7 determinations. Ranges give minimum and maximum values observed.* ●, *G1-arrested cells infected with Ad12 (input MOI ∿ 100 p.f.u./cell);* O, *mock-infected G1-arrested BHK21 cells.*

model and not in experiments with SV40. In mock-infected cells, no significant changes in cAMP concentrations were detected.

ACTIVITY OF ADENYLATE CYCLASE IN AD12-INFECTED BHK21 CELLS

The intracellular concentration of cAMP is controlled by two basic enzymatic activities, adenylate cyclase and cAMP phosphodiesterase. To obtain some information about the possible mechanism by which the concentration of cAMP decreases in Ad12-infected cells, activity of adenylate cyclase was determined at various times after infection (Figure 4). In infected cells, the first significant change of activity was detected at 7.5 h.p.i.; from 9 h.p.i. through 22 h.p.i., the observed enzymatic activity corresponded to about 50 percent of that in mock-infected cells. The obvious question that comes to mind is to determine whether an inhibitor is synthesized in infected cells. The obvious candidate for such a function is virus-induced T-antigen. T-antigen was purified (16) and tested for inhibitory activity. The experiments performed so far failed to detect any inhibitory effects on adenylate cyclase from rat astrocytoma (Brostrom and Raska, unpublished).

ACTIVITY OF cAMP PHOSPHODIESTERASE IN AD12-INFECTED CELLS

The activity of cAMP phosphodiesterase was determined with three different concentrations of cyclic AMP substrate (Table 4). It is evident that no significant change in the activity of the enzyme was observed until 13 h.p.i. with all three tested concentrations of the substrate in

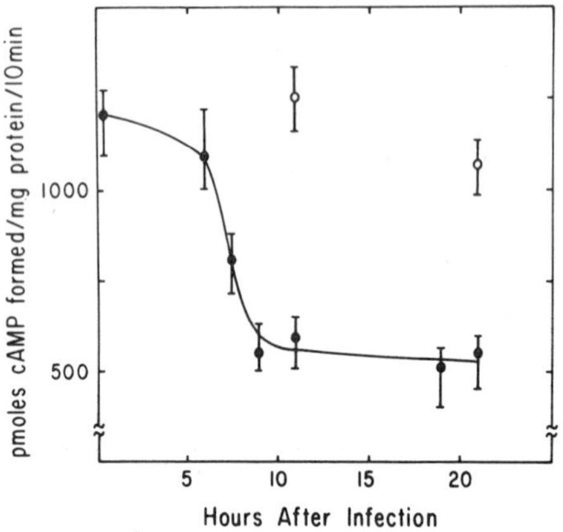

FIGURE 4. *Activity of adenylate cyclase in G1-arrested BHK21 cells infected with Ad12. The points represent averages of 3-4 determinations. Ranges give minimum and maximum values observed. ●, G1-arrested cells infected with Ad12 (input MOI ∿ 100 p.f.u./cell); O, mock-infected, G1-arrested cells.*

TABLE 4. *Activity of cyclic AMP phosphodiesterase in Ad12 infected BHK21 cells*

Source of Enzyme Activity	Phosphodiesterase Activity[a] (n mole per mg protein per 10 min[b]) 60 μ*M* Cyclic AMP	5 μ*M* Cyclic AMP	0.5 μ*M* Cyclic AMP
Gl-arrested cells	9.6	0.94	0.085
Infected cells (h after infection)			
5	10.1	ND[c]	ND
7	9.9	1.07	0.086
10	9.9	ND	0.081
13	9.4	0.93	0.075
19	6.5	0.66	0.049
22	7.8	0.46	ND
24	4.1	0.39	0.045

[a]*Assayed with 60.0 μM, 5.0 μM and 0.5 μM cyclic AMP as substrate.*
[b]*Values presented are averages of 2-4 determinations which varied not more than ± 10%.*
[c]*ND = not done.*

in the incubation mixture. At 19 h.p.i., the decrease in activity was detected in all cases. These results indicate that the cAMP phosphodiesterase activity is not profoundly affected by Ad12 infection. The decreases late after infection are not surprising in view of the observations that the synthesis of cAMP phosphodiesterase is regulated by intracellular concentrations of cAMP (1). The decreased activity could be interpreted as a result of reduced levels of cAMP in infected cells.

COMPARISON OF cAMP METABOLISM IN BHK21 CELLS AND BHK21 CELLS TRANSFORMED WITH AD12

Since the original observation of differences in cAMP metabolism in BHK21 cells and cells transformed by polyoma virus (4), a large volume of data on the differences in cAMP metabolism in normal and transformed fibroblasts accumulated [for review see (12)]. The observed differences are often controversial. It is apparent today that early generalizations are not supported by data available. One of the biggest problems, in our opinion, was that established fibroblastic cell lines were compared to "transformed" cell lines, but the actual relationship between the two was often conjectural. When we compared six different lines of "fibroblastic" BHK21 cells, we observed a number of differences in: saturation density of growth, cAMP metabolism, and yield of Ad2, G1-arrest in low serum, etc. We, therefore, decided to compare the one line of contact inhibited BHK21 cells used in our laboratory with Ad12 transformed cells derived from that line by Ad12 transformation *in vitro* (29). The transformed cells (line T637) show typical morphology, synthesize T-antigen and have lost "contact inhibition."

When the dibutyryl-cAMP alone or together with theophylline is added to the medium, the growth of cells is inhibited (Figure 5); removal of the drug results in prompt resumption of growth. Morphology of transformed cells changes after addition of dibutyryl-cAMP. The cells change from the round transformed morphology to somewhat elongated, fibroblast-like cells, but complete restoration of "normal" morphology reported for some transformed cell lines, was not observed and the treated cells never resembled the BHK21 cells from which they were originally derived. Removal of the drug resulted in a prompt reversion of morphology.

The immunofluorescence staining for T-antigen was not significantly changed by the addition of the drug (Sehulster, unpublished).

When both cell lines were seeded at a comparable density, the intracellular levels of cAMP in growing cells were similar (Figure 6). The dramatic difference, however, was detected at confluency (15) when a significant increase of cAMP level occurred in fibroblastic cells (Figure 6), while a time dependent decline in cAMP level was detected in transformed cells.

Addition of prostaglandins to the tissue culture medium

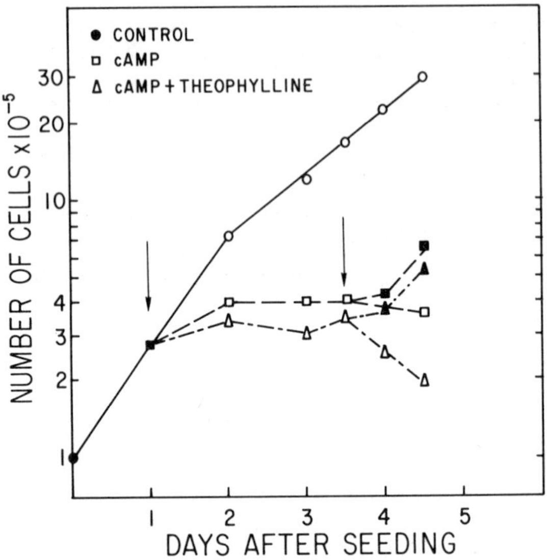

FIGURE 5. *Effects of cAMP and theophylline on growth of T637 cells in culture. 10^5 cells were plated in tissue culture dishes and their number determined at varying times thereafter. Dibutyryl-cAMP (□) and dibutyryl-cAMP + theophylline (Δ) were added to some cultures at 24 h after plating. The medium with drugs was replenished every 24 h. The second arrow indicates removal of the drugs from some cultures (■, ▲). O, control cells.*

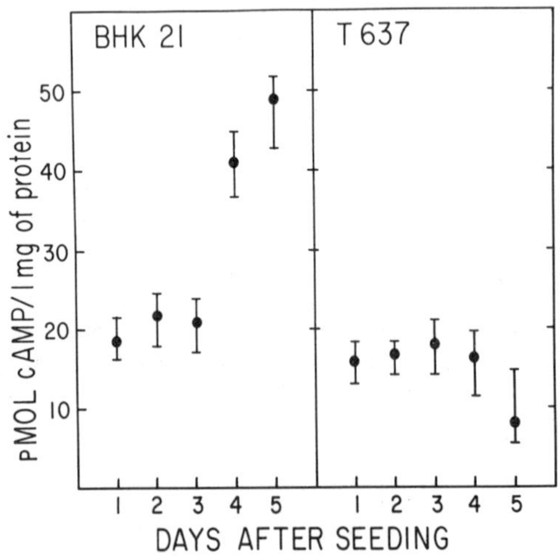

FIGURE 6. *Levels of cAMP in BHK21 cells and T637 cells at varying times after plating. The ranges represent results of five independent experiments.*

increases the intracellular concentration of cAMP in some cell lines (11). Results in Table 5 indicate that prostaglandin E_1 did not significantly increase the cAMP level in BHK21 cells. In transformed cells there was also no effect.

TABLE 5. *Cyclic AMP levels in BHK21 and T637 cells at various times after treatment with prostaglandin E_1*

Cell	Time After Addition (min)	cAMP Level (pmole/1 mg Protein) Control (0.5% Ethanol)	PGE_1
BHK21	0	18.5	18.5
	5	19.3	25.3
	30	17.3	22.2
	60	18.1	17.2
T637	0	16.2	16.2
	5	15.8	16.3
	30	16.3	17.1
	60	16.1	15.2

The polyoma transformed BHK21 cells were reported to have less adenylate cyclase activity than their parental fibroblasts (4). In our two cell lines, the activity of the enzyme was significantly lower in T637 cells. In heavy confluent cultures, a decrease of activity was detected in both cell lines (Figure 7). Adenylate cyclase activity in some cell lines can be stimulated by prostaglandins (13). When the activity of enzyme was measured in the absence of fluoride, the activity in transformed cells was still lower than in fibroblastic BHK21 cells. Prostaglandin E_1, however, did not stimulate the enzyme activity in either cell line (Figure 8). We detected the response to prostaglandin E_1 in only one of the six BHK21 lines that we tested (Raska, unpublished).

The activity of cAMP phosphodiesterase was tested in both cell lines (Table 6) at three levels of the cAMP substrate in growing and heavy cultures. At 0.6 μM and 6 μM concentrations, the activity was comparable in both cell lines in that it increased at confluency in both cell lines. At 60 μM concentration, the activity in growing cells was significantly higher in fibroblastic cells, but at confluency it decreased, while a reproducible increase was observed in heavy cultures of transformed cells (14).

DISCUSSION

The infection of G1-arrested BHK21 cells with oncogenic Ad12 leads to changes of cell regulatory mechanisms, release of cells from G1 arrest and, in most cells, to cell death. In relatively few cells, i.e., those that

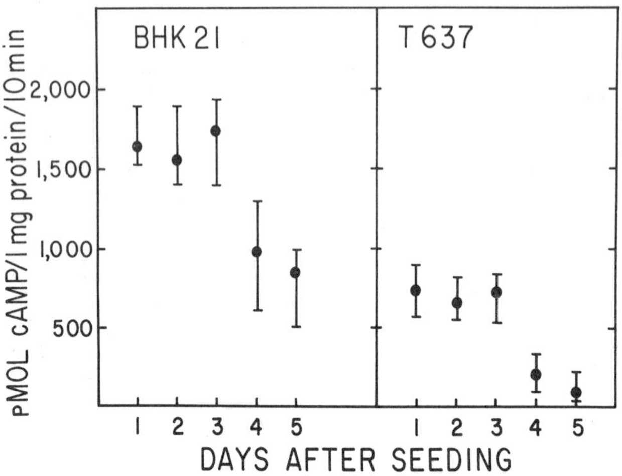

FIGURE 7. *Activity of adenylate cyclase in BHK21 cells and T637 cells at varying times after plating. The ranges represent maximum and minimum values obtained in four independent experiments.*

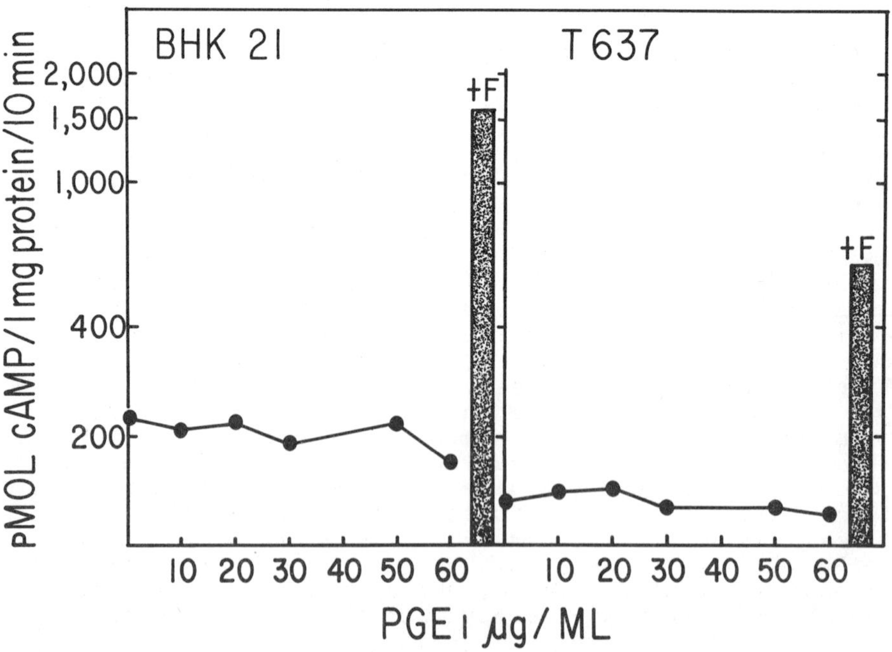

FIGURE 8. *Effects of prostaglandin* E_1 *on the activity of adenylate cyclase in BHK21 and T637 cells. The activity of the enzyme was determined in the absence of fluoride. Different concentrations of prostaglandin* E_1 *were added to the incubation medium.*

TABLE 6. *Activity of cyclic AMP phosphodiesterase in BHK21 and T637 cells**

		Phosphodiesterase Activity (nmole/mg Protein/10 min)		
Cell	State of Culture	50.0 μ*M* cAMP	6.0 μ*M* cAMP	0.6 μ*M* cAMP
BHK21	Growing	4.08	0.993	0.125
	Heavy	2.03	1.819	0.160
T637	Growing	1.90	1.44	0.051
	Heavy	2.57	2.2	0.11

**The enzymatic activity was determined as described in Table 4.*

survive the infection, establishment of transformed state occurs (28).

The infection is accompanied by profound changes in cAMP metabolism. The activity of adenylate cyclase and intracellular cAMP levels decrease after virus-specific products appear in the infected cells. Although active viral genome is required for these changes, cautious interpretation is required in view of the cell killing resulting from infection and a possibility that the changes may be of secondary nature.

It was previously claimed that one of the major adenovirus capsid proteins is phosphorylated (22) and it was shown that all adenovirus capsid proteins can be phosphorylated *in vitro* (30). Although phosphorylation of adenovirus virion protein components was recently modified (21), it was shown that in addition to virus structural proteins, some of the virus-specific nonstructured proteins in infected cells are also phosphorylated. Of these, the most interesting is the single-stranded DNA-binding protein (31), that is early in nature. Although synthesis of any of these proteins was not as yet shown in Ad12-infected BHK21 cells, their possible significance to changes observed in cAMP metabolism should be considered since adenylate cyclase activity in the cell may be regulated by phosphorylation (9).

The differences in cAMP metabolism between control BHK21 cells and BHK21 cells transformed with Ad12 are compatible with changes observed in abortively-infected cells. The T637 cells contain several Ad12 genomic equivalents, integrated into their genome (7) and most of the virus-specific mRNA sequences present in Ad12-infected BHK21 cells are detected in T637 cells (24).

Although one may infer that the changes of cAMP metabolism are a direct result of expression of viral function in abortively infected or transformed cells, the direct evidence is sought in experiments currently in progress.

ACKNOWLEDGEMENTS

The experimental work reported in this communication was supported by Grant CA-08851 from the National Cancer Institute and Grant DRG-1289 from the Damon Runyon-Walter Winchell Cancer Memorial Fund. I want to thank Dr. Margaret Brostrom, Department of Pharmacology, CMDNJ-Rutgers Medical School for reading this manuscript and helpful suggestions.

REFERENCES

1. D'Armiento, M., Johnson, G. S. and Pastan, I. 1972. Regulation of adenosine 3':5'-cyclic monophosphate phosphodiesterase activity in fibroblasts by intracellular concentrations of cyclic adenosine monophosphate. Proc. Nat. Acad. Sci. USA *69*:459-462.
2. Biron, K. K. and Raska, K., Jr. (1973). Effects of dibutyryl-cAMP on adenovirus replication in different cell lines. Virology *56*:383-386
3. Burger, M. M., Bombik, B. M., Breckenridge, B. M. and Sheppard, J. R. 1972. Growth control and cyclic alterations of cyclic AMP in the cell cycle. Nature New Biol. *239*:161-164.
4. Bürk, R. R. 1968. Reduced adenyl cyclase activity in a polyoma virus transformed cell line. Nature *219*:1272-1275
5. Doerfler, W. 1968. Nonproductive infection of baby hamster kidney cells (BHK-21) with adenovirus type 12. Virology *38*:587-606.
6. Doerfler, W. 1970. Integration of the deoxyribonucleic acid of adenovirus type 12 into the deoxyribonucleic acid of baby hamster kidney cells. J. Virol. *6*:652-666.
7. Fanning, E. and Doerfler, W. 1976. Adenovirus type 12 sequences in abortively infected and transformed hamster cells. In: *Molecular Biology of Adenoviruses*, European Molecular Biology Organization, p. 53, Cologne.
8. Hauschka, P. V., Everhart, L. P. and Rubin, R. W. 1972. Alteration of nucleoside transport of Chinese hamster cells by dibutyryl adenosine 3':5'-cyclic monophosphate. Proc. Nat. Acad. Sci. USA *69*:3542-3546.
9. Najjar, V. A. and Constantopoulos, A. 1973. The activation of adenylate cyclase. Mol. Cell Biochem. *2*:87-96.
10. Ortin, J. and Doerfler, W. 1975. Transcription of the genome of adenovirus type 12. I. Viral mRNA in abortively infected and transformed cells. J. Virol. *15*:27-35.

11. Otten, J., Johnson, G. S. and Pastan, I. 1972. Regulation of cell growth by cyclic adenosine 3',5'-monophosphate. J. Biol. Chem. *247*:7082-7087.
12. Pastan, I. H., Johnson, G. S. and Anderson, W. B. 1975. Role of cyclic nucleotides in growth control. Ann. Rev. Biochem. *44*:491-552.
13. Peery, C. V., Johnson, G. S. and Pastan, I. 1971. Adenyl cyclase in normal and transformed fibroblasts in tissue culture. J. Biol. Chem. *246*: 5785-5790.
14. Raska, K., Jr. 1973. Cyclic AMP in Gl-arrested BHK21 cells infected with adenovirus type 12. Biochem. Biophys. Res. Commun. *50*:35-41.
15. Raska, K., Jr. 1973b. Cyclic AMP metabolism in normal and adenovirus type 12 transformed BHK-21 cells. Abstracts Ann. Meeting Amer. Soc. Micro. *73*:195.
16. Raska, K., Jr., Biron, K. K. and Rubio, V. 1976. Purification and DNA binding of adenovirus type 12 tumor antigen. Proc. Amer. Assoc. Cancer Res. *17*: 45.
17. Raska, K., Jr., Rubio, V. and Biron, K. K. 1976. Double-stranded DNA binding of adenovirus type 12 tumor antigen. FEBS Lett. *63*:173-177.
18. Raska, K., Jr. and Strohl, W. A. 1972. The response of BHK-21 cells to infection with type 12 adenovirus. VI. Synthesis of virus-specific RNA. Virology *47*:734-742.
19. Raska, K., Jr., Strohl, W. A., Holowczak, J. and Zimmerman, J. 1971. The response of BHK-21 cells to infection with type 12 adenovirus. V. Stimulation of cellular RNA synthesis and evidence for transcription of the viral genome. Virology *44*: 296-306.
20. Rein, A., Carchman, R. A., Johnson, G. S. and Pastan, I. 1973. Simian virus 40 rapidly lowers cAMP levels in mouse cells. Biochem. Biophys. Res. Commun. *52*:899-904.
21. Russell, W. C., Blair, G. E. and Rekosh, D. 1976. Phosphorylation in adenovirus infection. In: *Molecular Biology of Adenoviruses*, European Molecular Biology Organization, p. 26, Cologne.
22. Russell, W. C., Skehel, J. J., Machado, R. and Pereira, H. G. 1972. Phosphorylated polypeptides in adenovirus-infected cells. Virology *50*:931-934.
23. Ryan, W. L. and Heidrick, M. L. 1968. Inhibition of cell growth *in vitro* by adenosine 3':5'-monophosphate. Science *162*:1484-1485.
24. Scheidtmann, K. H., Ortin, J., Greenberg, R.,Westphal, M. and Doerfler, W. 1976. Transcription maps of adenovirus type 12 DNA in infected and transformed cells. In: *Molecular Biology of Adenoviruses*, European Molecular Biology Organization, p. 8, Cologne.

25. Sheppard, J. R. 1971. Restoration of contact-inhibited growth to transformed cells by dibutyryl adenosine 3':5'-cyclic monophosphate. Proc. Nat. Acad. Sci. USA *68*:1316-1320.
26. Strohl, W. A. 1969a. The response of BHK-21 cells to infection with type 12 adenovirus. I. Cell killing and T antigen systhesis as correlated viral genome functions. Virology *39*:642-652.
27. Strohl, W. A. 1969b. The response of BHK-21 cells to infection with type 12 adenovirus. II. Relationship of virus-stimulated DNA synthesis to other viral functions. Virology *39*:653-665.
28. Strohl, W. A. 1973. Alterations in hamster cell regulatory mechanism resulting from abortive infection with an oncogenic adenovirus. Progr. Exp. Tumor Res. *18*:199-239.
29. Strohl, W. A., Rouse, H., Teets, K. and Schlesinger, R. W. 1970. The response of BHK-21 cells to infection with type 12 adenovirus. III. Arch. ges. Virusforsch. *31*:93-112.
30. Tao, M. and Doerfler, W. 1972. Phosphorylation of adenovirus polypeptides. Eur. J. Biochem. *27*:448-452.
31. Van der Vliet, P. and Levine, A. 1972. DNA-binding proteins specific for cells infected by adenovirus. Nature New Biol. *246*:170-174.
32. Zimmerman, J., Raska, K., Jr. and Strohl, W. A. 1970. The response of BHK-21 cells to infection with type 12 adenovirus. IV. Activation of DNA-synthesizing apparatus. Virology *42*:1147-1150.
33. Zimmerman, J. E., Jr. and Raska, K., Jr. 1972. Inhibition of adenovirus type 12 induced DNA synthesis in G1-arrested BHK-21 cells by dibutyryl adenosine cyclic 3':5'-monophosphate. Nature New Biol. *239*:145-147.

Part IV

CYCLIC NUCLEOTIDE METABOLISM AND THE REGULATION OF CELL GROWTH

MURAD
Cyclic GMP Metabolism in Transplantable Tumors

HADDOX et al.
Regulation of Cyclic GMP Metabolism in Splenic Cells

JUNGMANN et al.
Cyclic AMP-Mediated Nuclear Translocation of Cytoplasmic cAMP-Dependent Protein Kinases: Identity of the Nuclear and Cytoplasmic Enzymes

SHEPPARD, BURGHARDT, and TSIEN
Cyclic AMP Systems in Selected Cells of Erythroid and Lymphoid Origin

Cyclic GMP Metabolism in Transplantable Tumors

Ferid Murad

Although a substantial amount of data supports the view that diminished levels of cyclic AMP are in some manner associated with transformation and increased growth rates, some observations do not support this hypothesis (see review of Chlapowski *et al.* (1)). A notable exception has been the increased levels of cyclic AMP in hepatomas of various growth rates (Table 1) (2,3). While much less data currently exist, increased levels of cyclic GMP in transplantable tumors have correlated with increased growth rates (Table 1) (2-6). Increased levels of cyclic GMP are also observed in transformed fibroblasts (7-9), proliferating epidermis (10,11), lymphocytes (12), and plants (13).

The accumulation of either cyclic nucleotide in tissues is a function of its rate of formation by the cyclase system, rate of degradation by phosphodiesterases, and the rate of release or secretion from tissues into the extracellular compartment. I wish to review for you some relatively new data from this laboratory that indicate that guanylate cyclase in transplantable liver and kidney tumors has an altered subcellular distribution pattern, altered properties, and perhaps altered control mechanisms compared to normal liver and kidney (3-5,14).

Although many of the early studies with guanylate cyclase stressed the localization of the enzyme in supernatant fractions of homogenates, all tissues examined to date have significant amounts of guanylate cyclase activity associated with particular fractions (15,16). However,

TABLE 1. *In situ levels of cyclic AMP and cyclic GMP in normal and neoplastic rat tissues**

Tissue	pmoles/mg protein Cyclic AMP	Cyclic GMP
Normal liver	3.58	0.11
Liver of hepatoma bearing rats	3.50	0.09
Normal kidney cortex	8.01	0.32
Slow growing tumors		
Hepatoma 16	8.30	0.26
Hepatoma 9633F	5.11	0.39
Hepatoma 20	4.00	0.19
Hepatoma 21	6.30	0.50
Hepatoma 9633	6.64	0.51
Intermediate growing tumors		
Hepatoma 7800	11.00	0.56
Hepatoma 9121	9.94	0.38
Hepatoma 7316B	11.11	0.45
Kidney sarcoma (MK2)	5.10	0.70
Kidney adenoma (MK3)	4.50	0.55
Fast growing tumors		
Hepatoma 7288ctc	8.41	2.50
Hepatoma 9618A2	5.10	0.99
Hepatoma 5123tc	8.80	0.88
Hepatoma 3924A	9.86	22.40

**Mean values for cyclic nucleotide levels have been compiled from references 2-5.*

the distribution of activity between soluble and particulate fractions of homogenates is quite variable with different tissues and between species (Figure 1). Furthermore, much of the particulate and soluble activities in tissues appear to be latent since activities may increase with detergents, preincubation, and/or activation with nucleophilic agents (16-21).

Studies in this laboratory (15,16) and by Chrisman *et al.* (18) have demonstrated that many mammalian tissues possess at least two forms of guanylate cyclase. The properties of the soluble and particulate forms of the enzyme from rat heart (15) and lung (18) are summarized in Table 2. While these properties are also generally shared by the soluble and particulate forms of the enzyme in other mammalian tissues, there are some exceptions. For example, 3 *mM* Ca^{2+} will increase the soluble and decrease the particulate activities from heart (Figure 2) (15). With particulate preparations from lung Ca^{2+} will increase activity at low Mn^{2+} concentrations and decrease

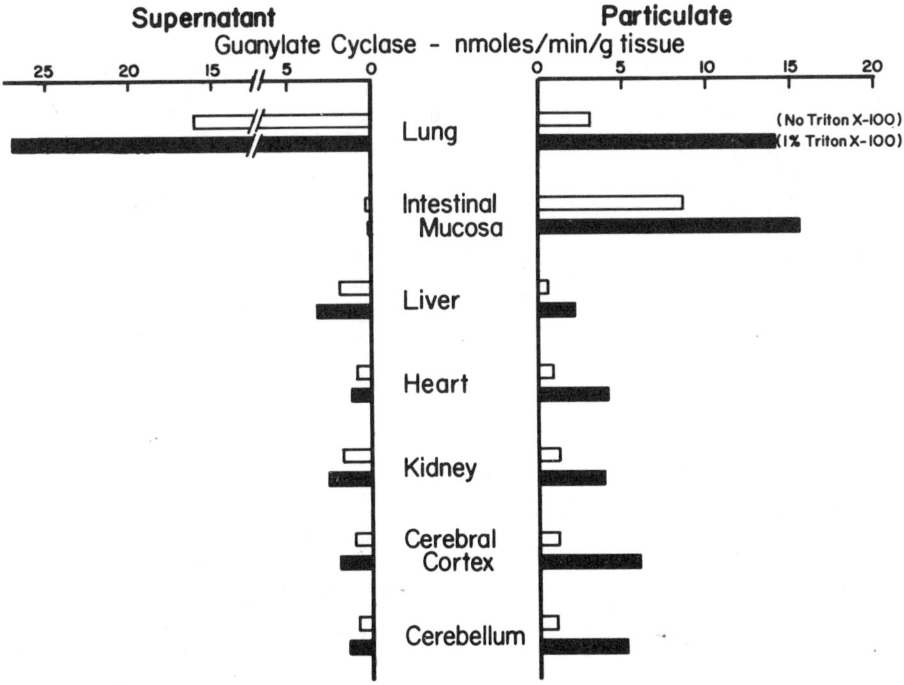

FIGURE 1. *Guanylate cyclase activity in 100,000 x g supernatant and particulate fractions of homogenates from various rat tissues and the effects of Triton X-100 (from reference 16).*

activity at Mn^{2+} concentrations greater than 2 *mM* (Figure 3). The particulate activity from rat intestinal mucosa is increased by Ca^{2+} at all Mn^{2+} concentrations examined (Figure 4). In spite of such exceptions the general properties summarized in Table 2 can be used as guidelines and markers for the soluble and particulate forms of the enzyme in many tissue preparations. Centrifugation of tissue homogenates at 105,000 x g is the simplest method of separating the two enzymes. Separation can also be achieved with gel filtration of detergent treated homogenates due to the differences in sizes of the enzymatic activities (Figure 5) (15,16,18).

The particulate form of guanylate cyclase in rat liver is associated with plasma membranes. Golgi and endoplasmic reticulum (22,23). Preparations of liver mitochondria and nuclei were not sufficiently clean to comment upon the location of guanylate cyclase in the latter fractions. The particulate enzyme activities associated with these various liver fractions had similar properties except that Ca^{2+} decreased the activity of plasma membrane and Golgi preparations and not that from microsomes (22). This isolated observation suggests that

TABLE 2. *Summary of the properties of soluble and particulate guanylate cyclase from rat heart and lung**

	Soluble	Particulate
Molecular weight	450,000	700,000 to 900,000
K_m for GTP	12 to 65 μ*M*	50 to 100 μ*M*
K_a for free Mn^{++}	0.2 m*M*	0.2 m*M*
Sites for metal-nucleotide complex	1	2 or more
Effect of Ca^{++}	Stimulates	Inhibits
Effect of EDTA or EGTA	No effect	No effect
Effect of maleate	Inhibits	Stimulates
Effect of Triton X-100	Stimulates (50-100%)	Stimulates (>300%)
50% inhibition by ATP	0.4 m*M*	>1 m*M*
50% inhibition by p-chloromercuri-phenyl sulfonic acid	0.1 to 0.01 μ*M*	1 to 10 μ*M*

**Compiled from references 15, 16, and 18.*

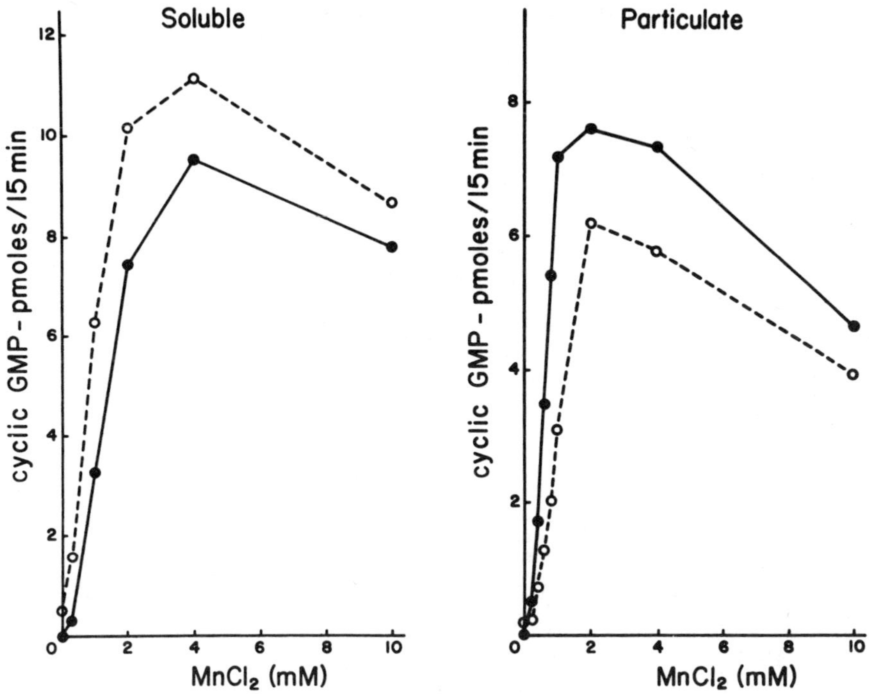

FIGURE 2. *Effects of Mn^{2+} and Ca^{2+} on soluble and particulate guanylate cyclase activities from rat heart. A 100,000 x g supernatant and particulate fraction from a rat heart homogenate was incubated with 1 mM GTP and the Mn^{2+} concentrations indicated. Some incubations also contained 3 mM Ca^{2+} (---O---) (from reference 15).*

the various particulate enzyme activities may also have different properties. White has reported guanylate cyclase activity associated with sarcoplasmic reticulum preparations from rat heart (24). While the apparent molecular sizes and properties of the soluble and particulate forms of the enzyme are different, there are no data to establish whether or not these activities are similar proteins.

The presence of two different enzyme activities in tissues suggests that each form of the enzyme could have different control mechanisms and perhaps different functions. In order to examine this hypothesis we studied guanylate cyclase in regenerating and fetal liver and hepatomas, since previous reports demonstrated increased cyclic GMP levels in proliferating fibroblast cultures (7-9), epidermis (10,11), hepatomas (2-6), lymphocytes (12), and plants (13). With regenerating rat liver we found an increase in particulate enzyme activity and a decrease in soluble activity (Figure 6) (14). The increase in particulate activity was associated with vari-

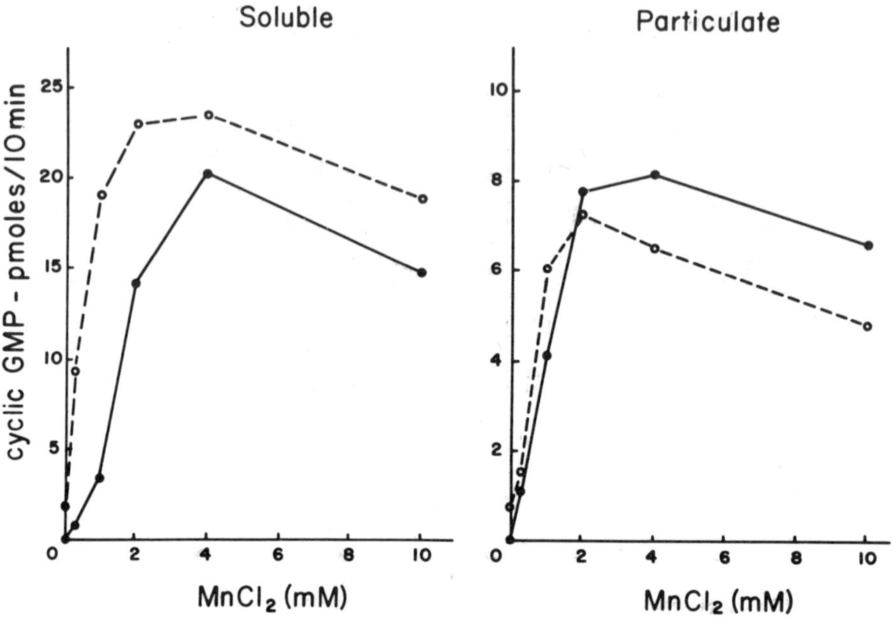

FIGURE 3. *Effects of Mn^{2+} and Ca^{2+} on soluble and particulate guanylate cyclase from rat intestinal mucosa. Conditions were similar to those in Figure 1.*

ous particulate fractions and could be prevented with cycloheximide (14). With fetal development in rats, liver particulate guanylate cyclase activity decreased and the soluble activity increased to approach adult values 2 to 3 weeks after birth (Figure 7) (14)

Although the mechanism(s) for the altered subcellular distribution of guanylate cyclase activity in regenerating and fetal liver is not known, these studies strongly suggest that each form of the enzyme is regulated independently of the other (14,16).

We also examined a third model of proliferating liver, transplantable Morris hepatomas with different growth rates (4,14). As summarized in Table 1, all of these tumors had increased *in situ* levels of GMP. While guanylate cyclase activities of homogenates of these tumors were less than normal liver, more of the homogenate enzyme activity was associated with particulate fractions and less with soluble fractions (Table 3). Similar observations were obtained with the transplantable renal tumors MK2 and MK3 (5). Some minor differences in the properties of the soluble and particulate activities were noted when tumor preparations were compared to preparations of normal liver (4) and kidney (5). Triton X-100 increased soluble and particulate guanylate cyclase activities from normal rat liver to a greater extent than with preparations of hepatoma 3924A and 9618A2 (4). The effects of

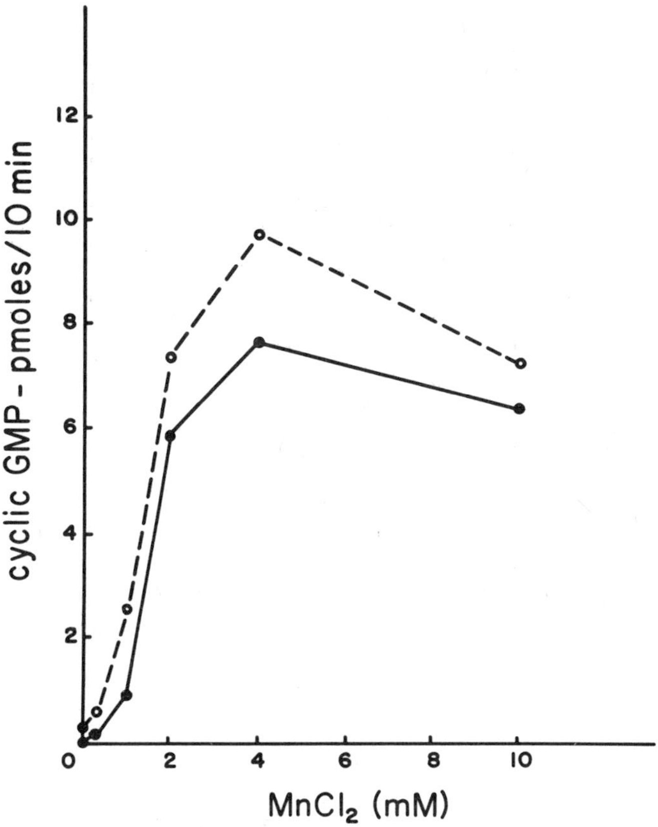

FIGURE 4. *Effects of Mn^{2+} and Ca^{2+} on particulate guanylate cyclase from rat intestinal mucosa. Conditions were similar to those in Figure 1.*

Triton X-100 with guanylate cyclase from renal tumors MK2 and MK3 were not different from those with normal rat kidney cortex (5). Soluble and particulate guanylate cyclase from rat liver can effectively use Mg^{2+} as a cofactor in the presence of suboptimal concentrations of Mn^{2+} (4,16,22). This was not the case with preparations from hepatoma 3924A and 9618A2 (4). The most striking difference with guanylate cyclase preparations from the tumors examined was the failure to increase enzyme activity with sodium azide (Table 4) (4,5). Sodium azide, among other nucleophilic agents, such as hydroxyl-amine and phenylhydrazine, can increase guanylate cyclase activity from a number of tissues, including liver and kidney (4,5,20,21,25). The ability of NaN_3 to activate guanylate cyclase is dependent upon the presence of a protein activator factor and the absence of a heat-labile inhibitor factor (20,21). Our current studies indicate that tumor preparations lack the protein activator factor

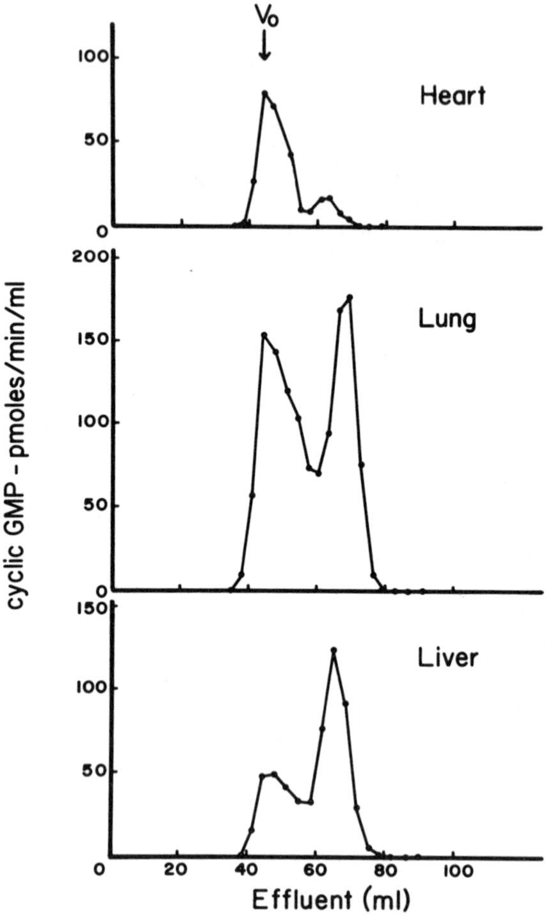

FIGURE 5. *Separation of two forms of guanylate cyclase from rat heart, lung and liver with Bio-Gel-P300 column chromatography. 105,000 x g supernatant fractions of homogenates treated with 1% Triton X-100 were applied to a Bio-Gel-P-300 column (2.6 x 34 cm). The first peaks of activity from each tissue eluted in the void volume and had properties of the particulate enzyme; the second peak had properties of the supernatant enzyme from homogenates (from reference 16).*

since addition of the partially purified liver activator factor to tumor preparations results in the appearance of a stimulatory effect of NaN_3.

The altered subcellular distribution of guanylate cyclase and its altered properties may in part explain the elevated *in situ* levels of cyclic GMP in these tumors (Table 1) (2-5). The homogenate activities of guanylate cyclase from hepatoma and renal tumor preparations are

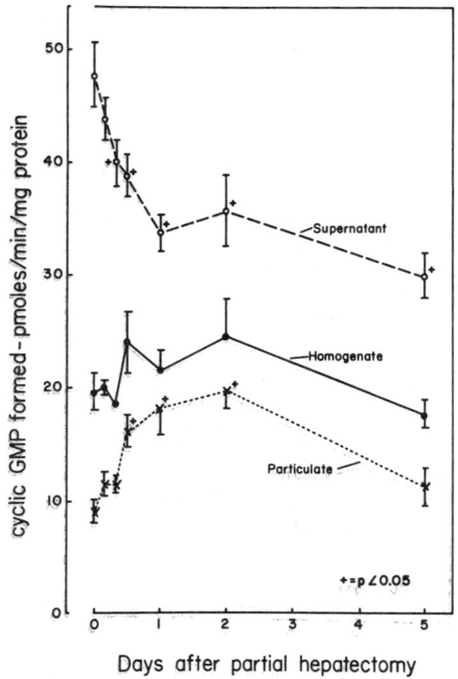

FIGURE 6. *Soluble and particulate guanylate cyclase activities in regenerating rat liver (from reference 14).*

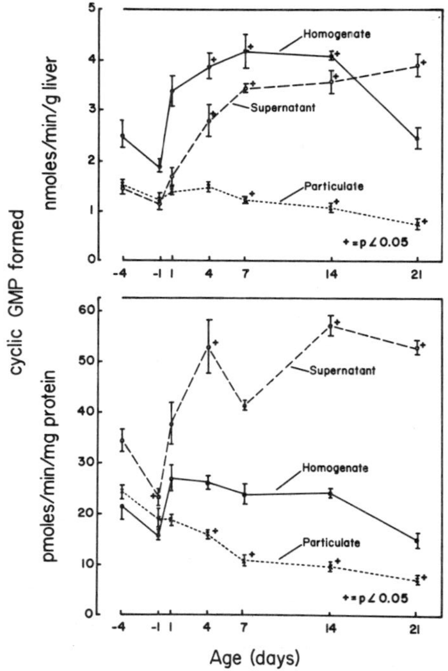

FIGURE 7. *Soluble and particulate guanylate cyclase activities in fetal and newborn rat liver (from reference 14).*

TABLE 3. *Subcellular distribution of guanylate cyclase activities in normal and neoplastic tissues from rat*. 105,000 x g supernatant and particulate fractions of homogenates were examined for guanylate cyclase activity.*

	Cyclic GMP Formed (nmoles/min/gm tissue)		Particulate: Soluble Ratio
	Particulate	Soluble	
Normal adult liver	0.6	2.5	0.24
Normal adult kidney	0.9	1.8	0.50
Slow growing tumors			
Hepatoma 20	1.2	0.5	2.40
Hepatoma 21	1.8	0.9	2.00
Intermediate growing tumors			
Hepatoma 9633F	1.3	0.5	2.60
Kidney sarcoma (MK2)	0.6	0.5	1.20
Kidney adenoma (MK3)	0.7	0.7	1.00
Fast growing tumors			
Hepatoma 3924A	1.3	0.4	3.25
Hepatoma 9618A2	0.8	0.6	1.33

**Compiled from references 4 and 5.*

TABLE 4. *Effect of sodium azide on guanylate cyclase activity*. 105,000 x g soluble and particulate fractions were preincubated with 1 mM NaN_3 for 10 minutes at 37°C prior to assay.*

Rat Tissue Fraction	Cyclic GMP Formed (pmoles/min/mg protein)		Ratio
	$-NaN_3$	$+NaN_3$	$+N_3/-N_3$
Normal liver (particulate)	5	27	5.4
Normal liver (soluble	42	281	6.7
Hepatoma 3924A (particulate)	51	54	1.1
Hepatoma 3924A (soluble)	23	27	1.2
Hepatoma 9618A2 (particulate)	33	38	1.2
Hepatoma 9618A2 (soluble)	23	25	1.1
Normal kidney cortex (particulate)	17	62	3.6
Normal kidney cortex (soluble	42	166	4.0
Renal tumor MK2 (particulate)	22	24	1.1
Renal tumor MK2 (soluble	38	32	0.8
Renal tumor MK3 (particulate)	14	13	0.9
Renal tumor MK3 (soluble)	23	25	1.1

**Compiles from references 4 and 5.*

about 50 to 75 percent of the activity from normal rat liver and kidney cortex (4,5). The proportionately greater quantities of homogenate activity in the particulate fractions, however, correlate with the increased levels of cyclic GMP observed in these tumors.

Equally important would be the rate of cyclic GMP hydrolysis by phosphodiesterase in these preparations. One laboratory working with preparations from transformed fibroblasts found decreased cyclic GMP hydrolysis (26). The increased levels of cyclic GMP in hepatoma 3924A and 9618A2 *in situ* are also observed in incubations of slices from these tumors[1]. While cyclic GMP is released from slices of liver and hepatoma 3924A and 9618A2 into the incubation medium, a decrease in the amount of cyclic GMP released cannot be offered as an explanation for the greater amounts of cyclic GMP observed in the tumors.

Rats bearing some of these transplantable tumors also have increased amounts of urinary cyclic GMP (Table 5) (27,28). When urinary cyclic GMP excretion is increased, as in rats bearing hepatoma 3924A, 9618A2 and renal tumor MK3, the excretion of cyclic GMP correlates with the tumor size (Table 5 and Figure 8). Treatment of animals bearing hepatoma 3924A with 5-fluorouracil or local irradiation to the tumor resulted in a decrease in tumor size and cyclic GMP excretion (Figure 8) (27). The source of the increased urinary cyclic GMP in these tumor bearing animals is presumably secretion from the tumors that results in increased plasma levels and renal filtration. However, this has not been established. Whatever the mechanism, the process appears to be specific for some types of tumors since not all tumors with elevated *in situ* levels of cyclic GMP are associated with increased urinary excretion of the nucleotide (Tables 1 and 5). With all of the tumor bearing animals examined the urinary excretion of cyclic AMP was not altered (27,28).

These recent studies suggest that increased urinary excretion of cyclic GMP might be expected in some varieties of tumors in patients. If this is the case, the excretion of cyclic GMP may be useful for diagnosing some tumors, and perhaps more likely, it could be useful in evaluating therapeutic responses in patients.

In summary, there are at least two forms of guanylate cyclase activity in mammalian tissues. These activities have different kinetic properties, molecular sizes and are separable with either centrifugation or gel filtration. The subcellular distribution and some properties of guanylate cyclase are altered in regenerating liver, fetal liver and all hepatomas and renal tumors that we have examined to date. The increased amounts of particulate enzyme and/or an alteration of its properties may account for the increased cyclic GMP levels observed in these transplantable tumors. It remains to be established whether or not these observations will be general findings in other tumors and proliferating tissues. Some, but not

[1]Criss, W. and F. Murad (unpublished observations).

TABLE 5. *Cyclic GMP in urine of normal and tumor bearing rats*.*

Tumor and Rat Strain	μmoles Cyclic GMP/gm Creatinine (Mean ± S.E.) Control or No Tumor Detectable	Diameter of Tumor (1-2 cm)	(2-3 cm)	(3-4 cm)	(4-6 cm)
Non-tumor bearing rats					
Sprague Dawley	1.5±0.07				
Buffalo	1.6±0.08				
ACI	1.4±0.06				
Hepatoma bearing rats					
20 (Buffalo)	1.6±0.06	1.6±0.06	2.6±0.09	3.2±0.10	3.6±0.10
21 (Buffalo)	1.4±0.07	1.4±0.07	3.3±0.09	2.8±0.09	2.9±0.11
9618A (Buffalo)	1.3±0.07	1.6±0.06	1.8±0.09	1.8±0.08	1.9±0.10
9633F (Buffalo)	1.4±0.05	1.8±0.06	1.6±0.08	1.8±0.07	1.5±0.08
3924A (ACI)	1.3±0.09	3.9±0.09[a]	10.4±0.20[b]	22.4±0.82[b]	41.9±2.81[b]
9618A2 (Buffalo)	1.4±0.09	1.9±0.09	2.6±0.10	3.3±0.12	5.6±0.23[a]
Kidney tumor bearing rats					
MK2 (Buffalo)	1.2±0.05	1.3±0.05	1.8±0.07	2.2±0.07	2.9±0.09
MK3 (Buffalo)	1.3±0.07	2.9±0.11[a]	9.7±0.46[b]	17.1±0.81[b]	32.7±1.62[b]

[a] $p < .05$
[b] $p < .01$
*Compiled from reference 28.

all, of these tumors are associated with increased urinary excretion of cyclic GMP. These latter findings could have useful diagnostic and prognostic value clinically with patients.

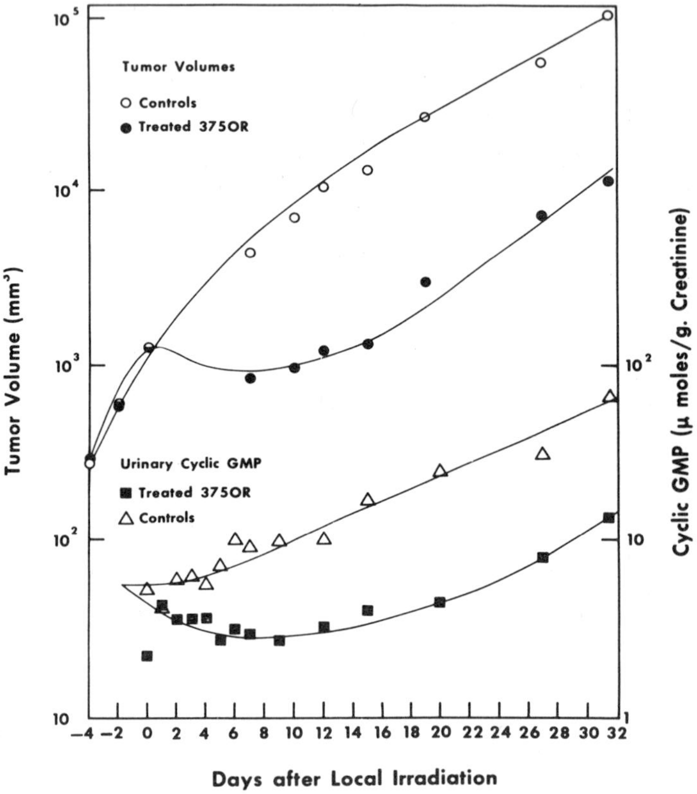

FIGURE 8. *Effects of local irradiation on urinary cyclic GMP levels and tumor size in rats bearing Morris hepatoma 3924A (modified from reference 27).*

REFERENCES

1. Chlapowski, F. J., Kelly, L. A. and Butcher, R. W. 1975. Cyclic nucleotides in cultured cells. Adv. Cyclic Nucleotide Res. *6*:245-338.
2. Thomas, E. W., Murad,F., Looney, W. B. and Morris, H. P. 1973. Adenosine 3',5'-monophosphate and guanosine 3',5'-monophosphate concentrations in Morris hepatomas of different growth rates. Biochim. Biophys. Acta *297*:564-567.
3. Criss, W. E. and Murad, F. Cyclic nucleotide metabolism in tumors. In: *Clinical Aspects of Cyclic Nucleotides*, L. Volicer, ed., Spectrum Publications (in press).

4. Criss, W. E., Murad, F., Kimura, H. and Morris, H.P. Properties of guanylate cyclase activity in adult rat liver and in a series of Morris hepatomas. Biochim. Biophys. Acta (in press).
5. Criss, W. E., Murad, F. and Kimura, H. 1976. Properties of guanylate cyclase from rat kidney cortex and transplantable kidney tumors. J. Cyclic Nucleotide Res. *2*:11-19.
6. Goldberg, M. L., Burke, G. C. and Morris, H. P. 1975. Cyclic AMP and cyclic GMP content and binding in malignancy. Biochem. Biophys. Res. Commun. *62*:320-327.
7. Rudland, P. S., Seeley, M. and Seifert, W. 1974. Cyclic GMP and cyclic AMP levels in normal and transformed fibroblasts. Nature *251*:417-419.
8. Moens, W., Vokaer, A. and Kram, R. 1975. Cyclic AMP and cyclic GMP concentrations in serum and density restricted fibroblast cultures. Proc. Nat. Acad. Sci. USA *72*:1063-1067.
9. Seifert, W. E. and Rudland, P. S. 1974. Possible involvement of cyclic GMP in growth control of cultured mouse cells. Nature *248*:138-140.
10. Voorhees, J. J., Stawiski, M., Duell, E. A., Haddox, M. K. and Goldberg, N. D. 1973. Increased cyclic GMP and decreased cyclic AMP levels in the hyperplastic, abnormally differentiated epidermis of psoriasis. Life Sciences *13*:639-653.
11. Voorhees, J., Kelsey, W., Stawiski, M., Smith, E. and Duell, E. Increased cyclic GMP and decreased cyclic AMP in rapidly proliferating epithelium of psoriasis. In: *The Role of Cyclic Nucleotides in Carcinogenesis*, J. Schultz and H. G. Gratzner, eds., pp. 325-373. Academic Press, New York, 1973.
12. Goldberg, N. D., Haddox, M. K., Estensen, R., White, J. G., Lopez, C. and Hadden, J. S. Evidence for a dualism between cyclic GMP and cyclic AMP in the regulation of cell proliferation and other cellular responses. In: *Cyclic AMP, Cell Growth and the Immune Response*, W. Braun, L. Lichtenstein, and C. Parker, eds., pp. 247-262. Springer-Verlag, New York, 1974.
13. Haddox, M. K., Stephenson, J. H. and Goldberg, N. D. 1974. Cyclic GMP in meristematic and elongating regions of bean roots. Fed. Proc. *31*:522.
14. Kimura, H. and Murad, F. 1975. Increased particulate and decreased soluble guanylate cyclase activity in regenerating liver, fetal liver, and hepatoma 3924A. Proc. Nat. Acad. Sci. USA *72*: 1965-1969.
15. Kimura, H. and Murad, F. 1974. Evidence for two different forms of guanylate cyclase in rat heart. J. Biol. Chem. *249*:6910-6919.

16. Kimura, H. and Murad, F. 1975. Two forms of guanylate cyclase in mammalian tissues and possible mechanisms for their regulation. Metabolism *24*:439-445.
17. Hardman, J. G., Chrisman, T. D., Gray, J. P., Suddath, J. L. and Sutherland, W. W. 1973. Guanylate cyclase: alteration of apparent subcellular distribution and activity by detergents and cations. Proc. Fifth Int. Cong. on Pharmacol. 143-145, 1973.
18. Chrisman, T. D., Garbers, D. L., Parks, M. A. and Hardman, J. G. 1975. Characterization of particulate and soluble guanylate cyclase from rat lung. J. Biol. Chem. *250*:374-381.
19. White, A. A. and Lad, P. J. 1975. Activation of the soluble guanylate cyclase from rat lung by preincubation. Fed. Proc. *34*:232.
20. Kimura, H. and Murad, F. 1975. Activation of guanylate cyclase from rat liver and other tissues by sodium azide. J. Biol. Chem. *250*:8016-8022.
21. Mittal, C. K., Kimura, H. and Murad, F. 1975. Requirement for a macromolecular factor for sodium azide activation of guanylate cyclase. J. Cyclic Nucleotide Res. *1*:261-269.
22. Kimura, H. and Murad, F. 1975. Localization of particulate guanylate cyclase in plasma membranes and microsomes of rat liver. J. Biol. Chem. *250*: 4810-4817.
23. Kimura, H. and Murad, F. 1975. Subcellular localization of guanylate cyclase. Life Sciences *17*: 837-844.
24. White, A. A. 1975. Guanylate cyclase activity in heart and lung. Adv. Cyclic Nucleotide Res. *5*: 353-373.
25. Kimura, H., Mittal, C. K. and Murad, F. 1975. Increases in cyclic GMP levels in brain and liver with sodium azide on activator of guanylate cyclase. Nature *257*:700-702.
26. Lynch, T. J., Tallant, F. A. and Cheung, W. J. 1975. Marked reductions of cyclic GMP phosphodiesterase activity in virally transformed fibroblasts. Biochem. Biophys. Res. Commun. *65*:1115-1122.
27. Murad, F., Kimura, H., Hopkins, H. A., Looney, W. B. and Kovacks, C. J. 1975. Increased urinary excretion of cyclic guanosine monophosphate in rats bearing Morris hepatoma 3924A. Science *190*:58-60.
28. Criss, W. E. and Murad, F. Urinary excretion of cyclic 3',5'-GMP and cyclic 3',5'-AMP in rats bearing transplantable liver and kidney tumors. Cancer Res. (in press).

Regulation of Cyclic GMP Metabolism in Splenic Cells

Mari K. Haddox, Leo T. Furcht, Mary E. Moser, Christine C. Bull, Janis H. Stephenson, and Nelson D. Goldberg

One focus of attention in our laboratory has been on exploring the role cyclic GMP may play in modulating lymphocyte function. Although this role has not yet been elucidated, the evidence uncovered to date indicates that cyclic GMP has a unique role which is distinctly different from that of cyclic AMP.

The suggestion that cyclic GMP may serve a role in promoting cell proliferation was based on studies with the human peripheral blood lymphocyte. These studies showed that minutes after the addition of the mitogens phytohemagglutinin or Concanavalin A, an increase in intracellular cyclic GMP was demonstrable (12). By using lectins with no agglutinating capacity this increase in the guanine nucleotide was dissociated from any change in cyclic AMP concentration. This initial observation has since been extended to rodent peripheral blood lymphocytes stimulated by Concanavalin A (32) and rodent B lymphocytes stimulated by lipopolysaccharide (28). Several other proliferative systems have also been shown to exhibit elevated levels of cyclic GMP after stimulation by mitogenic agents. These agents include serum, insulin (10,25), and phorbol ester (7,27).

These findings, however, constitute only indirect evidence for an involvement of cyclic GMP as an effector of the proliferative process. Numerous other cellular components and processes have been shown to be altered minutes after the addition of a mitogenic stimulus. It is not known whether all these metabolic changes are necessary for cell division to occur; whether some, although not absolutely required, serve to modulate the rate at which events in the proliferative process proceed; or whether some are merely related to the generalized changes effected in the cell membrane.

Another body of evidence which supports a positive involvement of cyclic GMP in regulating lymphocyte function derives from the demonstrations that some of the cellular events comprising the growth response can be enhanced by elevating intracellular cyclic GMP concentrations either by the addition of exogenous cyclic GMP (or 8-bromo or dibutyryl derivatives) or agents which stimulate the cellular accumulation of this cyclic nucleotide. In intact B lymphocyte systems the addition of cyclic GMP or 8-bromo cyclic GMP can be shown to directly stimulate DNA synthesis (28,29), RNA synthesis (29) and purine synthesis (4). In intact T cells the effect of the cyclic nucleotide is to augment the primary signal given by the mitogenic agent. Cholinergic agents, which elevate cyclic GMP concentration in lymphocytes, enhance the rate of DNA synthesis initiated by PHA (10) or a graft vs host reaction (26). In a similar manner, the cyclic GMP elevating agents, imidazole and levamisole, can be shown to augment PHA stimulated DNA synthesis (13). Cyclic GMP itself will enhance lymphoblast DNA synthesis (31), Con A-stimulated Ca^{++} uptake (9), and PHA-induced ornithine decarboxylase activity (18). In isolated nuclei of T lymphocytes, cyclic GMP can be shown to directly stimulate RNA synthesis (10), RNA polymerase (16), and acidic protein phosphorylation (15).

The answer to whether cyclic GMP is a required component in the sequence of events that occurs between the mitogenic signal and cell division awaits two important pieces of information still outstanding: one, the identification of the cellular component with which cyclic GMP interacts to modify the proliferative response and two, the elucidation of how cyclic GMP metabolism, specifically its biosynthesis, may be regulated by a mitogenic agent.

In this report we will present our findings regarding the regulation of cyclic GMP biosynthesis in a cell system consisting predominantly of lymphocytes. The evidence supports the concept that cellular events involving oxidation and reduction may serve as one mechanism for modulating guanylate cyclase activity in these cells. Several chemical agents with oxidative potential will be shown to promote an increased accumulation of cyclic GMP, and, in one case, the oxidant can be shown to also promote proliferation.

METHODS

Cell Preparation: Spleens, removed from cervically dislocated 6-8 week old BALB/c mice or ether anesthetized 250-300 g male Hartley guinea pigs, were homogenized gently in phosphate buffered saline (PBS, pH 7.4). After sedimentation of connective tissue the suspended cells were removed, the red blood cells lysed by suspension in ammonium chloride and the white cells washed by successive centrifugation and suspension in PBS. Cells were counted and adjusted to a concentration of 20 x 10^6 per ml PBS at 4°C for the periodate studies, 2 x 10^6 per ml RPMI 1640 at 37°C for the thiol and dehydroascorbic acid studies and 200 x 10^6 per ml PBS at 4°C for the guanylate cyclase studies. Suspensions for intact cell experiments were incubated at the indicated temperature 30 min prior to the addition of the agents to be tested.

Cyclic Nucleotide Determination: Samples were collected for the determination of cyclic nucleotides by adding perchloric acid to a final concentration of 1 *N*. Tracer amounts of [^{3}H]-cyclic GMP were added to each sample for determination of chromatographic recovery. The acid extracts were rapidly frozen and thawed three times and the denatured protein removed by centrifugation. The acid supernatants were stored at -70°C until they were processed further. The acid extracts were applied to 2 x 0.6 cm dry neutral alumina columns. The acid and media salts were washed from the columns with 10 ml distilled water. The cyclic nucleotides were then eluted with 5 ml of 200 m*M* NH_3 COOH, pH 6.0. This eluate was applied to a 2 x 0.6 cm distilled water washed AG 1-X8 anion exchange column (formate form). The salt was washed from the column with 10 ml distilled water, cyclic AMP eluted with 10 ml of 1 *N* formic acid and cyclic GMP eluted with 10 ml of 6 *N* formic acid. These eluates were evaporated to dryness.

Guanylate Cyclase Assay: Cell guanylate cyclase studies were performed with guinea pig splenic cells. After sonication to achieve 95 percent cell disruption the suspension was centrifuged for 60 min at 105,000 x g. The supernatant fraction was removed and the pellet resuspended by gentle homogenization in the original cell suspension volume of PBS. Cyclase activity was assayed by a modification of the method of Kimura and Murad (17). The standard rection mixture contained 50 m*M* Tris·HCl, pH 7.5, 1 m*M* GTP, 2 m*M* $MnCl_2$, 5 m*M* isobutylmethylxanthine, 15 m*M* creatine phosphate and 200 μg/ml creatine phosphokinase in a final volume of 30 μl. The reaction was initiated by the addition of 10 μl of the supernatant (20 μg protein) or particulate (40 μg protein) fraction and was incubated at 37°C. The reaction was terminated by the addition of 30 μl of 110 m*M* NaAc, pH 4, containing 11 m*M* EDTA and heated for 3 min at 90°C. Samples containing enzyme added during the boiling procedure served as controls. Cyclic GMP formed was assayed without purification by radioimmunoassay after it was determined that there was no cross reactivity of the reaction components in the antibody-antigen reaction. Polyethyleneimine thin layer chromatography of reaction mixtures in which either [^{3}H]-cyclic GMP or [^{3}H]-GTP had been present during the incubations demonstrated that under the cyclase assay conditions employed for both the supernatant and particulate fractions there was no measurable phosphodiesterase activity and 95 percent of the GTP substrate remained at the end of a 20 min incubation.

Cyclic Nucleotide Assays: Cyclic nucleotides were measured by a modification of the radioimmunoassay described by Steiner *et al.* (25). The modified procedure was fashioned after the innovation described by Cailla *et al.* (3), involving 2'-0-acylation of cyclic nucleotides. Evaporated column eluates were reconstituted in distilled water and guanylate cyclase reactions were diluted with water. Initial determinations were carried out with duplicate samples treated with phosphodiesterase prior to assay to establish that there were no components other than the cyclic nucleotide in the samples which would compete in the antibody-antigen reaction. Acylation of the samples was carried out by adding 2 λ of a mixture of triethylamine and acetic anhydride (2:1, v/v) to an 100 λ aliquot of sample or cyclic nucleotide standard. After vigorous mixing the reaction was incubated for 15 min at room temperature. The radioimmunoassay was performed by incubating the acylated samples in a reaction mixture containing 2'-0-succinyl cyclic nucleotide [^{125}I]-tyrosine methyl ester (25,000 cpm), anti-cyclic nucleotide antiserum (sufficient to bind 30% of the radioligand), 20 m*M* $CaCl_2$, 1.5 mg/ml γ-globulin and 50 m*M* sodium acetate, pH 4.75, in a final volume of 300 λ for 16-20 h at 4°C. Antibody-ligand complex was precipitated by the addition of 1.6 ml of 95 percent ethanol and centrifugation for 15 min at 1,800 x g. The supernatant was removed by aspiration and the radioactivity in the pellet measured in a γ spectrometer.

Proliferation: Cells in which the proliferative response was monitored were exposed to the periodate for 20 min, washed twice with PBS, and suspended at 2 x 10^6 cells/ml of RPMI 1640 containing fetal calf serum (5), penicillin (100 U/ml) and streptomycin (100 μg/ml). The cells were incubated in 5% CO_2 at 37°C for 48 h prior to a terminal pulse with 1 μCi [^{3}H]-thymidine for 3 h. Cultures were collected with a multiple automatic sample harvester and thymidine incorporation determined by liquid scintillation spectrometry.

RESULTS

The density at which splenic cells are suspended was found to have a profound effect on the cyclic nucleotide concentrations. As shown in Figure 1, as mouse splenic cells are suspended at higher densities the cyclic GMP levels decrease and the cyclic AMP levels increase. The guanine nucleotides responded similarly in guinea pig cell suspensions, while cyclic AMP levels appeared to be independent over the density range covered. Subsequent experiments showed that as the density was increased even further to 200 x 10^6 cells/ml, a density at which the cell volume is approximately 10 percent of the total suspension

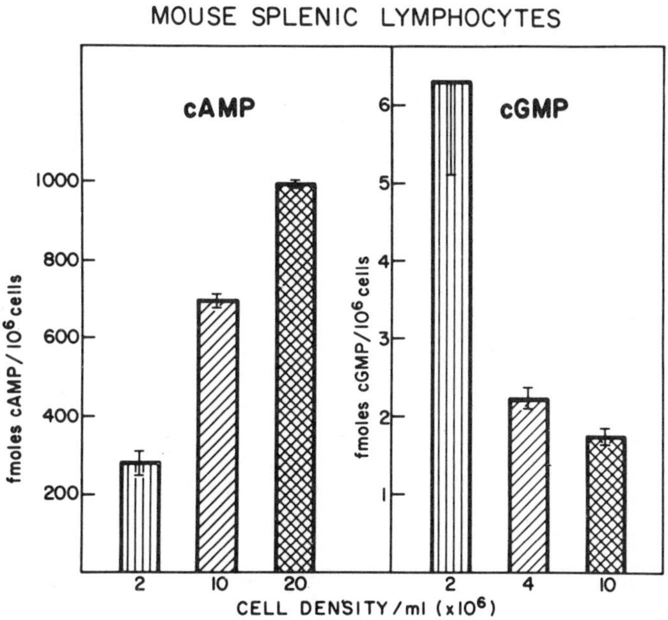

FIGURE 1. *Cell density-dependent nature of the cyclic nucleotide levels in mouse splenic cells. Cells were suspended at the densities indicated in RPMI 1640 and incubated for 30 min at 37°C.*

volume, the cyclic GMP levels fell to less than 0.8 femtomoles per 10^6 cells. Because of this sensitivity to density, the cyclic nucleotide levels for the intact cell experiments reported here are determined in the total cell suspension. Centrifugation of the cells in order to determine intracellular nucleotide concentrations could markedly effect both the basal and stimulated levels of the cyclic nucleotides and possibly obscure the regulatory events actually occurring. Other factors which we have found can markedly alter the basal cyclic nucleotide levels of these cells and have the potential of obscuring experimental results include temperature, suspension media component, cell preparation procedures, time of year and the nutritional state of the animal (Haddox, Stephenson, Moser and Goldberg; unpublished observations).

Novogrodsky has uncovered a novel means of stimulating lymphocyte proliferation by employing the oxidant periodate (19). Because of the definable chemical alterations this oxidant can produce in membrane components (20), we felt the study of this agent could provide new insights into the regulation of cyclic GMP metabolism in relation to the initiating mechanism of proliferation (14).

The effect of IO_4 on the concentrations of the cyclic nucleotide in suspensions of mouse splenic cells, maintained at 4°C in PBS, is shown in three separate experiments in Figure 2. The concentrations of cyclic GMP in

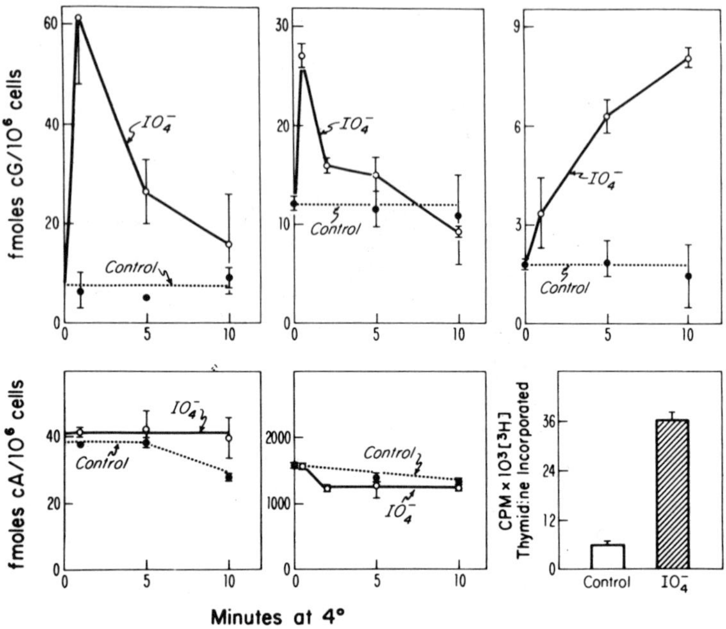

FIGURE 2. *Cyclic GMP and cyclic AMP levels and (^3H)-thymidine incorporation after exposure of mouse splenic cells to IO_4. The IO_4, freshly dissolved in PBS, was added to achieve a final concentration of 5 mM while the cell suspension was maintained at 4°C.*

the cells not exposed to the mitogen varied from animal to animal, ranging from 2.5 to 12.5 fmoles per 10^6 cells. In spite of the variation, treatment with IO_4 consistently caused an enhanced accumulation of cyclic GMP but did not alter the concentration of cyclic AMP. The increase in cyclic GMP concentration ranged from 2- to 10-fold and the temporal characteristics of the enhanced accumulation differ in each of the experiments shown. The IO_4-induced increases in cyclic GMP content were accompanied by a stimulation of lymphocyte proliferation as illustrated by the increased [^3H]-thymidine incorporated into cells cultured after exposure to IO_4 (Panel C).

The cyclic nucleotides in guinea pig splenic cells responded similarly upon exposure to IO_4; the cellular levels of cyclic GMP were increased but the cyclic AMP levels were unaffected by the oxidizing agent (Figure 3). The addition of $CaCl_2$ (0.2 m*M*) to the cation-free PBS media did not enhance the ability of IO_4 to promote cyclic GMP accumulation. The IO_4-induced increase in mouse splenic cell cyclic GMP was also not increased by the addition of extracellular cation. The ability of the oxidant to increase the guanine nucleotide in a Ca^{++}-independent manner is in contrast to the extracellular Ca^{++}-dependent nature of the stimulation reported for the plant lectin Concanavalin A (32).

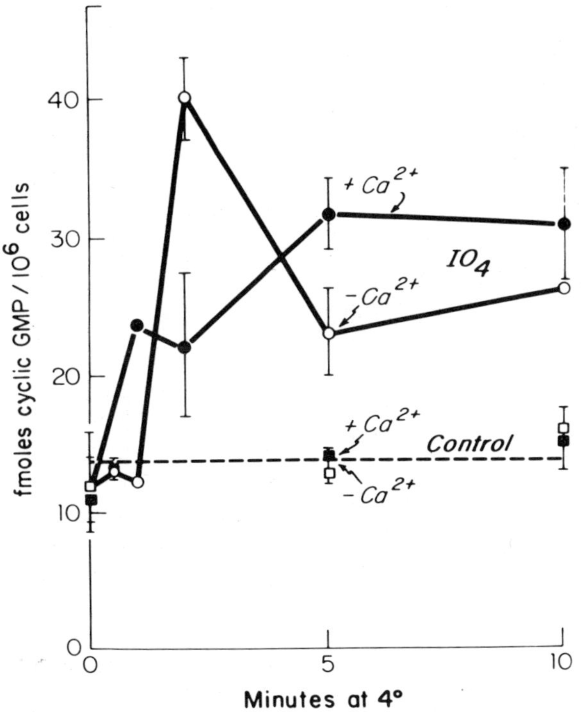

FIGURE 3. *Effect of IO_4 exposure on cyclic GMP levels of guinea pig splenic cells in the presence and absence of Ca^{++}. $CaCl_2$ (0.2 mM) was added to the cells suspended in PBS at 4°C for 10 min prior to the addition of the IO_4 (5 mM).*

Thirteen experiments were conducted during this study to examine the effects of IO_4 on cyclic GMP levels. In four of the experiments there was no detectable increase in cyclic GMP content. In the nine experiments in which an increase in cyclic GMP content was measurable (70-1000%) there was an increased rate of [^{3}H]-thymidine incorporation in the cells exposed to IO_4. However, in the four experiments in which there was no measurable increase in cyclic GMP, there was also no detectable increase in [^{3}H]-thymidine incorporation, indicating that the oxidative influence of IO_4 was ineffective in these experiments.

In order to further explore the relationship between the oxidative stimulation of proliferation and the increases in cyclic GMP levels the reducing agent cysteine was employed. The inclusion of this reducing agent during exposure of the cells to IO_4 prevented the IO_4-promoted increase in cyclic GMP content. The sulfhydryl reagent alone or together with IO_4 significantly lowered (50%) the concentration of this cyclic nucleotide (Figure 4). In the experiments with cysteine the increase in [^{3}H]-thymidine incorporation seen after exposure to IO_4 alone was

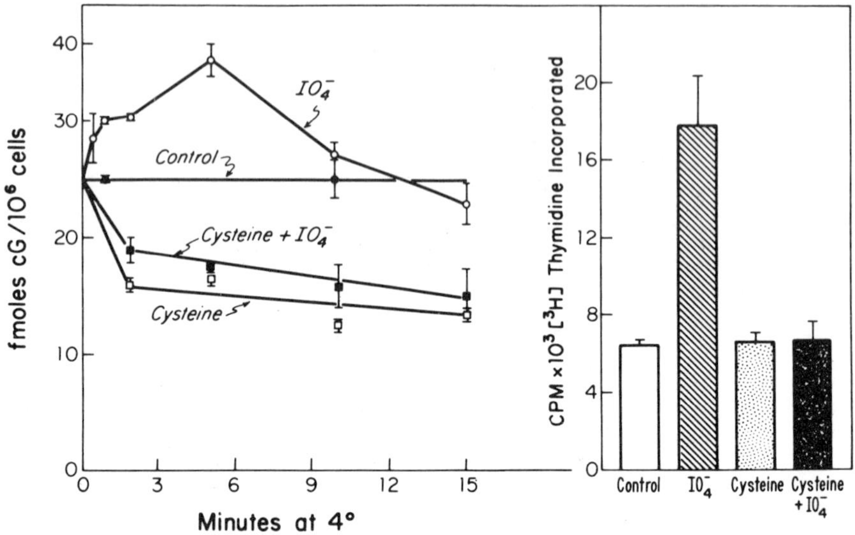

FIGURE 4. *Effect of cysteine on basal and IO_4-stimulated cyclic GMP levels of mouse splenic cells. Cysteine (1 mM) was added to the cells just prior to the addition of the IO_4 (1 mM).*

prevented in the cells exposed to IO_4 in combination with cysteine. There was no change in the cyclic AMP content after the addition of either the oxidizing or the reducing agent or both.

The ability of cysteine in these experiments to lower cyclic GMP levels appears to conflict with a suggested role for cyclic GMP as a positive effector of the proliferative process since the inclusion of reducing agents in lymphocyte cultures has been shown to augment the mitogenic response (2,8). When the effect of cysteine and another thiol, DTT, were tested on splenic cell cyclic nucleotide concentrations more extensively over a wide range of concentrations they were found to lower cyclic GMP levels in a dose dependent fashion (Figure 5) and slightly increase cyclic AMP levels. However, it should be noted that the conditions used for incubation of the cells in the experiment depicted in Figures 4 and 5 did not include the presence of serum or albumin, the serum component shown to be necessary to support mitogen-stimulated lymphocyte proliferation (24). When the splenic cells were suspended in a media containg 3 percent albumin which would support proliferation, the basal levels of cyclic GMP increased, while the concentration of cyclic AMP decreased as compared to cells suspended in culture media without albumin (Figure 6). Furthermore, the presence of albumin apparently "reversed" the effect of both reducing agents on the cyclic nucleotide levels. The addition of either cysteine or DTT to cells in albumin containing media elevated the cyclic GMP levels and lowered cyclic

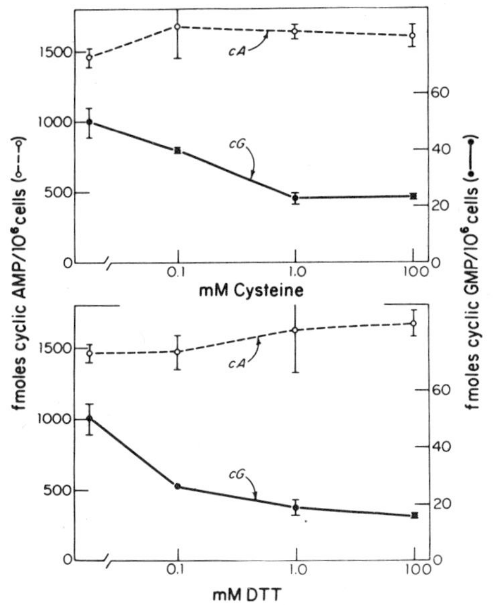

FIGURE 5. *Dose-dependent effect of cysteine and DTT on cyclic nucleotide levels in guinea pig splenic cells. The cells were incubated for 1 min at 37°C in the presence of the indicated concentrations of the reducing agents.*

AMP levels. Czech has shown in another suspended cell system that the addition of thiol reagents in the presence of albumin, which contains copper, results in the nonenzymatic generation of peroxide (5) through the following reaction:

$$2\ RSH + 2\ Cu^{++} \rightarrow RSSR + 2\ Cu^{++} + 2\ H^{++}$$

$$2\ H^{+} + 2\ Cu^{+} + O_2 \rightarrow 2\ Cu^{++} + H_2O_2$$

The biological actions of the thiols in this system have been postulated to be mediated by the peroxide formed. Thus, under conditions similar to those used to study the ability of reducing agents to augment proliferation (i.e., in the presence of albumin), conditions which are known to result in the generation of an oxidizing agent, the reducing agents can be shown to result in an increased accumulation of cyclic GMP.

Another compound possessing oxidizing potential which when added to splenic cell suspensions results in an enhanced accumulation of cyclic GMP is dehydroascorbic acid. The increase in cyclic GMP with this agent occurs in a concentration dependent manner and is accompanied by a decrease in cyclic AMP content (Figure 7). If *N*-ethylmaleimide (NEM), a sulfhydryl reagent which binds covalently to sulfhydryl groups and prevents their oxidation, is included in the incubation mixture the basal level of cyclic GMP

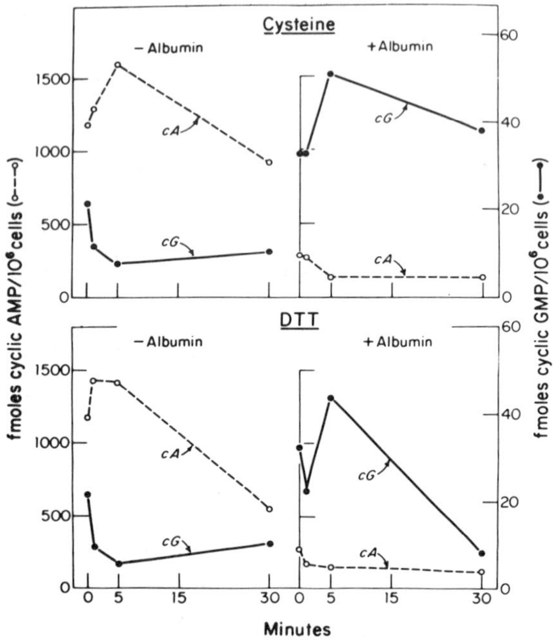

FIGURE 6. *Reversal of the effects of reducing agents effect on guinea pig splenic cell cyclic nucleotide content by albumin. The cells were preincubated at 37°C in the presence or absence of 3% bovine serum albumin for 30 min prior to the addition of the reducing agent (1 mM).*

is lowered and the dehydroascorbate-promoted increase of the cyclic nucleotide is blocked.

The observation which originally suggested the possibility that oxidants might be involved in promoting cyclic GMP accumulation by enhancing guanylate cyclase activity was the reported finding that soluble fractions of the enzyme from platelets (1) and lung (30) could undergo a time dependent, O_2 requiring "autoactivation" which is inhibitable by the reducing agent, DTT. Figure 8 shows that a similar activation process occurs with the soluble fraction of guanylate cyclase from splenic cells. The velocity of the enzyme prepared from cells obtained from guinea pig spleen increases 3-fold during 20 min of incubation at 37°C but remains constant if 1 m*M* DTT is included in the incubation mixture. The particulate enzyme, although it does not undergo an "autoactivation," also appears to be susceptible to modulation through a reductive mechanism. The addition of DTT to the cyclase incubation media results in an inhibition of the particulate guanylate cyclase activity (Figure 9).

Although a number of different agents which elevate intracellular cyclic AMP levels in a variety of cell systems have also been shown to activate adenylate cyclase (21), this is not the case with guanylate cyclase. Hor-

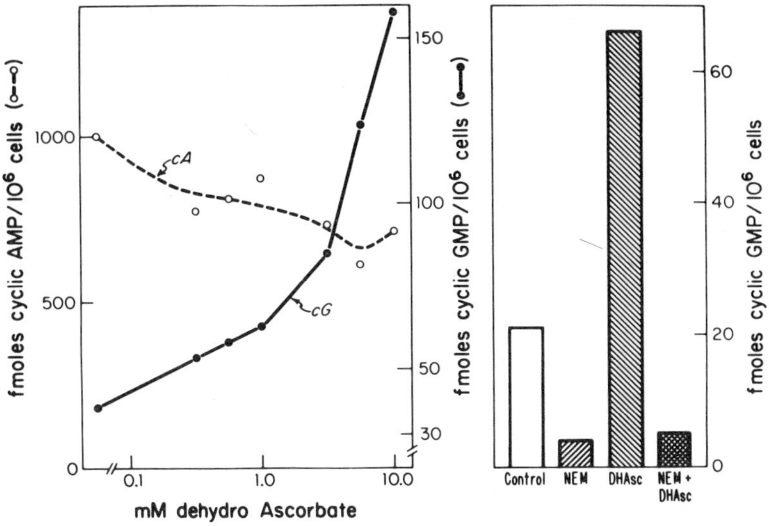

FIGURE 7. *Dehydroascorbic acid-promoted changes in guinea pig splenic cell cyclic nucleotides. (A) The cells were incubated at 37°C in the presence of the indicated concentration of dehydroascorbate for 1 min. (B) 10 mM NEM was added, where indicated, 5 min prior to the addition of dehydroascorbate (3 mM).*

mones which can elevate cyclic GMP levels in intact cells have not been shown to activate guanylate cyclase in the same cells after disruption. Nor has the addition of phytomitogens to broken lymphocyte cell systems been shown to result in increased guanylate cyclase activity.

However, dehydroascorbic acid, which we have found, stimulates cellular cyclic GMP accumulation in the whole splenic cell system, can also be shown to activate guanylate cyclase in the broken cell system. When the oxidized ascorbate is added to either the particulate or soluble guanylate cyclase preparations the activity of the enzyme is enhanced (Figure 10). Incubation with dehydroascorbic acid increases the velocity of the particulate enzyme from 35 to 66 pmoles/min/mg protein, the initial velocity of the soluble enzyme from 117 to 192 and the 20 min velocity of the soluble enzyme from 383 to 608 pmoles/min/mg protein. The stimulation of both the particulate and soluble enzyme is concentration dependent with respect to dehydroascorbic acid (Figure 11) with maximal stimulation occurring at 6 and 3 m*M*, respectively.

Although the dehydroascorbic acid promoted interaction that results in activation could derive its ability to promote oxidation of the enzyme (or a regulatory component) other mechanisms involving allosteric interaction of dehydroascorbic acid with the enzyme may be involved. As shown in Figure 12, the inclusion of 1 m*M* DTT in the guanylate cyclase reaction does not prevent the dehydroascorbic acid promoted activation. In fact, the addition

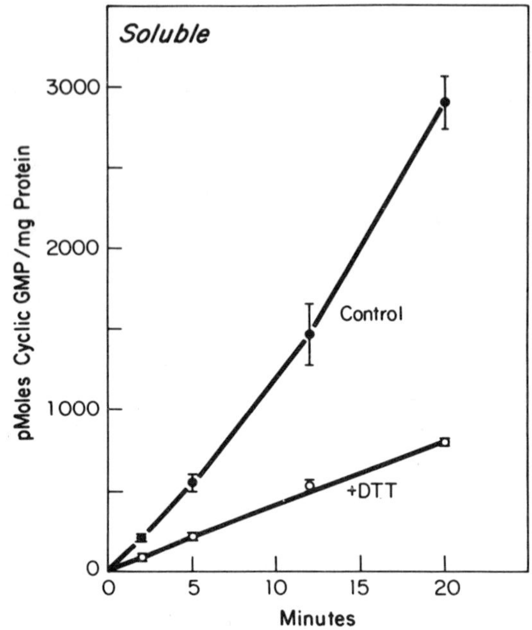

FIGURE 8. *"Autoactivation" of splenic cell soluble guanylate cyclase. The 104,000 x g supernatant fraction from splenic cells was incubated for the times indicated in the guanylate cyclase reaction in the presence or absence of 1 mM DTT.*

of the reducing agent appears to result in a greater dehydroascorbic acid-promoted activation of the enzyme because of the lowered control velocity that occurs in the presence of DTT (Figures 8 and 9).

DISCUSSION

This report presents evidence that guanylate cyclase activity can be modulated through cellular mechanism involving oxidation and reduction. The ability of IO_4 to stimulate an increase in cyclic GMP levels suggests that one mechanism for promoting cyclase activation may involve an alteration of the cell membrane. The modification of a specific membrane constituent, such as the sialic acid moiety indicated by the work of Novogrodsky (20), though an oxidation may be coupled to a system involved in regulating guanylate cyclase. The increases in cyclic GMP detectable after IO_4 are of a lesser magnitude than those reported to occur after stimulation of human or rodent peripheral blood lymphocytes with PHA or Con A. This may reflect the difference in the cell populations studied or a difference in the nature of the mitogenic stimulus However, under the conditions of mitogenic stimulation in the present experiments the changes occurred at 4°C, may have minimized the increases seen. Another factor to

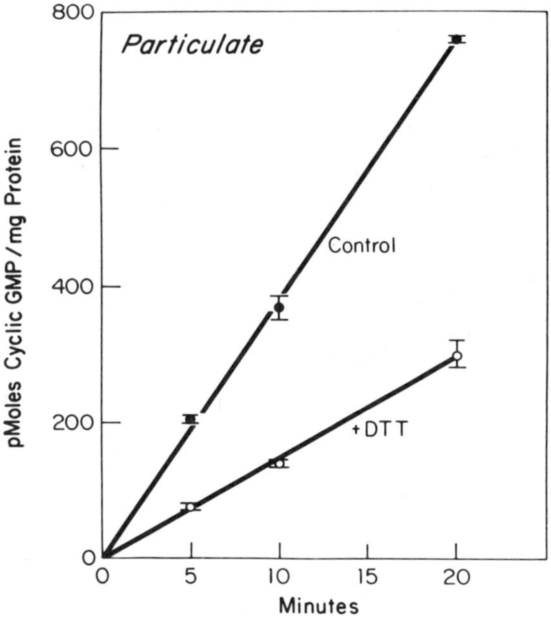

FIGURE 9. *DTT inhibition of splenic cell particulate guanylate cyclase. The 105,000 x g particulate fraction from splenic cells was incubated for the times indicated in the guanylate cyclase reaction in the presence or absence of 1 mM DTT.*

be considered is that the splenic cell suspensions used in these experiments, although consisting primarily of lymphocytes, contain other cells such as polymorphonuclear leukocytes and macrophages. Therefore, it is not possible to determine from the experiments reported here if the increases in cyclic GMP derive from an alteration induced in lymphocytes directly or secondarily, or whether they reflect an influence on another cell type. This question is becoming of greater concern in light of demonstrations of a macrophage requirement for IO_4 stimulation of lymphocytes (11) and of an involvement of cyclic GMP in the release of lymphocyte stimulating factors by macrophages (6).

The reciprocal effect on cyclic GMP levels of thiols and the peroxide-generating combination of albumin and thiols, and the demonstrated ability of the sulfhydryl reagent NEM to lower basal levels of cyclic GMP and block the dehydroascorbic acid-promoted increases suggest that a sulfhydryl-disulfide transformation may be involved in the control of cyclic GMP metabolism. This possibility is further supported by the reciprocal nature of the activation of guanylate cyclase through an oxygen-dependent autocatalytic reaction and the inhibition of this cyclase by the reducing agent DTT.

Dehydroascorbic acid, an agent with oxidizing

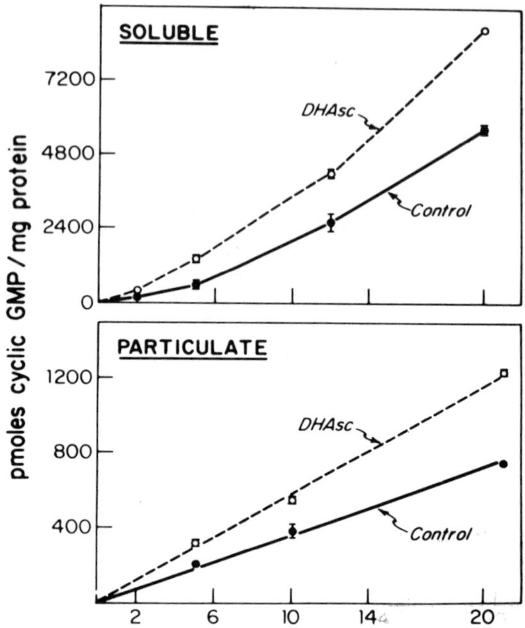

FIGURE 10. *Activation of splenic cell guanylate cyclase by dehydroascorbic acid. The particulate and soluble fractions of splenic cells were incubated in the guanylate cyclase reaction in the presence or absence of 1 mM dehydroascorbic acid.*

potential, elevates cyclic GMP levels in the intact cell and activates guanylate cyclase in the broken cell. However, this stimulation may involve a mechanism other than oxidation. The ability of dehydroascorbic acid to increase the soluble cyclase activity to a level higher than that resulting from the autoactivation suggests that several pathways for modulation of this enzyme may exist.

Since agents such as acetylcholine and imidazole can elevate lymphocyte cyclic GMP levels without promoting proliferation it is apparent that the increased cellular accumulation of this cyclic nucleotide is not the sole determinant in initiating the events in the proliferative process. However, there are three bodies of evidence which suggest that there is a positive involvement of cyclic GMP in lymphocyte biological regulation. The first is the effects of cyclic GMP-stimulating agents like acetylcholine and imidazole to potentiate the actions of mitogens. The second is the association in several types of lymphocyte systems between mitogenic stimulation and enhanced cyclic GMP accumulation. The third is the demonstrations that cellular events which are components of the growth response can be stimulated by the addition of exogenous cyclic GMP.

It would seem likely that cyclic GMP may represent an effector which, when properly integrated with changes

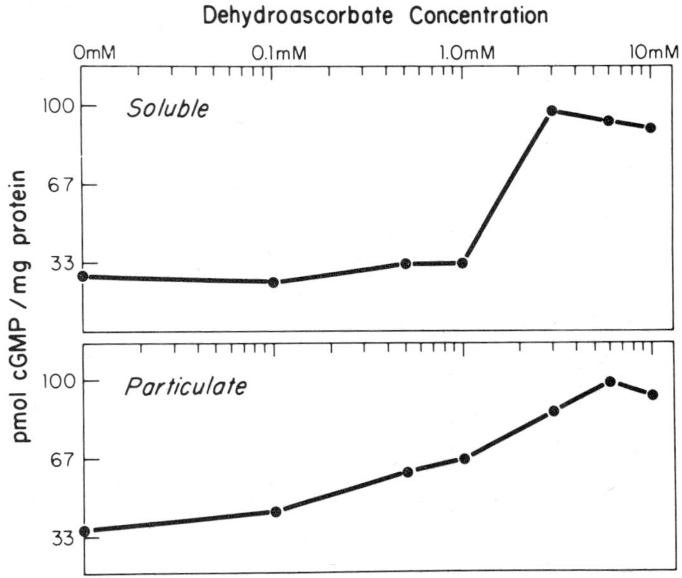

FIGURE 11. *Concentration-dependence of the activation of splenic cell guanylate cyclase by dehydroascorbic acid. The particulate and soluble fractions of splenic cells were incubated for 2 min in the guanylate cyclase reaction in the presence of the indicated concentration of dehydroascorbic acid*

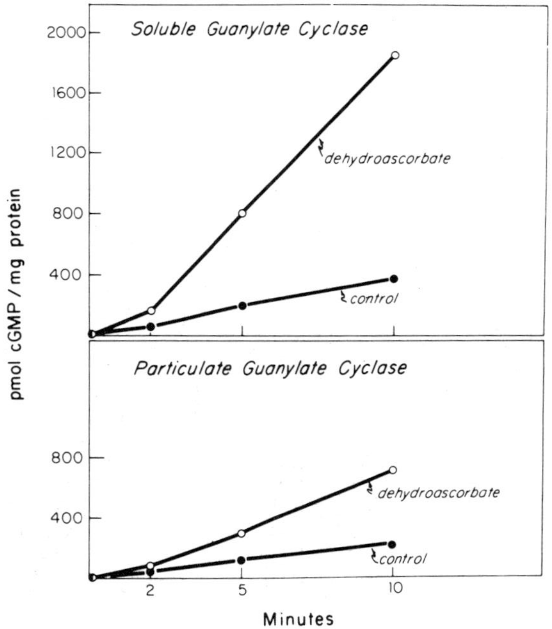

FIGURE 12. *Dehydroascorbic acid promoted activation of guanylate cyclase in the presence of DTT. The particulate and soluble fractions of splenic cells were incubated for the times indicated in the guanylate cyclase reaction containing 1 mM DTT in the presence or absence of 3 mM dehydroascorbic acid.*

in other metabolic components that are brought about by the mitogen, acts in parallel and/or in sequence to aid in promoting the commitment process and progress through the proliferative cycle. Modulation of cyclic GMP metabolism through discrete cellular events involving oxidation and reduction could then provide a mechanism by which this aspect of the regulation of the cellular growth process could be explored.

ACKNOWLEDGEMENTS

This research was supported by the following grants: NIH NS05979, ACS BC-166, HL 06314 and the Minnesota Leukemia Foundation.

REFERENCES

1. Bohme, E., Jung, R. and Mechler, I. Guanylate cyclase in human platelets. In: *Methods in Enzymology*, J. G. Hardman and B. W. O'Malley (eds.), Vol. 38C, Academic Press, New York, 1974, pp. 199-202.
2. Broome, J. D. and Jeng, M. W. 1973. Promotion of replication in lymphoid cells by specific thiols and disulfides *in vitro*. J. Exp. Med. *138*:574-592.
3. Cailla, H. L., Racine-Weisbuch, M. S. and DeLaage, M. A. 1973. Adenosine 3',5' cyclic monophosphate assay at 10^{-15} mole level. Anal. Biochem. *56*:394-407.
4. Chambers, D. A., Martin, D. W. and Weinstein, Y. 1974. The effect of cyclic nucleotides on purine biosynthesis and the induction of PRPP synthetase during lymphocyte activation. Cell *3*:375-380.
5. Czech, M. P., Lawrence, J. C. and Lynn, W. S. 1974. Evidence for electron transfer reaction involved in the Cu^{++}-dependent thiol activation of fat cell glucose utilization. J. Biol. Chem. *249*:1001-1006.
6. Diamantstein, T. and Ulmer, A. 1975. Stimulation by cyclic GMP of lymphocytes mediated by foluble factor released from adherent cells. Nature *256*:418-419.
7. Estensen, R., Hadden, J. W., Hadden, E. M., Touraine, F., Touraine, J. L., Haddox, M. K. and Goldberg, N. D. 1974. Phorbol myristate acetate: effects of a tumor promoter on intracellular cyclic GMP in mouse fibroblasts and as a mitogen on human lymphocytes. *The Cold Spring Harbor Conference on the Control of Proliferation in Animal Cells*, B. Clarkson and R. Baserga (eds.). Cold Spring Harbor, New York, pp. 627-634.
8. Fanger, M. W., Hart, D. A., Wells, J. V. and Nisonoff, A. 1970. Enhancement by reducing agents of the transformation of human and rabbit peripheral lymphocytes. J. Immun. *105*:1043-1045.

9. Freedman, M. H., Raff, M. C. and Gomperts, B. 1975. Induction of increased calcium uptake in mouse T lymphocytes by concanavalin A and its modulation by cyclic nucleotides. Nature *255*:378-382.

10. Goldberg, N. D., Haddox, M. K., Dunham, E., Lopez, C. and Hadden, J. W. 1974. The Yin Yang hypothesis of biological control: opposing influences of cyclic GMP and cyclic AMP in the regulation of cell proliferation and other biological processes. *The Cold Spring Harbor Conference on the Control of Proliferation in Animal Cells*, B. Clarkson and R. Baserga (eds.). Cold Spring Harbor, New York, pp. 609-625.

11. Greineder, D. K. and Rosenthal, A. S. 1975. The requirement of macrophage-lymphocyte interaction in T lymphocyte proliferation induced by generation of aldehydes on cell membranes. J. Immun. *115*: 932-938.

12. Hadden, J. W., Hadden, E. M., Haddox, M. K. and Goldberg, N. D. 1972. Guanosine 3',5' monophosphate: a possible intracellular mediator of mitogenic influences in lymphocytes. Proc. Nat. Acad. Sci. USA *69*:3024-3027.

13. Hadden, J. W., Coffey, R. G., Hadden, E. M., Lopez-Carrales, E. and Sunshine, G. H. 1975. Effects of levamisole and imidazole on lymphocyte proliferation and cyclic nucleotide levels. Cell. Immun. *20*:98-103.

14. Haddox, M. K., Furcht, L. T., Gentry, S. R., Moser, M. E., Stephenson, J. H. and Goldberg, N. D. 1976. Periodate-induced elevation of cyclic GMP levels in mouse and guinea pig splenic cells in association with mitogenesis. Nature, in press.

15. Johnson, E. and Hadden, J. 1975. Phosphorylation of lymphocyte nuclear acidic proteins: regulation by cyclic nucleotides. Science *187*:1198-1200.

16. Johnson, L. and Hadden, J. 1975. Cyclic GMP and lymphocyte proliferation: effects on DNA-dependent RNA polymerase I and II activities. Biochem. Biophys. Res. Commun. *66*:1498-1505.

17. Kimura, H. and Murad, F. 1974. Evidence for two different forms of guanylate cyclase in rat heart. J. Biol. Chem. *249*:6910-6916.

18. Mizoguchi, Y., Otani, S., Matsui, I. and Morisawa, S. 1975. Control of ornithine decarboxylase activity by cyclic nucleotides in the phytohemagglutinin induced lymphocyte transformation. Biochem. Biophys. Res. Comm. *66*:328-335.

19. Novogrodsky, A. and Katchalski, E. 1971. Induction of lymphocyte transformation by periodate. FEBS Lett. *12*:297-300.

20. Novogrodsky, A. 1975. Lymphocyte activation induced by modifications of surface membrane saccharides. In: *Immune Recognition*, Rosenthal, A. (ed.). Academic Press, New York, pp. 43-61.

21. Robison, G. A., Butcher, R. W. and Sutherland, E. W. 1971. *Cyclic cAMP*. Academic Press, New York.
22. Schumm, D. E., Morris, H. P. and Webb, T. E. 1974. Early biochemical changes in phytohemagglutinin-stimulated peripheral blood lymphocytes from normal and tumor-bearing rats. Eur. J. Cancer *10*: 107-113.
23. Seifert, W. E. and Rudland, P. S. 1974. Possible involvement of cyclic GMP in growth control of cultured mouse cells. Nature *248*:138-140.
24. Spieker-Polet, H. and Polet, H. 1976. Identification of albumin as the serum factor essential for the growth of activated human lymphocytes. J. Biol. Chem. *251*:987-992.
25. Steiner, A. L., Pagliara, A. S., Chase, L. R. and Kipnis, D. M. 1972. Radioimmunoassay for cyclic nucleotides. J. Biol. Chem. *247*:1114-1120.
26. Strom, T. B., Hirsh, M. S., Black, P. H., Carpenter, C. B., Phillips, S. M. and Merill, J. P. 1975. Modulation of GvH proliferation by cyclic nucleotides. Transpl. Proc. *7*:305-307.
27. Voorhees, J. J., Colburn, N. H., Stawiski, M., Duell, E. A., Haddox, M. and Goldberg, N. D. 1974. Imbalanced cyclic AMP and cyclic GMP levels in the rapidly dividing, incompletely differentiated epidermis of psoriasis. In: *The Cold Spring Harbor Conference on the Control of Proliferation in Animal Cells*, Clarkson, B. and Baserga, R. (eds.). Cold Spring Harbor, New York, pp. 635-648.
28. Watson, J. 1975. The influence of intracellular levels of cyclic nucleotides on cell proliferation and the induction of antibody synthesis. J. Exp. Med. *141*:97-111.
29. Weinstein, Y., Chambers, D. A., Bourne, H. R. and Melmon, K. L. 1974. Cyclic GMP stimulates lymphocyte nucleic acid synthesis. Nature *251*:352-353.
30. White, A. A. and Lad, P. J. 1975. Activation of the soluble guanylate cyclase from rat lung by preincubation. Fed. Proc. *34*:232.
31. Whitfield, J. F., MacManus, J. P., Rixon, R. H. and Gillan, D. J. 1973. The calcium-independent stimulation of thymic lymphoblast DNA synthesis by low cyclic GMP concentrations. Proc. Soc. Exp. Biol. Med. *144*:808-812.
32. Whitfield, J. F., MacManus, J. P., Boynton, A. L., Gillan, D. J. and Isaacs, R. J. 1974. Concanavalin A and the initiation of thymic lymphoblast DNA synthesis and proliferation by a calcium-dependent increase in cyclic GMP level. J. Cell. Physiol. *84*:445-458.

Cyclic AMP-Mediated Nuclear Translocation of Cytoplasmic cAMP-Dependent Protein Kinases: Identity of the Nuclear and Cytoplasmic Enzymes

Richard A. Jungmann, Mary L. Christensen, John S. Schweppe, Maija I. Mednieks, and Anna M. Spielvogel

The precise mechanism by which cyclic adenosine 3',5'-monophosphate (cAMP) exerts its regulatory effects on gene expression are largely unknown. The ubiquitous distribution of cAMP-regulated protein kinases in mammalian cells has led to the general notion that protein kinase-mediated phosphorylation of key cellular proteins underlies all or most of the effects of cAMP in eukaryotic cells (30). There has accumulated considerable suggestive evidence that cAMP-mediated activation of protein kinase and subsequent phosphorylative and functional modification of regulatory nuclear proteins may represent a key mechanism controlling gene transcription and induction of protein synthesis (1,14,16,20,25,28).

Although considerable protein kinase activity is found in the cytosol of most mammalian cells, nuclear protein kinase activity has been found to be remarkably insensitive to stimulation by cAMP in isolated intact nuclei. Thus, it is difficult to explain the cAMP-mediated phosphorylation of nuclear proteins *in vivo* through a direct action of cAMP on compartmentalized nuclear cAMP-dependent protein kinase. Our finding that ovarian nuclear protein kinase activity is gradually acquired during postnatal development of the rat, together with the fact that gonadotropin treatment of hypophysectomized rats increases the compartmentalized protein kinase activity of rat ovaries (9) has led us to propose that cytoplasmic cAMP-dependent protein kinase translocates to the nucleus as a consequence of cAMP stimulation (21). An intracellular redistribution of cAMP-dependent protein kinase activity, interpreted as intracellular redistribution and translocation, has repeatedly been observed after glucagon stimulation of rat liver (5,39), during transsynaptic induction of tyrosine hydroxylase (6,7), and after norepinephrine stimulation of glial tumor cells (42).

In support of the concept of translocation of cytoplasmic cAMP-dependent protein kinase to nuclear acceptor sites, we present evidence in this paper indicating that (a) nuclei of porcine ovaries possess a cAMP-dependent protein kinase which is identical with the cytoplasmic cAMP-dependent protein kinase; and that (b) nuclear translocation of cytoplasmic cAMP-dependent protein kinase does not arise as an artifact and is not due to contamination of nuclei by soluble cytosol protein kinase during the isolation of nuclei in aqueous media.

CYCLIC cAMP-MEDIATED INTRACELLULAR REDISTRIBUTION OF PROTEIN KINASE ACTIVITY IN CONFLUENT WI-38 FIBROBLASTS

Our initial studies have indicated that cAMP mediates the binding *in vitro* of purified cytosol cAMP-dependent protein kinase and cAMP-binding protein to nuclear and chromatin acceptor sites (19,21). The conclusions from these experiments were that intracellular cAMP may mediate the specific binding of cytoplasmic protein kinase to particulate fractions, particularly nuclei. Such an intracellular cAMP-mediated redistribution of kinase activity can, for example, be demonstrated in WI-38 fibroblast cells after their incubation with the cAMP-analogue 8-p-chlorophenylthio-cAMP. Figures 1 and 2 illustrate marked increases in nuclear and microsomal protein kinase activity with a concomitant decrease of cytosol and lysosomal-mitochondrial protein kinase activity in WI-38 cells after treatment with the cyclic nucleotide. The involvement of the cAMP-dependent but not the cAMP-independent protein kinase moiety in the intracellular redistribution can be demonstrated by testing the subcellular fractions for changes in protein kinase activity due to acquisition or loss of protein kinase catalytic subunit C with the aid of the heat-stable protein kinase inhibitor from rabbit skeletal muscle. The heat-stable inhibitor was shown to inhibit specifically cAMP-dependent protein kinase activity by directly interacting with the free catalytic subunit forming an inhibitor-subunit complex (2,3,50,51). When subcellular fractions of WI-38 cells were tested for catalytic subunit activity before and after 8-p-chlorophenylthio-cAMP treatment, the results of Figure 3 were obtained. Cyclic AMP-independent protein kinase activity measured in the presence of the inhibitor did not change in the subcellular fractions as a result of cyclic nucleotide treatment. Cyclic nucleotide-mediated increases of kinase activity in the nuclear and microsomal fractions, as well as the decreases of kinase activity in the cytosol and the lysosomal-mitochondrial fractions appear to involve only changes of catalytic protein kinase subunit.

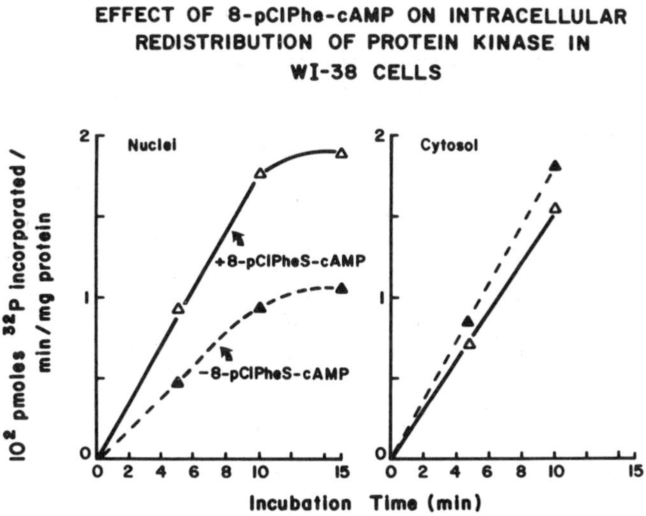

FIGURE 1. *WI-38 fibroblasts were grown to confluency. After incubation of the cells for 30 min without and with 10^{-7} M 8-p-chlorophenylthio-cAMP (ICN Pharmaceuticals), subcellular fractions were prepared and purified. Protein kinase activity of all fractions was assayed in the presence of 10^{-6} M cAMP with protamine as substrate. All experimental details concerned with the isolation and purification of subcellular fractions and protein kinase assay were described by Lee and Jungman (36).*

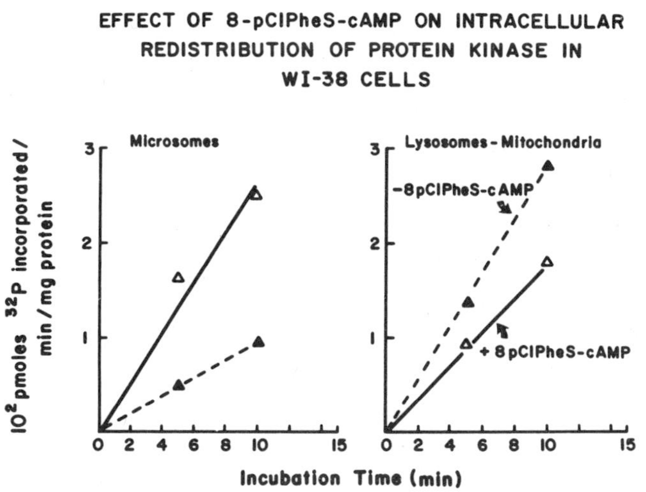

FIGURE 2. *See legend of Figure 1.*

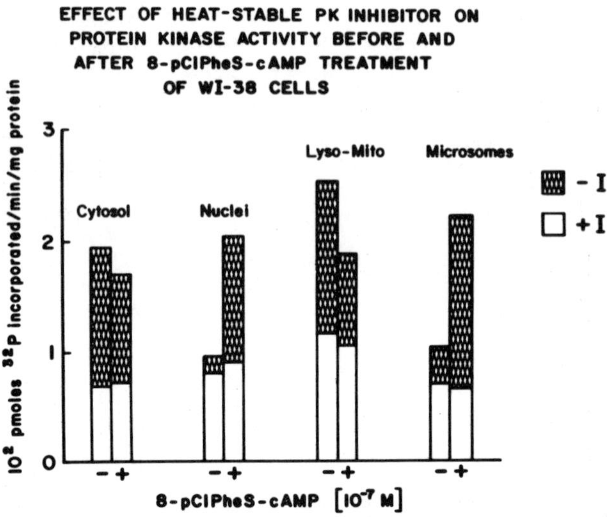

FIGURE 3. *Confluent WI-38 fibroblasts were incubated for 30 min without and with 10^{-7} M 8-p-chlorophenylthio-cAMP. After incubation, the cells were washed with buffer, recovered by centrifugation, and subcellular fractions were prepared. Protein kinase activity of all fractions was assayed with 10^{-6} M cAMP, protamine as substrate, in the absence and presence of saturating amounts of heat-stable protein kinase inhibitor as described by Lee et al. (37).*

DEMONSTRATION OF cAMP-DEPENDENT PROTEIN KINASE ACTIVITY IN PORCINE OVARY NUCLEI

The concept of translocation of cytoplasmic cAMP-dependent protein kinase to nuclear acceptor sites would be strongly supported, if it were possible to demonstrate (a) the presence of cAMP-dependent protein kinase activity in nuclei, and (b) the identity of cytoplasmic and nuclear cAMP-dependent protein kinases. The following results describe the identification of cAMP-dependent protein kinase activity in porcine ovary nuclei, and the identity of the nuclear cAMP-dependent protein kinase with the cytoplasmic enzyme.

DEAE-CELLULOSE CHROMATOGRAPHY AND SEPHADEX GIL FILTRATION OF PORCINE OVARY NUCLEAR PROTEIN KINASES

When purified porcine ovary nuclei were washed with 0.14 *M* NaCl and subsequently extracted with 0.5 *M* NaCl, about 85 percent of the total nuclear protein kinase activity was recovered in the extract (Table 1). Upon dialysis of the 0.5 *M* NaCl extract (fraction A) against 10 m*M* NaCl - 10 m*M* $MgCl_2$ - 3 m*M* dithiothreitol - 5% - glycerol - 20 m*M* Tris, pH 7.4, the majority of the extracted nuclear protein precipitated (fraction B). Analysis of fraction B indicated the presence of protein kinase activity which

TABLE 1. *Fractionation of porcine ovary nuclear protein kinase activity*

Fraction	Protein Kinase Activity			Recovery %		cAMP-Binding Activity		
	Protein (mg)	Total Units[1]	Specific Activity[2]	PS[3]	CA[4]	Total Units[5]	Specific Activity[6]	% Recovery
Nuclei	450	21,700	48.0	100	100	370	0.82	100
0.14 *M* NaCl (soluble fraction)	40	2,200	20.0	10	5	48	0.60	15
Fraction A (0.4 *M* NaCl extract before dialysis)	180	18,400	102.0	85	80	320	1.37	68
Fraction B (pellet after dialysis of 0.5 *M* NaCl extract)	157	2,400	15.5	11	52	-	N.D.[7]	-
Fraction C (soluble fraction after dialysis of 0.5 *M* NaCl extract)	18	11,700	650.0	54	16	165	9.2	47

Nuclear fractions were prepared as described in the text.

[1]*One unit = Amount of protein kinase activity which incorporates 1 pmole of* ^{32}P *from* $[^{32}P]$ *ATP in 1 min at 37°C with protamine as substrate in the presence of* 10^{-6} *M cAMP.*

[2]*Specific activity = pmoles of* ^{32}P *incorporated/min/mg protein with protamine as substrate.*

[3]*PS = Recovery of protein kinase activity measured with protamine as substrate.*

[4]*CA = Recovery of protein kinase activity measured with casein as substrate.*

[5]*One unit = Amount of binding activity which bind 1 pmole of c*$[^3H]$*AMP in 1 h at 2°C.*

[6]*Specific activity = pmoles of c*$[^3H]$*AMP bound per 1 h/mg protein.*

[7]*cAMP-binding activity not detectable.*

preferred casein as substrate. The protein kinase activity of fraction B was insensitive to cAMP stimulation, it did not bind cAMP, and it was not inhibited by saturating amounts of heat-stable protein kinase inhibitor from rabbit skeletal muscle (results not shown). We, therefore, concluded that the protein kinase activity of fraction B was primarily of the cAMP-independent variety, and we did not further characterize it.

The soluble nuclear protein fraction obtained after dialysis (Table 1; fraction C) contained the majority of the cAMP-dependent protein kinase activity. Protein kinase activity of fraction C was stimulated two-fold by 10^{-6} *M* cAMP, it was inhibited 50 percent by saturating concentrations of heat-stable inhibitor from rabbit skeletal muscle, and fraction C exhibited significant c[^{3}H]AMP-binding activity. We concluded that fraction C contained most of the nuclear cAMP-dependent protein kinase activity and proceeded further with the isolation and purification of that activity. As might be expected, most of the c[^{3}H]AMP-binding activity was recovered with the cAMP-dependent protein kinase in fraction C due to the binding of c[^{3}H]AMP to the regulatory subunit of the protein kinase.

Although previous workers have purified and separated nuclear protein kinases on DEAE-cellulose or phosphocellulose (11,12,27,45), chromatography of the soluble nuclear protein fraction C (Table 1) on DEAE-cellulose resulted in the elution of only one major peak of protein kinase and c[^{3}H]AMP-binding activity at an ionic strength of about 0.13 *M* NaCl (Figure 4). The fractions containing the

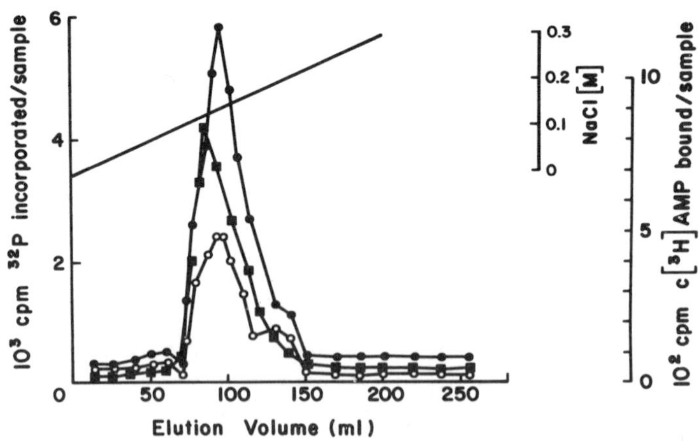

FIGURE 4. *DEAE-cellulose chromatography of porcine overy nuclear nonhistone protein and determination of protein kinase and c[^{3}H]AMP-binding activity. Porcine ovary nuclear nonhistone protein (Table 1; fraction C) was subjected to DEAE-cellulose chromatography and aliquots (0.05 ml) of each fraction were assayed for protein kinase and c[^{3}H]AMP-binding activity as described by Lee et al. (37). Protein kinase activity was assayed using protamine*

as substrate. O——O, protein kinase activity in the absence of cAMP; ●——●, protein kinase activity measured in the presence of 1 × 10^{-6} M cAMP; ■——■, c[^{3}H]AMP-binding activity; ——, NaCl [M].

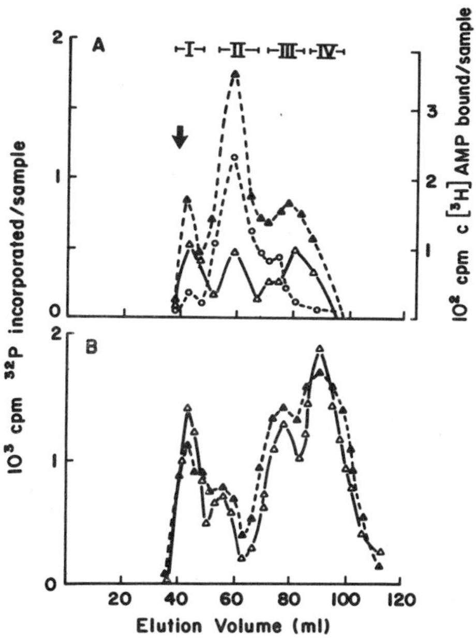

FIGURE 5. *A. Sephadex G-200 gel filtration of porcine ovary nuclear protein kinase activity obtained after DEAE-cellulose chromatography of ovary nuclear nonhistone protein. Fractions eluted between 80-130 ml from the DEAE-column (Figure 5) were concentrated by ultrafiltration in an Amicon model 202 ultrafiltration cell, and subjected to gel filtration on Sephadex G-200. Aliquots (0.05 ml) of each fraction were assayed for protein kinase and c[^{3}H]AMP-binding activity. Protein kinase activity was assayed with protamine as substrate without and with 1 × 10^{-6} M cAMP. Δ——Δ, protein kinase activity in the absence of cAMP; ▲---▲, protein kinase activity measured in the presence of cAMP; O---O, c[^{3}H]AMP-binding activity.*
B. Sephadex G-200 gel filtration of porcine ovary nuclear protein kinase activity after incubation with cAMP. Fractions eluted between 80-130 ml from the DEAE-column (Figure 4) were concentrated by ultrafiltration, incubated for 17 h at 4°C with 2 × 10^{-6} M cAMP, and subjected to gel filtration on Sephadex G-200. Aliquots (0.05 ml) of each fraction were assayed for protein kinase activity. Protein kinase activity was assayed using protamine as substrate without and with 1 × 10^{-6} M cAMP. ▲---▲, protein kinase activity in the presence of cAMP; Δ——Δ, protein

kinase activity without cAMP. The void volume of the column as determined by the elution of Blue Dextran 2000 is indicated by the arrow.

majority of the protein kinase activity that had initially been applied to the column (eluting between 80 and 130 ml (Figure 4) were pooled, concentrated by ultrafiltration, and subjected to gel filtration on Sephadex G-200. When the eluted fractions were assayed for protein kinase activity with protamine as substrate, three peaks of kinase activity were detected, protein kinase I, II, and III (Figure 5A). The activity of protein kinase II was stimulated about 3.5-fold by 10^{-6} *M* cAMP, and considerable c[^{3}H]AMP-binding activity was identified in the fractions with protein kinase II activity. Neither protein kinase I nor kinase III bound c[^{3}H]AMP, and they were only slightly (20%) stimulated by 10^{-6} *M* cAMP.

When the fractions containing protein kinase activity eluted from the DEAE-cellulose column (Figure 4; fractions eluting between 80-130 ml) were incubated for 2 h at 4°C in the presence of 5 x 10^{-6} *M* cAMP and subsequently subjected to Sephadex G-2000 gel filtration, the elution pattern shown in Figure 5B, was obtained. Protein kinases I and III were retained unchanged in their relative elution positions shown in Figure 5A. However, the majority of cAMP-dependent protein kinase II activity was absent from the elution profile after cAMP incubation. Instead, the elution profile of Figure 5B indicates the elution (between 86 and 100 ml) of a new peak of kinase activity, protein kinase IV. Protein kinase IV was not stimulated by 10^{-6} *M* cAMP and did not bind c[^{3}H]AMP, indicating the possibility that it was the catalytic subunit of cAMP-dependent protein kinase II.

To obtain additional confirmation that cAMP-dependent protein kinase II was composed of regulatory and catalytic subunits, the fractions exhibiting protein kinase II activity (Figure 5A; eluting between 55 and 65 ml) were pooled, incubated for 2 h at 4°C with 5 x 10^{-6} *M* c[^{3}H]AMP, and again filtered through Sephadex G-200. The elution pattern is presented in Figure 6. The majority of the protein kinase activity eluted between 82 and 95 ml at an elution position identical to protein kinase IV in Figure 5B. The eluted protein kinase IV was not stimulated by 10^{-6} *M* cAMP and presumably represented the free catalytic subunit which dissociated from the protein kinase II RC complex upon its interaction with c[^{3}H]AMP. The regulatory subunit, labeled with c[^{3}H]AMP, eluted between 70 and 80 ml (Figure 6) and did not exhibit protein kinase activity.

CYCLIC NUCLEOTIDE DEPENDENCY

To test their cyclic nucleotide dependency, protein kinases I, II, and III were assayed alone and in the presence of increasing concentrations of cAMP and cGMP with protamine as substrate (Figure 7). Under the standard

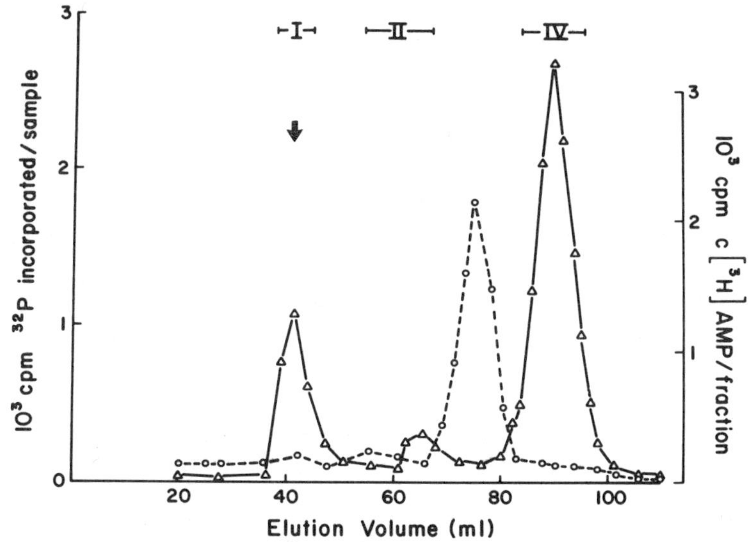

FIGURE 6. *Sephadex G-200 gel filtration of nuclear protein kinase II after incubation with c[^{3}H]AMP. Fractions eluted between 52 and 68 ml from the Sephadex column (Figure 5A) were concentrated by ultrafiltration, incubated for 2 h at 4°C with 2 × 10^{-6} M c ^{3}H AMP (10 μCi), and subjected to Sephadex G-200 gel filtration. Aliquots (0.05 ml) of each fraction were assayed for content of radioactivity (c[^{3}H]AMP), and protein kinase activity. Protein kinase activity was assayed with protamine in the absence of cAMP. Δ——Δ, protein kinase activity; O---O, tritium radioactivity. The void volume of the column as determined by the elution of Blue Dextran 2000 is indicated by the arrow.*

assay conditions, incorporation of ^{32}P-phosphate into protein was linear during the 15 min incubation period. There was only a slight or no effect on the activity of either protein kinase I or kinase III using concentrations of cyclic nucleotides as high as 10^{-4} *M*. In contrast, the activity of protein kinase II was maximally stimulated (six-fold) by 10^{-6} *M* cAMP. Higher concentrations of cGMP were needed to achieve the same degree of stimulation as that achieved with cAMP.

SUBSTRATE SPECIFICITY

Table 2 lists the relative rates of reaction of protein kinases I, II, and III with various protein substrates. Incubations were carried out for 5, 10, and 15 min and the reaction rates were calculated from the slopes of the linear rates. Each of the different substrates was present in equal concentration by weight. Of the various substrates tested, protamine was found to be the best substrate for protein kinases II and III. Histone F1 was the best substrate for protein kinase I.

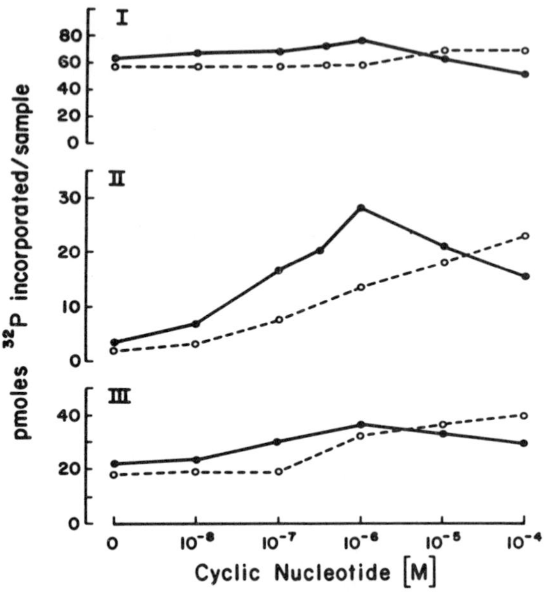

FIGURE 7. *Activity of porcine ovary nuclear protein kinases I, II, and III as a function of cAMP and cGMP concentration. The initial reaction rates of protein kinase I, II, and III obtained after Sephadex gel filtration (Figure 5A) were determined with protamine as substrate in the presence of the indicated concentrations of cAMP: ●——●, or cGMP: O---O. The values shown are the arithmetic means of three determinations.*

TABLE 2. *Substrate specificity of ovarian nuclear protein kinases I, II, and III*

Protein Substrate	^{32}P Incorporation (ρmoles/min)					
	I		II		III	
	- cAMP	+ cAMP	- cAMP	+ cAMP	- cAMP	+ cAMP
Histone Fl	17.2	20.3	1.7	4.7	2.8	3.4
Protamine	9.2	11.8	2.1	11.0	7.5	12.0
Casein	2.1	1.7	0.2	0.4	0.6	0.8

The phosphorylation reactions were carried out with 100 μg of substrate under standard assay conditions as described by Lee and Jungmann (36). Cyclic AMP was present where indicated at a concentration of 10^{-6} M. The nuclear protein kinases were prepared by extraction of ovary nuclei with 0.5 M NaCl, followed by DEAE-cellulose chromatography and Sephadex gel filtration of the extract as described in legend of Figure 5. The values shown are calculated from the slopes of the initial reaction rates.

Using an endogenous substrate, a porcine ovary nuclear nonhistone protein preparation, protein kinases I, II, and III exhibited remarkable substrate specificities as illustrated in Figure 8. In consideration of a recent

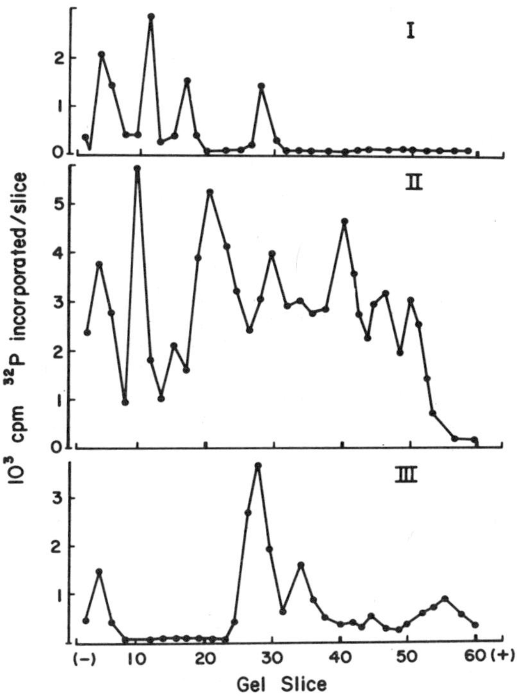

FIGURE 8. *Electrophoretic separation and ^{32}P-labeling patterns of porcine ovary nuclear nonhistone proteins after their phosphorylation by nuclear protein kinases I, II, and III. A porcine ovary nuclear nonhistone protein fraction, essentially free of endogenous protein kinase activity, was prepared by combining all fractions eluted from the DEAE-column (Figure 4) with the exception of those fractions exhibiting protein kinase activity (eluting between 70 and 140 ml). Pooled fractions were concentrated by ultrafiltration, dialyzed against 10 mM NaCl - 20 mM Tris - 10 mM $MgCl_2$ - 10 mM mercaptoethanol - 5% glycerol, pH 7.2, and adjusted to a protein concentration of 4 mg/ml. ^{32}P-labeling of the nuclear nonhistone protein was carried out by incubating 160 μg of nonhistone protein with protein kinases I, II, or III obtained after Sephadex gel filtration (Figure 5A) in the presence of 1 μM [γ-^{32}P]ATP (10 μCi) and 1 × 10^{-6} M cAMP in a total reaction volume of 0.12 ml. Incubation was carried out at 37°C for 15 min during which time incorporation of ^{32}P into protein proceeded linearly. The reaction was stopped by the addition of 10 μl of sodium dodecyl sulfate. Each sample was directly applied to the polyacrylamide gel and*

electrophoresis was carried out as described by Teng et al. (47). Radioactivity was determined after trichloroacetic acid-fixing of the protein, transverse slicing of the gel, and liquid scintillation counting. Autophosphorylation of nuclear nonhistone protein was determined after its incubation with [γ-^{32}P]ATP and cAMP in the absence of protein kinases I, II, or III. The cpm ^{32}P listed in the figure represent the cpm ^{32}P/gel slice obtained after phosphorylation of nonhistone protein in the presence of protein kinases I, II, or III minus the cpm ^{32}P/gel slice obtained after autophosphorylation of nonhistone protein without protein kinases I, II, or III.

report by Bylund and Krebs (4) which showed that denaturation of a substrate protein will increase its susceptibility for relatively unspecific phosphorylation by protein kinases, we have prepared undenatured nuclear nonhistone protein substrate by combining all nuclear protein eluted from the DEAE-cellulose column (Figure 4) with the exception of those fractions exhibiting protein kinase activity (eluting between 70 and 140 ml). The pooled protein represented about 85 percent of the total nuclear nonhistone protein initially applied to the DEAE column and exhibited only very low levels of endogenous protein kinase activity. After incubation of equal amounts of this nuclear nonhistone protein fractions with either protein kinase I, II, or III in the presence of [γ-^{32}P]ATP, the radioactive labeling patterns of the phosphorylated nuclear nonhistone proteins were analyzed after their separation by polyacrylamide gel electrophoresis in the presence of 0.1 percent sodium dodecyl sulfate. The ^{32}P-labeling patterns shown in Figure 8 clearly indicate a substrate-specific phosphorylation which was characteristic for each kinase. Close analysis of the patterns reveals that apparently only two of the nuclear proteins served as a common substrate for more than one kinase, e.g., protein in gel slices 2-4 was phosphorylated by all three kinases, and protein in gel slices 27-29 became phosphorylated by protein kinase I and III but not by kinase II.

K_m VALUES FOR ATP

When the effect of varying ATP concentrations on the protein kinase reactions was studied, it was found that protein kinases I, II, and III followed Michaelis-Menten kinetics for ATP (data not shown). The computed K_ms for ATP were 2.45 x 10^{-5} *M* for kinase I, 0.9 x 10^{-5} *M* for kinase II, and 1.9 x 10^{-5} *M* for kinase III.

EFFECT OF INHIBITOR PROTEIN ON NUCLEAR PROTEIN KINASES

A heat-stable inhibitor protein effective in inhibiting the activity of cAMP-dependent protein kinase has been described by Walsh *et al.* (2). The heat-stable inhibitor was shown to inhibit cAMP-dependent protein kinase activity by directly interacting with the free catalytic subunit forming an inhibitor-catalytic subunit complex (2,3,51). To evaluate the effect of the inhibitor, nuclear protein kinases I, II, and III, and the catalytic subunit, protein kinase IV, were assayed without and with cAMP in the presence of increasing concentrations of inhibitor protein isolated from rabbit skeletal muscle (Figure 9). The activity of protein kinase III was not inhibited and there was only a slight degree of inhibition of kinase I activity presumably due to the presence of some contamination with protein kinase II. Interaction of the heat-stable inhibitor with protein kinase II or with the catalytic subunit, protein kinase IV, resulted in a high degree of inhibition of kinase activity.

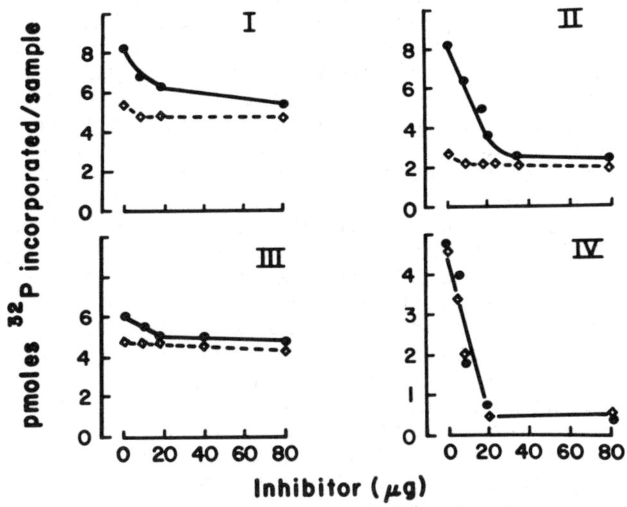

FIGURE 9. *Effect of heat-stable inhibitor protein on activity of nuclear protein kinases I, II, III, and IV. Protein kinases I, II, III, and IV were obtained after Sephadex gel filtration (Figures 5A and B) and assayed with protamine in the absence (◇--- ◇), or presence (●——●) of 1 × 10^{-6} M cAMP. Varying concentrations of heat-stable inhibitor protein partially purified from rabbit skeletal muscle (50) were added to the standard reaction mixtures. Incubation was carried out for 15 min at 37°C. The values shown are the arithmetic means of three determinations.*

COMPARISON OF CYTOPLASMIC AND NUCLEAR cAMP-DEPENDENT PROTEIN KINASES

To determine whether the nuclear cAMP-dependent protein kinase activity could have arisen as a result of contamination with the cytoplasmic cAMP-dependent protein kinase during isolation of nuclei, it was of interest to determine the degree of contamination by cytosol proteins occurring during homogenization of the tissue and subsequent isolation of nuclei. To test this possibility, porcine ovary slices were incubated with a mixture of uniformly ^{14}C-labeled amino acids under conditions allowing incorporation of amino acids into protein (17). After homogenization of the slices, a ^{14}C-labeled cytosol (105,000 x g supernatant fraction) was prepared which was incubated for 30 min at 4°C with nonradioactive nuclei isolated from ovaries that had not been labeled with radioactivity. When the nuclei were reisolated from the ^{14}C-labeled cytosol under conditions routinely used for the preparation and purification of nuclei, only statistically insignificant amounts of ^{14}C radioactivity remained associated with the nuclei (data not shown). Similarly, when the ^{14}C-labeled cytosol cAMP-dependent protein kinase preparation isolated by DEAE-cellulose chromatography (under conditions similar to those described in Figure 4) was incubated for 30 min at 4°C with unlabeled nuclei, insignificant amounts of radioactivity were recovered with the reisolated nuclei. These results strongly argue against a contamination of nuclei with cytoplasmic proteins.

Porcine ovary cytosol protein kinase activity exhibited DEAE elution characteristics identical to the nuclear protein kinase preparation shown in Figure 4, and both the nuclear and cytosol protein kinase activity eluted at an ionic strength of about 0.13 *M* NaCl (DEAE-elution profile for cytosol protein kinase not shown). When the cytosol and nuclear protein kinase preparations were preincubated for 17 h at 4°C with 2 x 10^{-7} *M* cAMP to achieve dissociation of the holoenzyme RC complex into the catalytic subunit C and the regulatory subunit R, it was found that subsequent DEAE-cellulose chromatography resulted in the elution of about 60-70 percent of the total applied protein kinase activity with the wash buffer (fractions eluting in the wash between 20-60 ml; Figures 10A and B). The nuclear as well as the cytosol protein kinase activity which eluted in the wash buffer was insensitive to stimulation by 10^{-6} *M* cAMP, did not bind c[^{3}H]AMP, and was inhibited 96 percent by saturating amounts of heat-stable protein kinase inhibitor from rabbit skeletal muscle, suggesting the presence of dissociated catalytic subunit C. Upon Sephadex G-200 gel filtration, the catalytic subunits of nuclear and cytosol protein kinase exhibited identical elution patterns (Figures 11A and B). The combined catalytic subunits from the nuclear and cytosol protein kinases eluted together and no separation of the kinase activities was observed (Figure 11C).

Additional comparative studies (not shown) on the catalytic subunits obtained after gel filtration on Sephadex G-200 (Figure 11) indicated that the activity of both subunits was inhibited 96 percent by the heat-stable inhibitor protein; both subunits exhibited an identical pH optimum of 7.5, and an identical K_m ATP of 0.8 x 10^{-5} *M*. Both protein kinases utilized the following substrates listed in the order of decreasing efficiency:

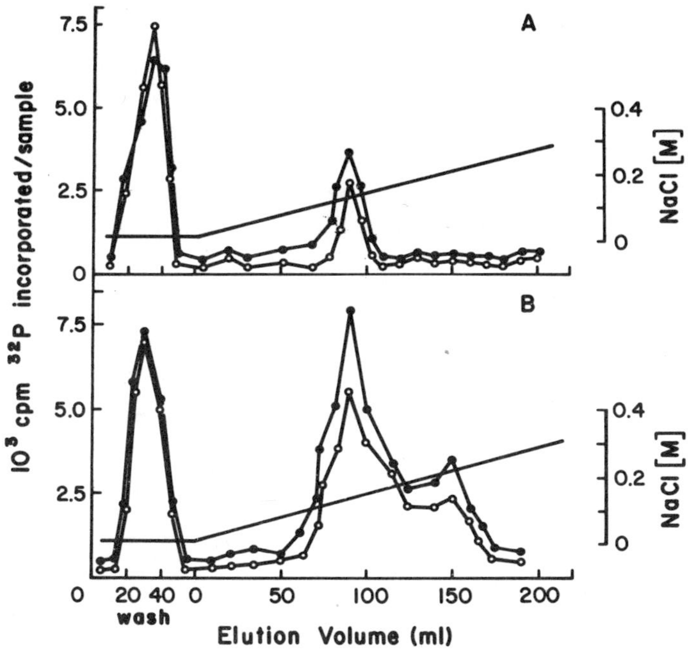

FIGURE 10. *DEAE-cellulose chromatography of porcine ovary nuclear nonhistone protein (panel A), and of porcine ovary cytosol protein (panel B) after their incubation with cAMP and determination of protein kinase activity. Nuclear nonhistone protein (fraction C - Table 1) and porcine ovary cytosol (105,000 x g supernatant fraction) were incubated for 17 h at 4°C with 2 × 10^{-6} M cAMP and subjected to DEAE-cellulose chromatography as described by Lee et al. (37). Aliquots (0.05 ml) of each fraction were assayed for protein kinase activity with protamine in the absence (O——O), and presence (●——●) of 10^{-6} M cAMP. ——, NaCl* [*M*].

protamine > histone F1 > phosvitin > casein

The apparent identity of the nuclear and cytosol protein kinase catalytic subunits was further confirmed after phosphorylation of undenatured endogenous nuclear nonhistone protein and analysis of the ^{32}P-labeling patterns by polyacrylamide gel electrophoresis. Figures 12A and B illustrate the similarity of the ^{32}P-labeling patterns of the nuclear nonhistone proteins after their phosphorylation by either the cytosol or the nuclear protein kinase catalytic subunit, indicating that both catalytic subunits exhibit identical substrate specificity toward endogenous nuclear nonhistone protein.

Porcine ovary cytosol and nuclear cAMP-dependent protein kinases also showed similar antigenic properties. Antiserum prepared against beef heart cAMP-dependent pro-

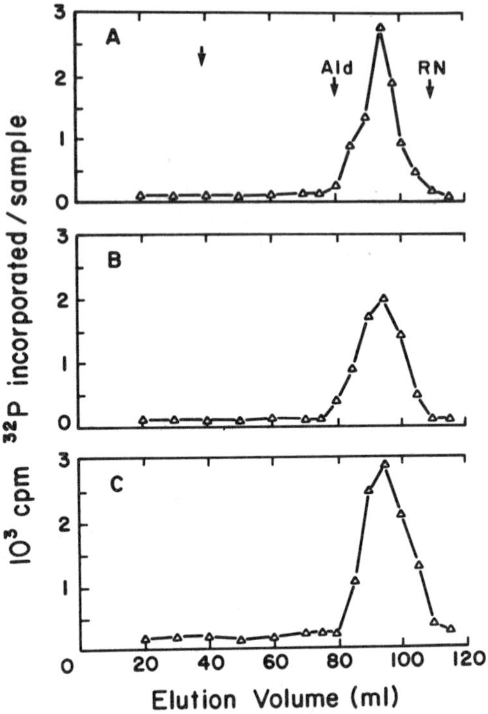

FIGURE 11. A. *Sephadex G-200 gel filtration of the catalytic subunit of ovary nuclear cAMP-dependent protein kinase. Fractions eluted between 20-60 ml (wash) from the DEAE-column (Figure 10A) were concentrated by ultrafiltration, and subjected to gel filtration.* B. *Sephadex G-200 gel filtration of the catalytic subunit of ovary cytosol cAMP-dependent protein kinase. Fractions eluted between 20-60 ml (wash) from the DEAE-column (Figure 10B) were concentrated by ultrafiltration, and subjected to gel filtration.* C. *Sephadex G-200 gel filtration of the combined catalytic subunits of ovary cytosol and nuclear cAMP-dependent protein kinases. Fractions eluted between 20-60 ml (wash) from the DEAE-columns (Figures 10A and B) of cytosol and nuclear protein kinase were combined, concentrated by ultrafiltration, and subjected to gel filtration. The combined fractions contained 9.7 units of cytosol and 7.3 units of nuclear protein kinase catalytic subunit. For definition of "unit" see legend of Table 1. Gel filtration was carried out as described in legend of Figure 5. Aliquots (0.05 ml) of each fraction were assayed for protein kinase activity without added cAMP. The void volume of the column as determined by the elution of Blue Dextran 2000 is indicated by the arrow. Standards used: Ald, aldolase; RN, ribonuclease.*

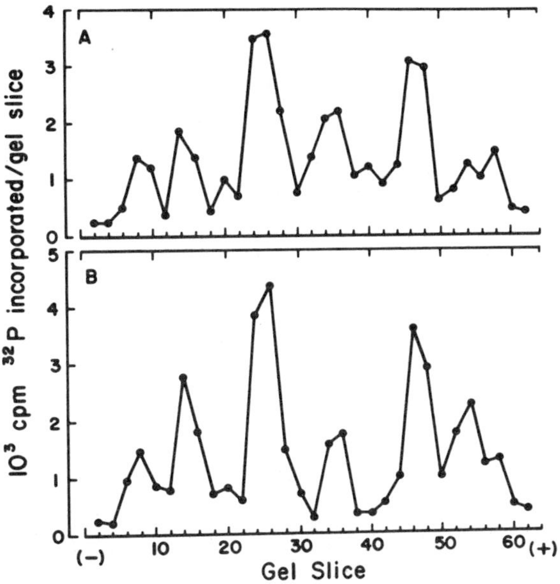

FIGURE 12. *Electrophoretic separation and ^{32}P-labeling patterns of porcine ovary nuclear nonhistone proteins after their phosphorylation by the catalytic subunit of cytosol (panel A), and of nuclear (panel B) cAMP-dependent protein kinase. Protein kinase-free nuclear nonhistone protein substrate was prepared as described in legend of Figure 8. ^{32}P-labeling of the nuclear nonhistone protein was carried out by incubating [γ-^{32}P]ATP, 160 μg of nonhistone nuclear protein with the cytosol or nuclear protein kinase catalytic subunit obtained after Sephadex gel filtration (Figures 11A and B). For further experimental details see legend of Figure 8. The cpm ^{32}P/gel slice plotted are corrected data and were obtained after subtraction of the cpm ^{32}P incorporated/gel slice contributed by the autophosphorylation of the nuclear nonhistone protein and subtraction of the cpm ^{32}P/gel slice contributed by the autophosphorylation of the catalytic subunits from the total cpm ^{32}P/gel slice determined after incubation of nonhistone protein with the catalytic subunits.*

tein kinase had relatively low titers but high affinity for the antigen and inhibited both the cAMP-dependent protein kinase from ovary cytosol as well as protein kinase II from ovary nuclei to a similar degree (Table 3). No inhibition of the nuclear cAMP-independent protein kinase I by the antiserum was observed.

To determine whether the nuclear cAMP-dependent protein kinase II activity could have arisen as a result of contamination with the cytoplasmic cAMP-dependent protein kinase during isolation of nuclei in the aqueous medium, we have prepared nuclei by a nonaqueous method. The method consists in breakage of lyophilized cells and

TABLE 3. *Inhibition of protein kinase activity with anti-cAMP-dependent protein kinase antiserum*

Protein Kinase Preparation	^{32}P Incorporated (pmoles/min) Normal Serum	Anti-serum	Inhibition %
Beef heart	22.8	6.6	82.3
(Sigma Chemical Co.)	45.2	13.8	72.7
Porcine ovary cytosol	9.8	1.4	85.2
	18.6	2.1	89.8
Protein kinases from nuclei isolated by the aqueous method			
Nuclear protein kinase I	6.8	6.4	5.9
Nuclear protein kinase II	2.1	0.6	70.7
	10.7	1.6	86.8
Protein kinases from nuclei isolated by the nonaqueous method			
Nuclear protein kinase I	5.9	5.9	-
Nuclear protein kinase II	23.9	3.5	85.4
	35.7	4.1	88.6

Beef heart cAMP-dependent protein kinase was obtained from Sigma Chemical Co. Porcine ovary cytosol cAMP-dependent protein kinase was obtained after DEAE-cellulose chromatography of an ovary cytosol 105,000 x g supernatant fraction. Nuclear protein kinases I and II were obtained after DEAE-cellulose chromatography and Sephadex gel filtration (Figure 5) of nuclear extracts obtained either from nuclei isolated by the aqueous method, or by the non-aqueous method as described in the text. Enzyme-antiserum incubation media contained the following: PBS (pH 7.4) containing protein kinase, 0.1 ml; and either 0.1 ml of normal rabbit serum or 0.1 ml of antiserum at a 1:2 dilution. Incubation was carried out for 1 h at 23°C and subsequently for 18 h at 4°C. After incubation, the mixture was centrifuged for 10 min at 20,000 x g and 0.02 ml aliquots of the supernatant fraction were assayed for protein kinase activity. The phosphorylation reactions were carried out with 100 µg of protamine as substrate and in the presence of 10^{-6} M cAMP under standard assay conditions as described by Lee et al. (37).

separation of nuclei by centrifugation in various mixtures of cyclohexane-carbon tetrachloride. Nuclei prepared by this method are essentially free of non-nuclear contaminants as indicated by the absence of several marker enzymes such as glucose-6-phosphatase and succinate-cytochrome c reductase (44). When 0.5 *M* NaCl extracts were prepared from these nuclei and analyzed by Sephadex G-200 gel filtration, elution patterns were obtained that were identical to the elution pattern shown in Figure 5A. Additionally, the presence of cAMP-dependent protein kinase II in nuclei isolated by the nonaqueous method was clearly identified by its inhibition caused by interaction with anti-cAMP-dependent protein kinase antiserum (Table 3).

TRANSLOCATION OF CYTOPLASMIC cAMP-DEPENDENT PROTEIN KINASE TO THE NUCLEUS

Previous comparative studies on nuclear protein kinases in our laboratory have revealed that nuclei from the ovaries of neonatal pigs (2-4 days old) contain no cAMP-dependent protein kinase II activity, although cAMP-dependent protein kinase activity could be identified in the 105,000 x g supernatant fraction of these ovaries. To determine whether or not a translocation of cytoplasmic protein kinase to the nucleus could be achieved as a consequence of cAMP action, cell suspensions prepared from the ovaries of neonatal pigs were incubated for 15 min in the absence and presence of 10^{-7} *M* 8-p-chlorophenylthio-cAMP. Untreated and treated ovarian cells were subsequently lyophilized, nuclei were isolated in nonaqueous media, and nuclear extracts were analyzed by Sephadex gel filtrtion. Comparison of the protein kinase elution profiles illustrated in Figures 13A and B reveals the presence of cAMP-dependent protein kinase II in nuclei from neonatal ovaries incubated with 8-p-chlorophenylthio-cAMP (Figure 13B), whereas nuclei from untreated neonatal ovaries lack protein kinase II (Figure 13A).

The translocation of cytoplasmic cAMP-dependent protein kinase to the nucleus as a result of 8-p-chlorophenylthio-cAMP action was also confirmed by the data of Table 4. Antiserum against cAMP-dependent protein kinase inhibited 45 percent of the protein kinase activity in extracts of nuclei from 8-p-chlorophenylthio-cAMP-treated cells, whereas no inhibition of kinase activity was observed in nuclear extracts from untreated cells. Isolated nuclear protein kinase II from the nuclei of ovarian cells incubated with the cAMP analogue was markedly inhibited by the antiserum.

DISCUSSION

Studies designed to elucidate the sequence of events of cAMP action in hormonal target cells point to the nucleus as a major site of cAMP action and to the ability of cAMP to influence transcription. The effect of cAMP upon transcription may be mediated through its action upon cAMP-dependent protein kinases (35) and the subsequent phosphorylation and functional modification of chromosomal proteins that play an important role in the regulation of transcription (1,14,15,20,25,28).

Recent studies showed that multiple forms of cAMP-dependent protein kinases are present in the cytosol of most mammalian tissues (35). Yet, despite some indirect

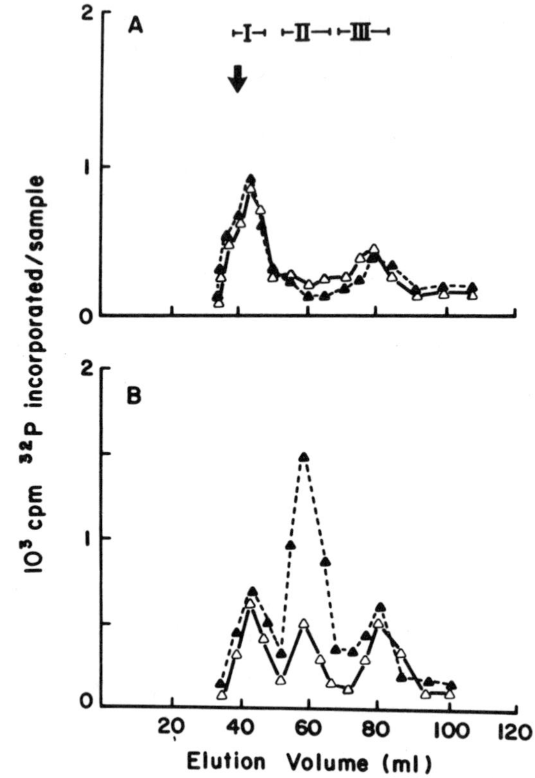

FIGURE 13. *Sephadex G-200 gel filtration of nuclear protein kinase activity from neonatal porcine ovaries before and after their incubation with 8-p-chlorophenylthio-cAMP. Cell suspensions prepared from the ovaries of neonatal pigs (3-4 days old) were incubated for 15 min at 37°C in 0.05 M sodium phosphate buffer (pH 7.4) - 0.14 M NaCl in the absence or presence of 10^{-7} M 8-p-chlorophenylthio-cAMP. After incubation, the cells were recovered by centrifugation, quick-frozen in liquid nitrogen, and lyophilized. Nuclei were isolated in nonaqueous media as described by Siebert (44). Nuclei were subsequently suspended in and extracted with 0.5 M NaCl. The extract was dialyzed against 10 mM NaCl - 10 mM $MgCl_2$ - 5% glycerol - 3 mM dithiothreitol - 20 mM Tris, pH 7.4. The dialyzed fractions were subjected to DEAE-cellulose chromatography which resulted in the elution of one major peak of protein kinase activity. The combined fractions containing all protein kinase activity were concentrated by ultrafiltration, and subjected to Sephadex G-200 gel filtration. The eluted fractions were assayed for protein kinase activity as described in legend of Figure 5. Panel A: protein kinase activity of nuclei from untreated ovarian cells; Panel B: protein kinase activity of nuclei from ovarian cells incubated with 8-p-chlorophenylthio-cAMP. Δ——Δ, protein*

kinase activity assayed without cAMP; ▲---▲, protein kinase measured in the presence of 10^{-6} M cAMP. The void volume of the column as determined by the elution of Blue Dextran 2000 is indicated by the arrow.

TABLE 4. *Inhibition of nuclear protein kinase activity from neonatal porcine ovaries with anti-cAMP-dependent protein kinase antiserum*

Protein Kinase Preparation	^{32}P Incorporated (pmoles/min) Normal Serum	Antiserum	Inhibition %
0.5 *M* NaCl extract of nucleic from untreated ovaries	6.5	6.7	-
	13.8	13.6	-
0.5 *M* NaCl extract of nuclei from ovaries incubated with 8-p-chlorophenylthio-cAMP	18.6	4.7	45.4
	21.6	11.7	45.8
Nuclear protein kinase II	25.7	3.6	86.0
	36.5	4.7	87.1

Cell suspensions prepared from the ovaries of neonatal pigs were incubated without and with 10^{-7} M 8-p-chlorophenylthio-cAMP. After incubation, the cells were recovered by centrifugation, quick-frozen, and lyophilized Nuclei were isolated in nonaqueous media, and extracted with 0.5 M NaCl. The dialyzed extracts were either tested for protein kinase activity or subjected to Sephadex G-200 gel filtration to isolate nuclear protein kinase II. For experimental details refer to legend of Figure 13. Enzyme-antiserum incubations and protein kinase assays were carried out as described in legend of Table 3.

evidence suggesting a cAMP-mediated activation of nuclear cAMP-dependent protein kinase and subsequent phosphorylation of nuclear proteins (18,32-34,38,49), attempts to stimulate protein kinase activity in intact nuclei with cAMP have been largely unsuccessful (8,10,11,19,22-24,29, 31,36,40,41,45,46). The demonstrated inability of cAMP to activate nuclear protein kinases and our previous observation that cAMP mediated the binding of cytoplasmic cAMP-dependent protein kinase to nuclear acceptor sites *in vitro* (19) have led us to propose that hormones, mediated through cAMP, cause the nuclear translocation of catalytic and regulatory subunits of the dissociated cytoplasmic cAMP-dependent protein kinase (21) which in turn leads to cAMP-mediated phosphorylation of nuclear proteins.

Most previous attempts to fractionate nuclear protein kinases have been unsuccessful in identifying nuclear cAMP dependent protein kinases in either their dissociated or undissociated form, and the presently available evidence seems to suggest that nuclear protein kinases present in rat liver and rat mammary gland are of the cAMP-independent variety (10-13,40,45,46). Desjardins *et al.* (11) identified low levels of cAMP-dependent protein kinase in rat liver nuclei but did not exclude the possibility that the presence of kinase activity was caused by cytoplasmic contamination. There exists, however, some convincing evidence for the presence of cAMP-dependent protein kinase in nuclei of several tissues. The work of Kish and Kleinsmith (27) suggests the presence of multiple protein kinases in beef liver nuclei and demonstrated that the various nuclear kinases differed considerably in their response to cAMP. In a recent paper, Johnson *et al.* (16) described the isolation of two cAMP-dependent protein kinases from extensively purified calf thymus nuclei. Rikans and Ruddon (41) obtained evidence for the presence of dissociated catalytic subunits in rat liver nuclei. These workers were able to demonstrate the inhibition of a nuclear protein kinase (K IV) by a cAMP-binding protein and by the regulatory subunit of rabbit muscle cAMP-dependent protein kinase suggesting that protein kinase IV was a catalytic subunit isolated from rat liver nuclei. Additionally, Schlepper and Knippers (43) identified two nuclear cAMP-dependent protein kinases in nuclei of rat C_6 glioma cells, Ehrlich ascites tumor cells, and mouse L cells. These nuclear protein kinases were shown to be similar to the cytoplasmic cAMP-dependent enzyme.

Our results presented here establish the presence of a cAMP-dependent form of protein kinase in porcine ovary nuclei. Convincing evidence for the identity of the ovarian nuclear cAMP-dependent protein kinase with the ovarian cytosol cAMP-dependent kinase was obtained in several ways. Protein kinase II from nuclei isolated in aqueous as well as in nonaqueous media and cytosol cAMP-dependent protein kinase exhibited identical elution characteristics on DEAE-cellulose and Sephadex G-200 indicating that both kinases are of similar molecular size and possess similar ionic charge. Both kinases exhibited an identical K_m for ATP, showed similar substrate specificity for endogenous nuclear nonhistone protein, and revealed similar antigenic properties.

We consider the possibility that nuclear cAMP-dependent protein kinase may have represented cytoplasmic contamination as highly unlikely for the following reasons: nuclei which had been incubated with a ^{14}C-labeled ovary cytosol did not retain ^{14}C-labeled cytosol contaminants; repeated treatment of nuclei with isotonic 0.14 *M* NaCl did not solubilize significant amounts of protein kinase; solubilization of nuclear cAMP-dependent protein kinase activity could be accomplished only with 0.5 *M* NaCl under conditions required for the solubilization of the majority of nuclear proteins, indicating that the nuclear cAMP-dependent protein kinase is strongly attached to chromatin; a relatively large amount of the total nuclear protein kinase activity was recovered as cAMP-dependent protein kinase II (32% recovery from nuclei isolated in aqueous media, and 39 % recovery from nuclei isolated in nonaqueous media) even after extensive washing of the nuclei with 0.14 *M* NaCl. Finally, and most convincingly, we were able to demonstrate the presence of cAMP-dependent protein kinase II in porcine ovary nuclei that had been isolated in nonaqueous media which precludes any exchange of water-soluble components between nucleus and cytoplasm. Nuclear cAMP-dependent protein kinase II was clearly identical with the cytoplasmic cAMP-dependent protein kinase based on the identity of their gel filtration elution profiles and antigenic properties.

Previous attempts to detect cAMP-dependent protein kinase in rat liver nuclei (10,11, 40,45,46) or rat mammary gland nuclei (12) may have been unsuccessful due to the relatively large quantity of nuclear cAMP-independent protein kinase activity obscuring small amounts of the cAMP-dependent enzyme; by the observed instability of isolated nuclear cAMP-dependent protein kinase (A. Spielvogel and R. A. Jungmann, unpublished observation); and by the lack of documentation of the presence of dissociated catalytic subunits according to all of the criteria given by Traugh *et al.* (48). It is also conceivable that appreciable quantities of cAMP-dependent protein kinase (or the dissociated catalytic subunit) can be detected only at certain times during hormonal stimulation of RNA synthesis and after the cAMP-mediated translocation and compartmentalization of cytoplasmic cAMP-dependent protein kinase at nuclear sites.

Using experimental conditions under which nonspecific binding of the dissociated catalytic protein kinase subunit to nuclei is prevented, Keely *et al.* (26) failed to demonstrate cAMP-mediated nuclear translocation of cytoplasmic cAMP-dependent protein kinase in beef heart preparations. Based on their results, Keely *et al.* suggested that binding of the catalytic subunit to nuclear fractions may be an artifact observed in aqueous solution at low salt concentrations and may lack physiological significance. Avoiding the use of aqueous media for the isolation of nuclei, we have obtained results which clearly indicate that a cAMP-mediated increase of nuclear cAMP-dependent protein kinase activity can be demonstrated and that the nuclear cAMP-dependent protein kinase acquired by nuclei after cyclic nucleotide treatment of intact ovarian cells is identical with the cytoplasmic protein kinase. Since the experimental conditions chosen rule out an unspecific contamination of nuclei with cytoplasmic components, it appears a distinct possibility that cAMP-mediate nuclear translocation of cytoplasmic cAMP-dependent protein kinase is a phenomenon of biological significance and represents a prerequisite in the acquisition of nuclei of the dissociated cAMP-dependent protein kinase. However, as indicated by preliminary studies, nuclear translocation of cytoplasmic cAMP-dependent protein kinase does not appear to occur in all instances of increases of intracellular cAMP levels or cAMP activation (dissociation) of cytoplasmic protein kinase. Rather, nuclear translocation of cytoplasmic cAMP-dependent protein kinase seems to occur only at times of cAMP-mediated stimulation of nuclear protein phosphorylation and nuclear RNA synthesis.

ACKNOWLEDGEMENTS

This research was supported by Grant No. GB-41387 from the National Science Foundation, and in part by the Research and Education Fund, Northwestern Memorial Hospital. Mary L. Christensen is a Fellow of the National Cancer Institute, NIH.

REFERENCES

1. Allfrey, V. G., Inoue, A., Karn, J., Johnson, E. M., Good, R. A. and Hadden, J. W. *Ciba Foundation Symposium on the Structure and Function of Chromatin*. Ciba Foundation Symposium 28, Amsterdam, Elsevier, 1974, pp. 199-206.
2. Ashby, C. D. and Walsh, D. A. 1972. Characterization of the interaction of a protein inhibitor with adenosine 3',5'-monophosphate-dependent protein kinases. I. Interaction with the catalytic subunit of the protein kinase. J. Biol. Chem. *247*:6637-6642.
3. Ashby, C. D. and Walsh, D. A. 1973. Characterization of the interaction of a protein inhibitor with adenosine 3',5'-monophosphate-dependent protein kinases. II. Mechanism of action with the haloenzyme. J. Biol. Chem. *248*:1255-1261.
4. Bylund, D. B. and Krebs, E. G. 1974. Effect of denaturation on the susceptibility of proteins to enzymic phosphorylation. J. Biol. Chem. *250*:6355-6361.
5. Castagna, M., Palmer, W. K. and Walsh, D. A. 1974. Nuclear protein-kinase activity in perfused rat liver stimulated with dibutyryl-adenosine cyclic 3':5'-monophosphate. Eur. J. Biochem. *55*:193-199.
6. Costa, E. Kurosawa, A. and Guidotti, A. 1976. Proc. Nat. Acad. Sci. USA *73*:in press.
7. Costa, E., Chuang, D. M., Guidotti, A. and Uzunov,P. Cyclic 3',5' adenosine monophosphate dependent molecular mechanisms in the trans-synaptic induction of tyrosine hydroxylase. In: *Chemical Tools in Catecholamine Research*, Almgren, O., Carlsson, A. and Engel, I. (eds.), Vol. II. North-Holland Publishing Co., Amsterdam, 1975, pp. 283-292.
8. Dastugue, B., Tichonicky, L. and Kruh, J. 1974. Multiple forms of protein kinase in a liver cell. II. Nuclear kinase and cytosol phosvitin kinase. Biochimie *56*:491-497.

9. DeAngelo, A. B., Schweppe, J. S., Jungmann, R. A., Huber, P. and Eppenberger, U. 1975. Ovarian cyclic adenosine monophosphate-dependent protein kinase activity: ontogeny and effect of gonadotropin. Endocrinology *97*:1509-1520.

10. Desjardins, P. R., Lue, P. F., Liew, C. C. and Gornall, A. G. 1972. Purification and properties of rat liver nuclear protein kinases. Can. J. Biochem. *50*:1249-1259.

11. Desjardins, P. R., Liew, C. C. and Gornall, A. G. 1975a. Rat liver nuclear protein kinases. Can. J. Biochem. *53*:354-363.

12. Desjardins, P. R., Mendelson, I. M. and Anderson,K. M. 1975b. Relationship between RNA polymerase and protein kinase activities in rat mammary gland nuclei. Can. J. Biochem. *53*:591-598.

13. Gamo, S. and Lindell, T. J. 1974. Presence of two protein kinases in highly purified rat liver nucleioli. Life Sci. *15*:2179-2187.

14. Johnson, E. M. and Allfrey, V. G. 1972. Differential effects of cyclic adenosine 3',5'-monophosphate on phosphorylation of rat liver nuclear acidic proteins. Arch. Biochem. Biophys. *152*:786-794.

15. Johnson, E. M., Karn, J. and Allfrey, V. G. 1974. Early nuclear events in the induction of lymphocyte proliferation by mitogens. J. Biol. Chem. *249*:4990-4999.

16. Johnson, E. M., Hadden, J. W., Inoue, A. and Allfrey, V. G. 1975. DNA binding by cyclic adenosine 3', 5'-monophosphate dependent protein kinase from calf thymus nuclei. Biochemistry *14*:3873-3884.

17. Jungmann, R. A., Schweppe, J. S. and Lestina, F. A. 1970. Studies on adrenal histones. J. Biol. Chem. *245*:4321-4326.

18. Jungmann, R. A. and Schweppe, J. S. 1972. Mechanism of action of gonadotropin. J. Biol. Chem. *247*: 5535-5542.

19. Jungmann, R. A., Hiestand, P. C. and Schweppe, J. S. 1974a. Mechanism of action of gonadotropin. IV. Cyclic adenosine monophosphate-dependent translocation of ovarian cytoplasmic cyclic adenosine monophosphate-binding protein and protein kinase to nuclear acceptor sites. Endocrinology *94*:168-184.

20. Jungmann, R. A., Hiestand, P. C. and Schweppe, J. S. 1974b. Adenosine 3':5'-monophosphate-dependent protein kinase and the stimulation of ovarian nuclear ribonucleic acid polymerase activities. J. Biol. Chem. *249*:5444-5451.

21. Jungmann, R. A., Lee, S. G. and DeAngelo, A. B. Translocation of cytoplasmic protein kinase and cyclic adenosine monophosphate binding protein to intracellular acceptor sites. In: *Advances in Cyclic Nucleotide Research*, Drummond, G. I., Greengard, P. and Robison, G. A. (eds.), Vol. 5. Raven Press, New York, 1975, pp. 281-306.

22. Kamiyama, M. and Dastugue, B. 1971. Rat liver non-histone proteins: correlation between protein kinase activity and activation of RNA synthesis. Biochem. Biophys. Res. Commun. *44*:29-36.
23. Kamiyama, M., Dastugue, B. and Kruh, J. 1971. Action of phosphoproteins and protein kinase from rat liver chromatin on RNA synthesis. Biochem. Biophys. Res. Commun. *44*:1345-1350.
24. Kamiyama, M., Dastugue, B., Defer, N. and Kruh, J. 1972. Liver chromatin non-histone proteins partial fractionation and mechanism of action on RNA synthesis. Biochim. Biophys. Acta *277*:576-583.
25. Karn, J., Johnson, E. M., Vidali, G. and Allfrey, V.G. 1973. Differential phosphorylation and turnover of nuclear acidic proteins during the cell cycle of synchronized HeLa cells. J. Biol. Chem. *249*:667-677.
26. Keely, S. L., Jr., Corbin, J. D. and Park, C. R. 1975. On the question of heart cAMP-dependent protein kinase. Proc. Nat. Acad. Sci. USA *72*:1501-1504.
27. Kish, V. M. and Kleinsmith, L. J. 1974. Nuclear protein kinases. J. Biol. Chem. *249*:750-760.
28. Kleinsmith, L. J., Allfrey, V. G. and Mirsky, A. E. 1966a. Phosphorylation of nuclear protein early in the course of gene activation in lymphocytes. Science *154*:780-781.
29. Kleinsmith, L. J., Allfrey, V. G. and Mirsky, A. E. 1966b. Phosphoprotein metabolism in isolated lymphocyte nucleic. Proc. Nat. Acad. Sci. USA *55*:1182-1189.
30. Kuo, J. F. and Greengard, P. 1969. Cyclic nucleotide-dependent protein kinases. IV. Widespread occurrence of adenosine 3',5'-monophosphate-dependent protein kinase in various tissues and phyla of the animal kingdom. Proc. Nat. Acad. Sci. USA *64*:1349-1355.
31. Langan, T. A. A phosphoprotein preparation from liver nuclei and its effect on the inhibition of RNA synthesis by histones. In: *Regulation of Nucleic Acid and Protein Synthesis*, Koningsberger, V.V. and Bosch, L. (eds.), Vol. 10. Elsevier, Amsterdam, 1969, pp. 233-242.
32. Langan, T. A. 1968. Histone phosphorylation: stimulation by adenosine 3',5'-monophosphate. Science *162*:579-580.
33. Langan, T. A. 1969. Phosphorylation of liver histone following the administration of glucagon and insulin. Proc. Nat. Acad. Sci. USA *64*:1276-1283.
34. Langan, T. A. 1971. Cyclic AMP and histone phosphorylation. Ann. N. Y. Acad. Sci. *185*:166-180.
35. Langan, T. A. Protein kinases. In: *Advances in Cyclic Nucleotide Research*, Greengard, P. and Robison, G. A. (eds.), Vol. 3. Raven Press, New York, 1973, pp. 99-153.

36. Lee, P. C. and Jungmann, R. A. 1975. Ontogeny of cyclic AMP-dependent protein phosphokinase during hepatic development of the rat. Biochim. Biophys. Acta *399*:265-276.
37. Lee, P. C., Radloff, D., Schweppe, J. S. and Jungmann, R. A. 1976. Testicular protein kinases: characterization of multiple forms and ontogeny. J. Biol. Chem. *251*:914-921.
38. Mallette, L. E., Neblett, M., Exton, J. H. and Langan, T. A. 1973. Phosphorylation of lysine-rich histone in the isolated perfused rat liver. J. Biol. Chem. *248*:6289-6291.
39. Palmer, W. K., Castagna, M. and Walsh, D. A. 1974. Nuclear protein kinase activity in glucagon-stimulated perfused rat livers. Biochem. J. *143*:469-471.
40. Ruddon, R. W. and Anderson, S. L. 1972. Presence of multiple protein kinase activities in rat liver nuclei. Biochem. Biophys. Res. Commun. *46*:1499-1508.
41. Rikans, L. E. and Ruddon, R. W. 1973. Role of 3',5'-cyclic AMP in the control of nuclear protein kinase activity. Biochem. Biophys. Res. Commun. *51*:387-394.
42. Salem, R. and DeVellis, J. 1976. Protein kinase (PK) activity and cAMP-dependent protein phosphorylation in subcellular fractions after norepinephrine (NE) treatment of glial cells. Fed. Proc. *35*:296.
43. Schlepper, J. and Knippers, R. 1975. Nuclear protein kinases from murine cells. Eur. J. Biochem. *60*:209-220.
44. Siebert, G. Isolation of Nuclei. In: *Methods in Cancer Research*, Busch, H. (ed.), Vol. II. Academic Press, New York, 1967, pp. 287-301.
45. Takeda, M., Yamamura, H. and Ohga, Y. 1971. Phosphoprotein kinases associated with rat liver chromatin. Biochem. Biophys. Res. Commun. *42*:103-110.
46. Takeda, M., Matsumura, S. and Nakaya, Y. 1974. Nuclear phosphoprotein kinases from rat liver. J. Biochem. *75*:743-752.
47. Teng, C. S., Teng, C. T. and Allfrey, V. G. 1971. Studies of nuclear acidic proteins. J. Biol. Chem. *246*:3597-3609.
48. Traugh, J. A., Ashby, C. D. and Walsh, D. A. 1974. Criteria for the classification of protein kinases. Methods. Enzymol. *38*:290-299.
49. Turkington, R. W. and Riddle, M. 1969. Hormone-dependent phosphorylation of nuclear proteins during mammary gland differentiation *in vitro*. J. Biol. Chem. *244*:6040-6046.
50. Walsh, D. A., Ashby, C. D., Gonzales, C., Calkins, D., Fischer, E. H. and Krebs, E. G. 1971. Purification and characterization of a protein inhibitor of adenosine 3',5'-monophosphate-dependent protein kinases. J. Biol. Chem. *246*:1977-1985.

51. Walsh, D. A. and Ashby, C. D. 1973. Protein kinases: aspects of their regulation and diversity. Rec. Progr. Horm. Res. *29*:329-333.

Cyclic AMP Systems in Selected Cells of Erythroid and Lymphoid Origin

H. Sheppard, C. R. Burghardt, and W. H. Tsien

During the 1960s, interest was developing in our laboratories in the interaction of hormones and pharmacological agents with cell membranes. Hemolysis of rat erythrocytes produced by hypotonicity could be modulated by a large number of drugs but these effects could not be correlated with their physiological profiles (26). The effects of hypotonicity were enhanced by some drugs, inhibited by others or first inhibited and then enhanced by another group. The hemolysis produced by the detergent, deoxycholate, was modulated by fewer compounds and usually in the direction of enhancement. Hemolysis produced by the reaction of digitonin with membrane cholesterol was most resistant to modulation by drugs.

It was recognized, however, that as an end-point, hemolysis was an all-or-none effect and that it might be possible to detect more subtle changes in the membrane by monitoring the action of some membrane bound enzymes. Our attention focused on the adenylate cyclase system because it was located mainly in the plasma membrane, was responsive to hormones and represented a model of mechano-chemical coupling at the molecular level. The enthusiasm with which we approached this aspect of the problem was dampened by the generalization that only nucleated erythrocytes possessed an adenylate cyclase (11). This conclusion was based on finding a cyclase in nucleated avian erythrocytes but not in the non-nucleated canine variety. It was recalled, however, that the dog erythrocyte had a high internal Na concentration and, therefore, lacked a Na-K

ATPase for pumping Na out of the cell. It was reasoned, therefore, that the dog erythrocyte might be a poor representative of the non-nucleated variety since a deficiency in one membrane-bound enzyme using ATP as the substrate might be accompanied by deficiencies in other such enzymes using the same substrate. To answer this question we looked for an adenylate cyclase in the Na-pump containing erythrocytes of the rat, mouse and human as well as the Na-pump deficient cells of the dog and cat.

CONVERSION OF ATP TO CYCLIC AMP AND OTHER NUCLEOTIDES

The adenylate cyclase assay used was based on a method developed by Krishna *et al.* (12) and measured the conversion of ^{14}C-ATP to cyclic AMP in the presence of a large unlabeled pool of cyclic AMP during a 30 min incubation period at 37°C. Using paper chromatography, the cyclic AMP and other hydrolytic products of ATP were separated and counted (19). In Table 1, NaF and norepinephrine (NE) at concentrations of 10 m*M* and 1 m*M*, respectively, markedly increased cyclic AMP with the ghosts of the rat and mouse erythrocytes but not with those of the dog, cat or human. The small apparent increase in cyclic AMP obtained with the last mentioned erythrocytes in the presence of NaF was due to the production of a substance which resembled xanthine in its chromatographic properties but which could be separated from carrier xanthine in one chromatographic system.

Though the dog and cat erythrocytes lacked a Na-K ATPase, it is apparent from Table 1 that they contained appreciable Mg-ATPase or phosphohydrolase activity, which might be somewhat weaker than that seen with the other varieties. In all cases, NaF inhibited the breakdown of ATP, while NE probably did not. It is of interest that none of these erythrocytes ghosts possessed much 5'-nucleotidase activity despite the fact that this enzyme is considered to be a good marker of plasma membranes.

These studies have clearly demonstrated the presence of a hormone sensitive adenylate cyclase in non-nucleated erythrocytes which, with the exception of the human variety, could be correlated with the presence of a Na-K ATPase. In addition, the erythrocytes lacking a Na pump did possess a Mg-ATPase at approximately the same level of activity seen with those in possession of such a pump. Lastly, the ghosts of all these species appeared to be deficient in the membrane marker enzyme, 5'-nucleotidase.

The possibility of a functional link existing between the Na-K ATPase and the adenylate cyclase was suggested by Sheppard (29). In addition, the transport of Na by turkey erythrocytes (6,18) and K and water by the duck erythrocyte (17) has been shown to be altered by the action of beta agonists on the adenylate cyclase and that this altered cation transport could have resulted from a cyclic AMP induced phosphorylation of the plasma membrane (18).

The adenylate cyclase of the rat erythrocyte was

TABLE 1. *The metabolism of ATP by mammalian erythrocyte ghosts.*

Species	Stimulant	Percent ^{14}C Found in: Cyclic AMP	ATP	ADP	AMP	Adenosine
Mouse	None	0.017	28.5	43.8	27.7	0.096
	F	0.498	34.3	46.0	18.9	0.058
	NE	0.110	24.8	44.8	30.0	0.093
Rat	None	0.018	38.2	35.4	25.7	0.112
	F	0.298	53.0	30.8	15.8	0.048
	NE	0.104	46.0	32.2	21.2	0.089
Human	None	0.013	31.5	48.6	19.9	0.035
	F	0.035	67.2	24.2	8.6	0.023
	NE	0.012	40.6	39.3	20.1	0.043
Dog	None	0.018	48.4	33.7	17.6	0.048
	F	0.045	57.2	30.2	12.2	0.039
	NE	0.015	51.4	32.1	16.4	0.053
Cat	None	0.018	57.0	30.2	12.3	0.077
	F	0.031	70.6	19.5	5.7	0.021
	NE	0.018	60.2	28.6	11.2	0.066

examined in more detail to obtain a profile of the hormone receptors which existed in the membrane. After testing a large number of substances which were known to stimulate the cyclases of other tissues, only two were active with the rat erythrocyte. The membranes of the ghosts contained receptors for prostaglandins of the E type and catecholamines of the beta type (20). While the ability of dopamine to stimulate the adenylate cyclase was first thought to be indicative of the existence of a special dopamine receptor, it was soon demonstrated that the dopamine was actually acting on the beta receptor (21).

It was not long before we and others wondered about the role of cyclic AMP in erythrocyte function and why the cyclase was coupled to a beta receptor. Apparently cyclic AMP does not play an important role in the gas-exchange function of erythrocytes since dogs and cats liver perfectly well without it. It was concluded that the cyclase was a "remnant of a more glorious past," where it was probably of some importance for the maturation of a progenitor cell. Thus, beta agonists and prostaglandins probably played some role in erythropoiesis but the nature of that role was unclear. In any case, this ideal could be tested by examining the cyclase activity of the more immature cells, the reticulocytes. Rather than produce a reticulocytosis by artificial means we took advantage of an available supply of neonate rats which were known to have reticulocytes comprising over 90 percent of the circulating erythroid cells. It was disappointing to find no great difference between the activity of the neonate reticulocyte and the adult erythrocyte. Our disappointment was even greater when it was soon shown that reticulocytosis produced by acetylphenylhydrazine was associated with much higher adenylate cyclase activity and that it was positively correlated with the percentage of circulating reticulocytes (7,16). The adenylate cyclase could also be demonstrated in human (10,16) and rabbit (1) erythroid cells which contained a high percentage of reticulocytes. It was becoming increasingly clear that adenylate cyclase was present in the erythrocyte progenitor cells of many, if not all, mammalian species. It would be of interest if this generalization was found to be true for even the dog and cat erythrocytes.

It was difficult to understand why the reticulocytes of the neonate rats did not possess activities greater than those seen in the adult animal. A rationalization was sought in the observations from several laboratories that adenylate cyclase activities of a number of tissues of the rat increased to a maximum and decreased during the period of maturation of the newborn animal (3,4). A similar phenomenon was seen by Sheppard and Burghardt (22) during a study of the adenylate cyclase of the erythrocytes of 11-, 25-, 74- and 372-day-old rats. As seen in Table 2, maximum stimulation over basal was noted in the 25-day-old rats with both NaF and isoproterenol followed by a fall to lower levels by 74 days of age. A similar fall during the post-weanling period was noted by Quiring *et al.* (16) and this was correlated with a decrease in reticulocyte counts. That phosphohydrolase and phosphodiesterase activities also fell (22) indicated that this phenomenon was not restricted to the cyclase nor just the membrane but was probably related to the age of the circulating cell.

TABLE 2. *The effect of NaF (1 mM) and isoproterenol (50 μM) on the adenylate cyclase preparations of erythrocytes from rats which were 11, 25, 74, and 372 days old*

	Stimulated/Basal			
Agent	11	25	74	372
NaF	3.7	5.2	1.9	2.3
ISO	2.0	3.7	1.4	2.2

CYCLIC AMP PHOSPHODIESTERASE (PD)

Cells containing an adenylate cyclase would be expected to contain a PD which would degrade the cyclic AMP to 5'-AMP. This proved to be the case for the ghost-free hemolysates not only of the rat and the mouse but also for the dog, cat and human (19). The PD activity was measured by incubating the enzyme preparations with ^{3}H-cyclic AMP and separating the labeled products by paper chromatography. Over 90 percent of the activity was in the ghost-free hemolysate but it could not be demonstrated conclusively that the 10 percent remaining with the ghost was an integral part of the membrane.

The enzyme of the rat was then used to test a number of interesting pharmacological agents for their ability to act as inhibitors. One of them, Ro 20-1724 (dl-4-(3-butoxy-4-methoxybenzyl)-2-imidazolidinone) had a surprisingly high potency, producing 50 percent inhibition at 0.1 μM (27). However, when this compound was tested with the PDs from other tissues its potency decreased almost two orders of magnitude (28). This great difference in potency was not seen with other classes of PD inhibitors.

Sometimes the erythrocyte preparation lost its sensitivity to Ro 20-1724 and it was noticed that this occurred after it was stored in the freezer. After following many false leads it was found that the preparation had to be frozen slowly to produce these changes and, therefore, could have been altered by the increase in ionic strength which occurs during slow freezing. Happily, when the ionic strength was increased with NaCl and the preparation stored overnight in the cold without freezing, the enzyme activity increased but the sensitivity to Ro 20-1724 decreased appreciably (23). It was also shown that 2 *M* NaCl would immediately and reversibly increase both the hydrolytic activity and inhibitor sensitivity. This was then followed by a slow decrease in inhibitor sensitivity which was irreversible. These immediate and slowly developing effects of hypertonic NaCl were studied with a number of different PD inhibitors and the results can be seen in Table 3. The immediate increase in the potencies of the inhibitors was noted in all cases but the subsequent loss in sensitivity was seen only with Ro 20-1724 and diazepam. It can also be noted that the other inhibitors, SQ-20006, papavarine and theophylline, were also less potent than the other two. We have examined a large number of inhibitors and while many structural analogues act like the representatives reported here, no unifying hypothesis could be developed which would satisfactorily explain the loss of potency.

A number of studies were undertaken to understand the mechanism responsible for the salt induced changes. It was soon learned that the enzyme initially had an apparent molecular weight of 250,000-300,000, that dithiothreitol could reduce this by 50 percent with no change in activity or inhibitor sensitivity and that storage in salt or the freezer reduced the apparent molecular weight to one quarter its original size as it reduced the inhibitor sensitivity. It was subsequently learned that the delayed effect of 2 *M* NaCl could be mimicked by 1 m*M* $CaCl_2$ without a time-delay and that the effects of NaCl were probably due to Ca as an impurity (24). Both and Na and Ca effects appear to require the presence of at least one and probably other macromolecular components. The thought occurred to us that the reduced sensitivites of the PDs from other

TABLE 3. *Effect of 2 M NaCl on the hydrolytic activity and inhibitor sensitivity of the rat erythrocyte cyclic AMP phosphodiesterase*

Inhibitor	Conc.	Preinc. Time	% Inhibition	
			Control	2 *M* Nacl
	(μ*M*)	(h)		
Ro 20-1724	0.1	0	51.8	69.1
		24	59.3	0
Theophylline	500.0	0	44.6	60.4
		24	45.5	46.8
Papavarine	10.0	0	35.9	60.9
		24	33.8	65.5
SQ-20006	10.0	0	58.7	79.9
		24	57.8	60.4
Diazepam	5.0	0	56.9	73.8
		24	61.2	3.5

tissues could result from Ca being liberated during the homogenization process. However, when we prepared homogenates or lysates in the presence of EGTA no increased sensitivities were obtained. On the chance that a Ca dependent protease was involved, we included protease inhibitors but to no avail. It remains for future studies to determine the significance of the Ca effects. We are still left with the agonizing question concerning the uniqueness of the erythrocyte PD. That it is not simply an artifact resulting from the aging of the erythrocytes is suggested by the studies with the rats of different ages. The sensitivities to Ro 20-1724 was the same (22) for all samples even though they contained different amounts of the younger reticulocytes (16).

It should be kept in mind that the erythrocyte enzyme had a low K_m of 5 μ*M* and acted only weakly, if at all, on cyclic GMP. The data presented here suggested that the enzyme was composed of four subunits and that two must remain dimerized to be sensitive to Ro 20-1724. The existence of the tetramer depended on the integrity of disulfide bonds but the dimer apparently did not. These conclusions have been conceptualized in Figure 1. All four subunits are considered to be identical in Model A and to contain an internal disulfide bond. As this disulfide bond is broken the arms which it joined now spread apart forcing the two vertical dimers to separate. In Model B, the tetramer is viewed as being composed of two identical dimers composed of different subunits. The two dimers could be held together by one bridging disulfide bond. In both models the dimers are held together in some manner which can be altered by a Ca dependent mechanism. Until purified enzyme preparations can be obtained it will not be possible to distinguish between these two models.

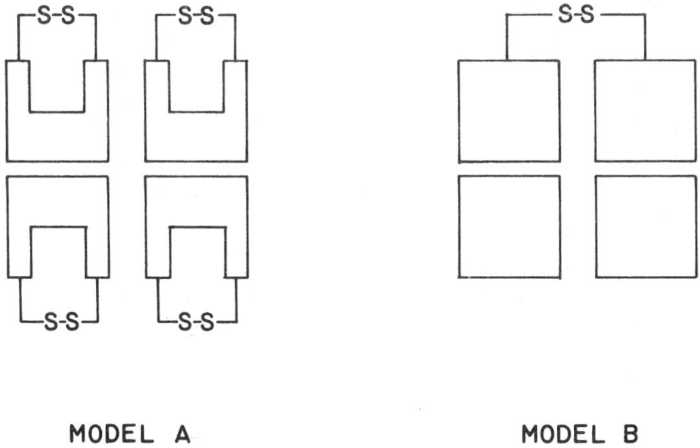

FIGURE 1. *Hypothetical models of the organization of the cyclic AMP phosphodiesterase of the rat erythrocyte.*

THE CYCLIC CAMP BINDING PROTEIN

In addition to the hormone sensitive adenylate cyclase and an unusual cyclic AMP PD, the rat erythrocyte contains a protein kinase and its regulatory subunit, the cyclic AMP binding protein. The binding protein could be detected in the ghost-free hemolysate at pH 4.0 by the membrane filtration method of Gilman (8), but not at pH 7.4 by the charcoal exclusion method of Brown *et al.* (5) and/or Sheppard and Tsien (25). With the 105,000 x g supernatant of an isotonic KCl homogenate of rat skeletal muscle, the cyclic AMP protein complex could be detected by both methods. By introducing 0.8 *M* KCl, however, the cyclic AMP protein complex could be detected at pH 7.4 by both methods. As far as is known, this masked binding activity at pH 7.4 is unique to the erythrocyte and its cause is unknown.

MOUSE THYMUS CELL PREPARATIONS

Though the rat erythrocyte is an interesting cell its limitations are many. The adenylate cyclase is affected by few hormones, its PD and binding proteins have unique properties, the function of the cyclic AMP is unknown and it lacks a cyclic GMP system. In considering a number of other easily obtainable mammalian cells which do not have these limitations, we turned to the mouse thymus cell because of its role in immunological responses. Cyclic AMP in intact cells was elevated by isoproterenol and PGE_1 (2) and by adenosine and Ro 20-1724 (31), while the cyclase of membrane preparations (9,13) could be stimulated by isoproterenol and PGEs. Using the cyclase assay, no stimulation was obtained with histamine, glucagon, ACTH, bovine growth hormone, prolactin, thyroid stimulating hormone or insulin. In our hands, only histamine in addition to isoproterenol, PGEs and adenosine could elevate cyclic AMP in the intact cells. Among the PD inhibitors, Ro 20-1724 was quite potent.

Short-term cultures of these thymus cells actively incorporate ^{3}H-thymidine (Tdr) into DNA and we were interested in knowing whether any of the substances which elevated cyclic AMP would modify this process. It can be seen in Table 4 that following a 2 h preincubation with

TABLE 4. *The effect of various agents on the incorporation of ^{3}H-Tdr into the cold acid soluble (CAS) and hot acid soluble (HAS) extracts of thymus cells during a 30 min pulse at 37°C after a 2 h preincubation with the test agents at 37°C*

Test Agent		% Inhibition	
(μ*M*)		CAS	HAS
Isoproterenol	(10)	+6.5	+22.1
	(1)	7.6	6.2
Ro 20-1724	(100)	20.5	24.1
IBMX	(100)	15.1	15.7
Papavarine	(100)	70.1	93.3

the test agents, Ro 20-1724, IBMX (3-isobutyl-1-methylxanthine) and papavarine at 0.1 m*M* inhibited the appearance of ^{3}H in both the cold acid soluble (CAS) and hot acid soluble (HAS) extracts during a 15 min incubation at 37°C. Papavarine was most active in this respect. Isoproterenol, on the other hand, increased the ^{3}H in the HAS but not the CAS extract. Following a 10 min preincubation, Ro 20-1724 was more active, while isoproterenol was most frequently inactive. Tdr incorporation into DNA appeared to be a function of the intracellular concentration of thymidine triphosphate and this depended on the concentration of Tdr in the medium (14). The incorporation into DNA was very temperature dependent and was negligible at 18°C although appreciable transport and phosphorylation occurred (14). We found that essentially no incorporation into DNA occurred if the incubation was carried out for 30 min in an ice bath and since over 90 percent of the label was phosphorylated, the rate limiting step was probably the transport into the cell. Under these conditions and with no preincubation period, isoproterenol was inactive at 1 μ*M* and 10 μ*M*, while Ro 20-1724, IBMX and papavarine inhibited the transport of Tdr (Table 5). Measurements of cyclic AMP in parallel cultures showed that none of the PD inhibitors significantly elevated cyclic AMP, while isoproterenol increased it over 30-fold. These data suggest that the effects on transport and phosphorylation could not be correlated with measurable changes in cyclic AMP. In addition, further work with isoproterenol showed that the stimulation which was obtained did not result from its action on the beta receptor since its effect was not blocked by propranolol, was not stereospecific and was not reversed by removing the medium and washing twice.

TABLE 5. *The effect of various agents on the CAS-^3H content and cyclic AMP levels following a 30 min incubation in an ice bath with ^3H-Tdr*

Test Agent (μ*M*)		% Inhibition	Cyclic AMP x Basal
Isoproterenol	(10)	0	32.7
	(1)	0	32.4
Ro 20-1724	(100)	66	0.6, NS
IBMX	(100)	53	-0.3, NS
Papavarine	(100)	99	0.2, NS

NS = Not statistically significant.

The effects of papavarine shown here with thymocytes agreed with the findings of Plagemann and Sheppard (15) using Novikoff rat hepatoma cells. Based on an extension of these studies to other cell types, Sheppard *et al.* (30) concluded that these effects were not related to elevations of cyclic AMP. Our results extended the list of inhibitors to include Ro 20-1724 and IBMX and showed that the effects were not related to measurable increases in cyclic AMP. At present, the mechanism responsible for a decreased transport of thymidine is not known but the effects of Ro 20-1724 can be overcome by time and increased concentrations of Tdr.

Once again it has been demonstrated that care must be exercised in interpreting data related to the incorporation of labeled nucleosides into nucleic acids.

REFERENCES

1. Babu, C. R., Azhar, S. and Murti, C. R. K. 1975. Loss of epinephrine stimulated synthesis of cyclic adenosine 3':5' monophosphate during maturation of rabbit and human reticulocytes. Med. Biol. *53*: 148-155.
2. Bach, M. A. 1975. Differences in cyclic AMP changes after stimulation by prostaglandins and isoproterenol in lymphocyte subpopulations. J. Clin. Invest. *55*:1074-1081.
3. Baer, H. P. and Hahn, P. 1971. Development of rat liver adenyl cyclase. Can. J. Biochem. *49*:85-89.
4. Bitensky, M. W., Russell, V. and Blanco, M. 1970. Independent variation of glucagon and epinephrine responsive components of hepatic adenyl cyclase as a function of age, sex and steroid hormones. Endocrinology *86*:154-159.
5. Brown, B. L., Albano, J. D. M., Elkins, R. P., Sghergi, A. M. and Tampion, W. 1971. A simple and sensitive saturation assay method for the measurement of adenosine 3',5'-cyclic monophosphate. Biochem. J. *121*:561-562.
6. Gardner, J. D., Klaeveman, H. L., Bilezikian, J. P. and Aurbach, G. D. 1974. Stimulation of sodium transport in turkey erythrocytes by cyclic 3',5'-AMP. Endocrinology *95*:499-507.
7. Gauger, D., Palm, D., Kaiser, G. and Quiring, K. 1973. Adenyl cyclase activities in rat erythrocytes during stress erythropoiesis: localization of the enzyme in the reticulocytes. Life Sci. *13*: 31-40.
8. Gilman, A. G. 1970. A protein binding assay for adenosine 3',5'-cyclic monophosphate. Proc. Nat. Acad. Sci. *67*:305-312.
9. Huang, Y. C., Duquesnoy, R. J. and Kemp, R. G. 1973. Properties of adenylate cyclase from mouse thymus. Int. J. Biochem. *4*:79-88.
10. Kaiser, G., Quiring, K., Gauger, D., Palm, D., Becker, H. and Schoeppe, W. 1974. Occurrence of adenyl cyclase activity in human erythrocytes. Blut *29*:115-122.
11. Klainer, L. M., Chi, Y.-M., Friedberg, S. L., Rall, T. W. and Sutherland, E. W. 1962. Adenyl cyclase. IV. The effects of neurohormones on the formation of adenosine 3',5'-phosphate by preparations from brain and other tissues. J. Biol. Chem. *237*:1239-1243.
12. Krishna, G., Weiss, B. and Brodie, B. B. 1968. A simple, sensitive method for the assay of adenyl cyclase. J. Pharmacol. Exp. Ther. *163*:379-385.
13. Makman, M. H. 1971. Properties of adenylate cyclase of lymphoid cells. Proc. Nat. Acad. Sci. USA *68*: 885-889.

14. Plagemann, P. G. W. and Erbe, J. 1972. Thymidine transport by cultured Novikoff hepatoma cells and uptake by simple diffusion and relationship to incorporation into deoxyribonucleic acid. J. Cell Biol. *55*:161-178.
15. Plagemann, P. G. W. and Sheppard, J. R. 1974. Competitive inhibition of the transport of nucleosides, hypoxanthine, choline and deoxyglucose by theophylline, papavarine and prostaglandins. Biochem. Biophys. Res. Commun. *56*:869-875.
16. Quiring, K., Gauger, D., Kaiser, G. and Palm, D. 1973. Adenyl cyclase activity in non-nucleated red blood cells: evidence for its localization in the reticulocytes. Experientia *29*:526-527.
17. Riddick, D. H., Kregenow, F. M. and Orloff, J. 1971. The effect of norepinephrine and dibutyryl cyclic adenosine monophosphate on cation transport in duck erythrocytes. J. Gen. Physiol. *57*:752-766.
18. Rudolph, S. A. and Greengard, P. 1974. Regulation of protein phosphorylation and membrane permeability by β-adrenergic agents and cyclic adenosine 3':5'-monophosphate in the avian erythrocyte. J. Biol. Chem. *249*:5684-5687.
19. Sheppard, H. and Burghardt, C. 1969. Adenyl cyclase in non-nucleated erythrocytes of several mammalian species. Biochem. Pharmacol. *18*:2576-2578.
20. Sheppard, H. and Burghardt, C. R. 1970. The stimulation of adenyl cyclase of rat erythrocyte ghosts. Mol. Pharmacol. *6*:425-429.
21. Sheppard, H. and Burghardt, C. R. 1971. The effect of alpha, beta, and dopamine receptor-blocking agents on the stimulation of rat erythrocyte adenyl cyclase by dihydroxphenethylamines and their β-hydroxylated derivatives. Mol. Pharmacol. *7*:1-7.
22. Sheppard, H. and Burghardt, C. R. 1973. Age-dependent changes in the adenylate cyclase and phosphodiesterase activity of rat erythrocytes. Biochem. Pharmacol. *22*:427-429.
23. Sheppard, H. and Tsien, W. H. 1974. The effect of slow freezing and hypertonic NaCl on the hydrolytic activity of rat erythrocyte cyclic AMP phosphodiesterase and its sensitivity to the inhibitor, D,L-4-(3-butoxy-4-methoxy-benzyl)-2-imidazolidinone. Biochim. Biophys. Acta *341*:489-496.
24. Sheppard, H. and Tsien, W. H. 1975a. Alterations in the hydrolytic activity, inhibitor sensitivity and molecular size of the rat erythrocyte cyclic AMP phosphodiesterase by calcium and hypertonic sodium chloride. J. Cyclic Nucleotide Res. *1*:347-242.
25. Sheppard, H. and Tsien, W. H. 1975b. The effect of pH and hypertonic KCl on the collection of the cyclic AMP-protein complex from rat skeletal muscle and erythrocyte extracts. Life Sci. *16*:139-148.
26. Sheppard, H., Tsien, W. H. and Burghardt, C. 1969. Effect of drugs on the hemolysis of rat erythrocytes. Biochem. Pharmacol. *18*:2215-2223.

27. Sheppard, H. and Wiggan, G. 1971. Analogues of 4-(3-,4-dimethoxybenzyl)-2-imidazolidinone as potent inhibitors of rat erythrocyte adenosine cyclic 3',5'-phosphate phosphodiesterase. Mol. Pharmacol. *7*:111-115.
28. Sheppard, H., Wiggan, G. and Tsien, W. H. 1972. Structure-activity relationships for inhibitors of phosphodiesterase from erythrocytes and other tissues. Adv. Cyclic Nucleotide Res. *1*:103-112.
29. Sheppard, J. R. 1976. 25th Annual Bureau of Biological Research Conference, Rutgers University, New Brunswick, N. J.
30. Sheppard, J. R. and Plagemann, G. W. 1975. Cyclic AMP, membrane transport and cell division. I. Effects of various chemicals on cyclic AMP levels and rate of transport of nucleosides, hypoxanthine and deoxyglucose in several lines of cultured cells. J. Cell Physiol. *85*:163-172.
31. Zenser, T. V. 1975. Formation of adenosine 3',5'-monophosphate from adenosine in mouse thymocytes. Biochim. Biophys. Acta *404*:202-213.

Part V

CYCLIC NUCLEOTIDES AND MECHANISMS OF GROWTH REGULATION

COFFINO et al.
Growth Regulation of Cyclic AMP in Mouse Lymphoma Cells

SHEPPARD, LELIEVRE, and PARAF
Membrane Enzyme Studies Using Drug Resistant Cell Lines

Growth Regulation of Cyclic AMP in Mouse Lymphoma Cells

Philip Coffino, Joe W. Gray, Irma Lemaire, and Paul A. Insel

We have examined the effect of cyclic AMP on the growth of the S49 lymphoma cell line. This work demonstrates that in these cells cyclic AMP rapidly causes a dose-dependent prolongation of transit time through the G_1 phase of the cell cycle. The growth inhibitory effect is mediated by cyclic AMP dependent protein kinase. Our results render implausible those models of cell cycle regulation which assert that fluctuation of cyclic AMP levels is a necessary requirement for cell growth. The modulation of growth by cyclic AMP in this system can be described in terms of a simple model of cell proliferation. Finally, we described a class of mutants specifically resistant to growth inhibition by cyclic AMP.

CYCLIC AMP BLOCKS S49 CELLS IN G_1

The S49 cell line was induced in Balb/c mice and adapted to suspension culture by Horibata and Harris (8). The cells grow in Dulbecco's modified Eagle's medium with a doubling time of 16-18 h. The S49 line has been maintained in continuous culture for several years. The cells are presumed to be of thymic origin since they are positive for Thy 1.1 surface antigen. Cholera toxin or the beta-adrenergic agonist isoproterenol produces an elevation of intracellular cyclic AMP levels in the cells (3). Addition of these agents or the dibutyryl derivative of cyclic AMP to a growing culture results first in the inhibition of proliferation and subsequently, after three to four days, in cell lysis (5). Figure 1 shows the effect of dibutyryl cyclic AMP on proliferation of an exponentially growing cell population. Maximally effective concentrations of drug, here 10^{-4} *M* dibutyryl cyclic AMP along with 2 x 10^{-4} *M* theophylline (a phosphodiesterase inhibitor) produce cessation of growth. The arrest occurs after about one cell generation, when the cell density is twice that present at the time of drug addition. Lower drug concentrations slow the growth rate in a dose-dependent manner after one generation time.

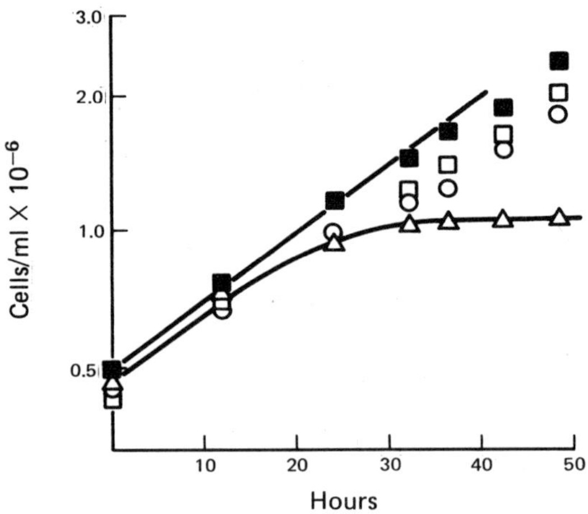

FIGURE 1. *Effect of bidutyryl cyclic AMP on the growth of S49 cells treated with theophylline, 0.2 mM, alone (■) or theophylline at that concentration together with dibutyryl cyclic AMP, 0.1 mM (Δ), 0.03 mM (O) or 0.01 mM (□) (6).*

To determine the cell cycle phase where this inhibition occurs, we used the flow cytometer (FCM) at the Lawrence Livermore Laboratory (15). This device can determine the population distribution of DNA content per cell, and hence the cell cycle distribution, of a cell population. It does so by measuring the fluorescence, at rates up to 1,000 per second, of individual cells fixed and then stained with acriflavine, a fluorescent dye that binds to DNA. The result of such an analysis for S49 cells in exponential growth is shown in Figure 2A. Cell number is plotted on the vertical axis against channel number, which is linearly related to DNA content per cell, on the horizontal axis. The prominent peak on the left corresponds to cells in G_1, the pre-DNA replicative phase of the cycle. At a channel number twice this are post-DNA replicative cells that are either in G_2 or mitosis. Distributed in intermediate channels are cells in S that have begun DNA synthesis. By appropriate curve fitting methods the total population can thus be resolved into fractions in the G_1, S and G_2 + M phases of the cycle. This information, the population doubling time and the growth fraction allow one

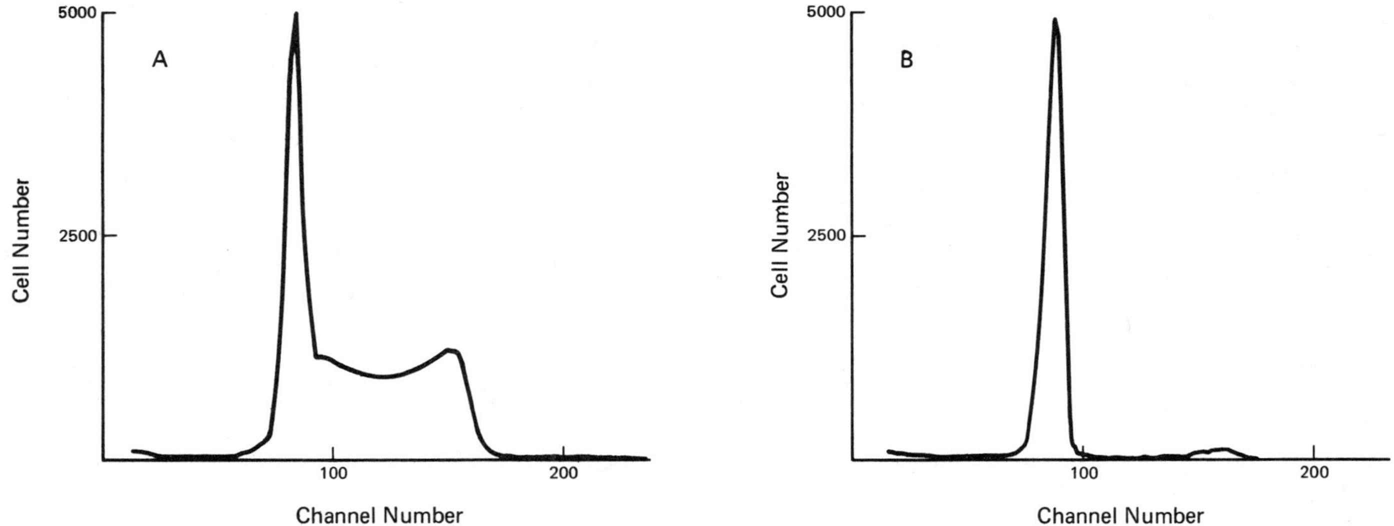

FIGURE 2. *Flow microfluorimeter analysis of S49 cells (A) in exponential growth and (B) 28 h after treatment with dibutyryl cyclic AMP (0.3 mM) and Ro 20-1724 (0.03 mM).*

to calculate mean cycle phase durations. For S49 cells in exponential growth, 30 percent of cells are in the G_1 and the mean duration of G_1 is about 2 h.

Flow cytometer analysis of cells treated for one doubling time with dibutyryl cyclic AMP shows a strikingly different distribution. More than 90 percent of cells are in G_1 (Figure 2A). At this time, cell proliferation has virtually ceased. Kinetic analysis (Figure 3A) reveals that the G_1 block is rapidly established after drug addition and that the proportion of cells in G1 increases progressively with time as cells initially in later parts of the cycle return to G_1. Simulation of the block using a mathematical model (Figure 3B) confirms this interpretation.

Three additional points may be noted. First, the block, even at maximally effective drug concentrations, is not absolute. A small fraction of cells is not in G_1. If a secondary mitotic block is produced with colcemid, it can be shown that these are cycling cells that have escaped the G_1 block induced by cyclic AMP (6). Second, the degree of effectiveness of the block can be regulated by the concentration of dibutyryl cyclic AMP. This can be shown, for instance, by determining the fraction of cells in G_1 after treatment with different drug concentrations when a new equilibrium state has been attained. Third, there are no important points of cyclic AMP mediated growth inhibition outside of G_1 in this system. This is apparent from the failure of dibutyryl cyclic AMP to affect early points in the growth curve, the good match to FCM kinetic data of a model that assumes a block only in G_1, and by the demonstration that cells that escape from the block have approximately normal rates of traverse through the remainder of the cycle (6).

KINASE IS REQUIRED FOR GROWTH INHIBITION

Is the G_1 block we have demonstrated specific for cyclic AMP? In S49 cells, cholera toxin (30 ng/ml) raises cyclic AMP levels, slows cell growth and increases the proportion of cells in the G_1 phase. These effects are potentiated by inhibitors of phosphodiesterase activity such as theophylline or Ro 20-1724.

In addition, a genetic approach has been used to examine the specificity and mechanism of cyclic AMP action. The cytolytic effect of cyclic AMP has enabled us to select mutants that are resistant to killing (5). More than 400 mutants selected for resistance to dibutyryl cyclic AMP have been isolated and examined. All mutant clones have abnormal cyclic AMP dependent protein kinase activity. This enzyme mediates the regulatory effects of cyclic AMP in glycogen synthesis and degradation and in lipolysis. Kuo and Greengard (10) have proposed that in animal cells the kinase acts as the universal effector of cyclic AMP action. Observations in diverse systems, including our own, support this hypothesis. We have used the kinase mutants to examine the specificity of growth regulation by

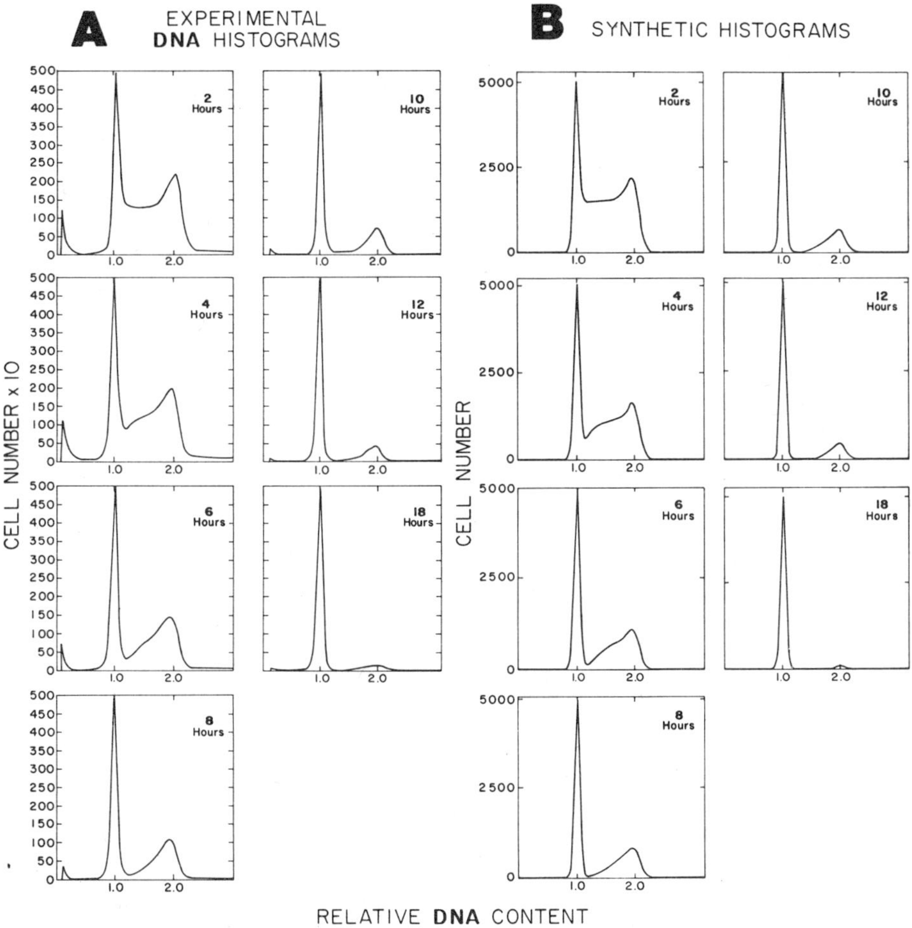

FIGURE 3. *Kinetics of change in cell cycle distribution (A) after treatment of S49 cells with dibutyryl cyclic AMP (0.1 mM) and theophylline (0.2 mM) and (B) computer simulation generated by perturbing an exponentially growing population only by imposing a block in G_1 (6).*

cyclic AMP and to determine the role of kinase in this process.

Kinase mutants are of three types (9). One, termed kin^-, is completely devoid of cyclic AMP stimulating kinase activity. A second type, K_m, is stimulated to half maximal activity only at 10-100-fold higher concentrations of cyclic AMP than is required in the parent cells. A third type, V_{max}, apparently has a normal cyclic AMP concentration requirement for half maximal activation, but possesses levels of kinase activity half or less of that in parent cells (Figure 4A). If the growth inhibition by dibutyryl cyclic AMP is, indeed, cyclic AMP, specific and kinase mediated, then the dose response relationship for

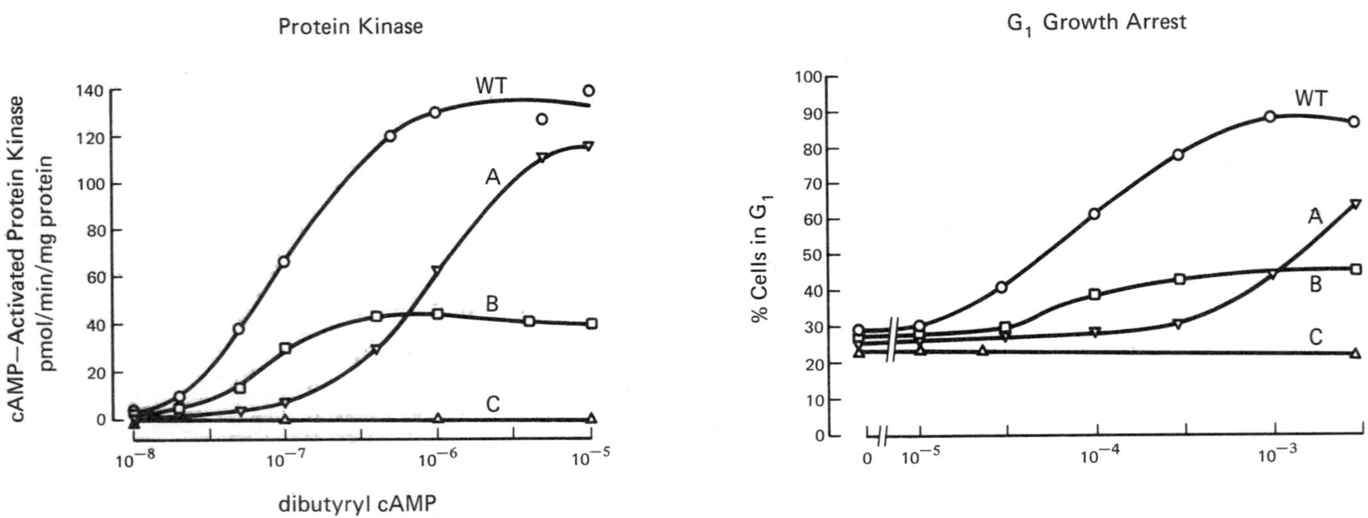

FIGURE 4. *(Left) - Cyclic AMP stimulated histone phosphorylation by cytosol extracts and (Right) - percent of cells in G_1 phase after 24 h treatment with dibutyryl cyclic AMP. WT, A, B and C designate respectively cells that are wild-type parental, Km mutant, V_{max} mutant, and kin$^-$ mutant. See Insel et al. (9) for experimental details.*

G_1 arrest in parent cells and each type of mutant should reflect the pattern of kinase activation by cyclic AMP determined *in vitro*. This is the case (Figure 4B). When the fraction of cells in G_1 is determined 24 h after treatment with different concentrations of dibutyryl cyclic AMP, K_m mutants require higher concentrations for maximal arrest and V_{max} mutants arrest to a lesser degree at maximal concentrations. Dibutyryl cyclic AMP at 3×10^{-3} *M*, the highest concentration tested, has no effect on the proliferation or cell cycle distribution of kin^- mutants. This last class of mutants is thus completely unresponsive to cyclic AMP, whether generated by addition of an exogenous analog or endogenously by cholera toxin. The resistance to killing by cyclic AMP occurs because the mutants lack the normal intracellular receptor for cyclic AMP that is required for growth inhibition.

Temporal fluctuations of cyclic AMP levels have been reported in synchronized cells (1,4,11,12). In particular, low levels are most commonly associated with mitotic cells. It has been proposed, in part on the basis of these findings, that changes in cyclic AMP trigger the cycle and serve as the regulatory mechanism for its orderly progression. In all reports in the literature, cycle-associated changes in intracellular cyclic AMP do not exceed 3-fold. In contrast, cyclic AMP levels in kin^- cells treated with cholera toxin are increased more than 50-fold at 90 min (2) and more than 1000-fold at 24 h (Insel, unpublished). Yet these cells continue to cycle normally, as do kin^- cells treated continuously for many generations with a concentration of dibutyryl cyclic AMP that is 10-fold higher than that needed to produce maximal growth inhibition in wild-type cells. Therefore, cells can cycle normally in the absence of the only known receptor for the cyclic AMP signal and can do so even when the mean level of that signal varies over an enormous range. It is difficult to imagine how kin^- cells could detect and respond to small fluctuations in cyclic AMP levels if the mean level is so markedly increased; the signal to noise ratio would be very low. We, therefore, conclude that such fluctuations are non-essential for cell growth.

While it can thus be argued that the reported small *fluctuation* in cyclic AMP levels associated with progression through the cycle cannot be required for cell growth, it remains possible that *minimal* levels might be necessary for optimum growth. If so, this positive regulation must work through some recognition system different from the predominant kinase. In addition, this required tonic level of cyclic AMP must be equal to or less than 20 percent of that present in parent S49 cells, since adenylate cyclase deficient S49 mutants with such a reduced intracellular level (3; Insel, unpublished) still grow normally.

MODELS FOR GROWTH REGULATION

A normal nondividing somatic cells has not yet replicated its DNA and hence has G_1 DNA content. To distinguish such a noncycling cell from a cycling one and to broaden the definition of the cell cycle, the idea of a G_0 state was introduced. Postmitotic cells could remain in the cycle or exit from G_1, coming to rest in G_0, perhaps to reenter the replicative cycle at some later time. This provides a two-state model with a decision point in G_1 and accommodates the observation that a cell's ability to divide or not divide is subject to experimental manipulation. This formulation has been refined and extended (7,14) to include the findings that cell growth rate is a continuous rather than a two-state variable, that cell cycle phase durations have dispersions about a mean and that multiple control points distributed throughout the cycle are possible. Such models are generally probabilistic in nature, so that regulatory events are depicted as affecting a time dependent probability controlling transition from one cell cycle state to the next. Mean phase durations are, thus, a function of transition probabilities at control points.

A simplified form of the more general model can be used to describe the data concerning cyclic AMP and S49 cells. A single point of the G_1 phase is designated a special control site, A. This is the point where cyclic AMP exerts its regulatory function. The rest of the cell cycle is designated B, which is of fixed duration. B comprises all of the cycle except a part of G_1. Each postmitotic cell enters A for an indeterminate period before returning to B, still in G_1. The exit from A occurs according to some constant probability P, the fraction of cells that leave A per unit time. When P is large, cells are almost always in the B state and the mean cycle time approaches the duration of B. P can be reduced by experimental manipulation; when P is small, cells are almost all in A, have G_1 phase DNA content and are effectively noncycling. The growth of cells can thus be described by P and the duration and dispersion of B. We have chosen our nomenclature to be consistent with that proposed by Smith and Martin (13) in their model of growth regulation.

It is evident that this kind of model provides a good qualitative description of S49 cell growth in the presence of different concentrations of dibutyryl cyclic AMP. Under basal conditions, P is large, all cells cycle rapidly and uniformly, G_1 is 2 h in duration, and the fraction of cells in G_1 is about 0.3. At high concentrations of drug, P becomes smaller in a dose-dependent manner. At maximally effective drug concentrations, P is sufficiently small that more than 0.9 of the cells are in G_1 and no growth is apparent. Yet, as we have seen, the block is not absolute and those cells that escape it transit the remainder of the cycle (B state) with normal kinetics. While this model of functional dependency of P on cyclic AMP is attractive and plausible, it remains to be determined whether it can provide a good quantitative description of the negative regulatory effect of cyclic AMP on S49 cell growth.

BIOCHEMICAL BASIS FOR GROWTH REGULATION

Model building of this kind has descriptive value but leaves unresolved the question of what biochemical mechanisms underlie growth regulation. We conclude from our work that cyclic AMP can modulate cell growth by kinase activation. The nature of the kinase substrate involved in this regulation is unknown, as is the manner in which phosphorylation affects its activity. The existence and role, if any, of subsequent steps are also unknown, as is the question of whether cyclic AMP has a unique role or is only one among several modulators of growth.

In an effort to resolve these questions, we have begun to isolate new classes of mutants with specific defects, distal to kinase, in cyclic AMP mediated growth inhibition. To do so we screened several hundred clones resistant to dibutyryl cyclic AMP for their kinase activity, but found none with normal kinase. We then reasoned that if cell death and growth inhibition work through distinct mechanisms, a mutant lacking the growth function but retaining the death function would not survive our single-step selection procedure. We, therefore, first selected "deathless" mutants (D^-) by growing parental cells for four days in dibutyryl cyclic AMP to kill parental cells and adding during the final 17 h cytosine arabinoside to kill kinase defective, nonarrested mutants. The survivors were recloned and two clones had the desired phenotype. They were growth arrested by cholera toxin and dibutyryl cyclic AMP but were not killed. Even after a week, normal growth resumed when inhibiting drugs were removed. Assays of cyclic AMP dependent kinase in cell lysates showed identical activity in parental and D^- cells.

The D^- cells were then cloned in the presence of dibutyryl cyclic AMP to select, in a second step, mutants resistant to growth inhibition (G^-). Two of these clones have now been tested for protein kinase activity and found to be indistinguishable from parental cells. Our present understanding of these mutants is indicated in Table 1. Kin^- mutants lack kinase activity but have a

TABLE 1. *Schematic representation of defects in kin⁻, D⁻ and G⁻ mutants*

	Protein Kinase	Sub_D	Sub_G
kin^-	-	+	+
D^-	+	-	+
G^-	+	-	-

normal complement of kinase substrates. The D^- mutant may be deficient in a kinase substrate activated by phosphorylation and required for death. Alternatively, the defect could be a structural alteration in the kinase itself that has affected its affinity for some, but not all, substrates, or it could be in a step beyond phosphorylation. The G^- mutant express a second defect which affects growth regulation and which can be similarly interpreted. By detailed comparison of these mutant and parental cells, we hope to discern the biochemical mechanisms underlying growth regulation.

ACKNOWLEDGEMENTS

This work was supported by National Science Foundation Grant BMS 75-06764, National Institute of Health Grants GM01791 and GM196496, by U. S. Energy Resources Development Agent Contract #W-7405-ENG-48, U. S. Public Service Health Grant 5R0114533, and University of California Cancer Research Coordinating Committee Grant 75SFg. Irma Lemaire is a Fellow of the Medical Research Council of Canada.

REFERENCES

1. Borgna, J.-L. and Mousseron-Canet, M. 1974. Le 3',5' cAMP au course du cycle cellulaire chez le cellules HeLa. C. R. Acad. Sci. Paris D. *279*:1479-1482.
2. Bourne, H. R., Coffino, P. and Tomkins, G. M. 1975a. Somatic genetic analysis of cyclic AMP action: characterization of unresponsive mutants. J. Cell Physiol. *85*:611-621.
3. Bourne, H. R., Coffino, P. and Tomkins, G. M. 1975b. Selection of a variant lymphoma cell deficient in adenylate cyclase. Science *187*:750-751.
4. Burger, M. M., Bombik, B. M., Breckenridge, B. McL. and Sheppard, J. R. 1972. Growth control and cyclic alterations of cyclic AMP in the cell cycle. Nature New Biol. *239*:161-163.
5. Coffino, P., Bourne, H. R. and Tomkins, G. M. 1975a. Somatic genetic analysis of cyclic AMP action: characterization of unresponsive mutants. J. Cell Physiol. *85*:603-615.
6. Coffino, P., Gray, J. W. and Tomkins, G. M. 1975b. Cyclic AMP. A nonessential regulator of the cell cycle. Proc. Nat. Acad. Sci. USA *72*:878-882.
7. Gray, J. W. 1974. Cell cycle analysis from computer synthesis of deoxyribonucleic acid histograms. J. Histochem. Cytochem. *22*:642-653.
8. Horibata, K. and Harris, A. W. 1970. Mouse myelomas and lymphomas in culture. Exp. Cell Res. *60*:61-69.
9. Insel, P. A., Bourne, H. R., Coffino, P. and Tomkins, G. M. 1975. Cyclic AMP-dependent protein kinase: pivotal role in regulation of enzyme induction and growth. Science *190*:896-898.

10. Kuo, J. F. and Greengard, P. 1969. Cyclic nucleotide-dependent protein kinases. IV. Widespread occurrence of adenosine 3',5'-monophosphate-dependent protein kinase in various tissues and phyla of the animal kingdom. Proc. Nat. Acad. Sci. USA *64*: 1349-1353.

11. Seifert, W. E. and Rudland, P. S. 1974. Possible involvement of cyclic GMP in growth control of cultured mouse cells. Nature *248*:138-141.

12. Sheppard, J. R. and Prescott, D. M. 1972. Cyclic AMP levels in synchronized mammalian cells. Exp. Cell Res. *75*:293-301.

13. Smith, J. A. and Martin, L. 1973. Do cells cycle? Proc. Nat. Acad. Sci. USA *70*:1263-1267.

14. Takahashi, M. 1968. Theoretical basis for cell cycle analysis. II. Further studies on labelled mitosis wave method. J. Theoret. Biol. *18*:195-204.

15. Van Dilla, M. A., Steinmetz, L. L., Davis, D. T., Calvert, R. N. and Gray, J. W. 1974. High-speed cell analysis and sorting with flow systems: biological applications and new approaches. I.E.E.E. Trans. Nucl. Sci. *NS-21*:714-723.

Membrane Enzyme Studies Using Drug Resistant Cell Lines

J. R. Sheppard, L. LeLievre, and A. Paraf

The interaction of extracellular molecular effectors with the plasma membrane generates physiological signals which activate the biochemical processes that result in cell proliferation and cell specialization. For example, the postulated mechanisms whereby lymphocytes are activated by antigens (1,2), lectins (3,4), or hormones (5) all depend on an initial membrane mediated event. This membrane generated signal stimulates cellular transport and macromolecular synthesis, eventually culminating in cell division.

The plasma membrane is the cellular site which controls whether a cell will continue to grow or remain in a specialized, quiescent state (6-8), and it has been proposed that in a variety of mammalian cell types, a set of metabolically unrelated biochemical reactions coordinately respond to changes in the environment (9,10). This "pleiotypic response" may act via the second messenger, cyclic AMP (11,12) directly implicating the enzyme adenylate cyclcase (E.C.4.6.1.1) and we hypothesized that other membrane enzymes such as the Na^+/K^+ ATPase (E.C.3.6.1.3) might also participate in membrane regulatory phenomena (6). Thus, we sought to test whether any relationship existed among the membrane associated enzymes adenylate cyclase and the Na^+/K^+ ATPase.

Comparison of thermosensitive (13,14) and drug resistant (15,16) mutants are useful in biochemical investigations of cell function. We used this approach in studying the plasma membrane properties of several clonal lines selected for resistance to ouabain and dibutyryl cyclic AMP (But_2-cAMP). With these drug resistant and wild type cell lines, we have investigated the modulation of adenylate cyclase and Na^+/K^+ ATPase by the cardiac glycoride ouabain.

MATERIALS AND METHODS

The parental cell lines, ME_2 (for murine epithelioid) and MF_2 (murine fibroblastic), which were derived from the MOPC 172 murine plasmocytoma are considered the wild type cells. Each of these two cell lines has been tested for their respective sensitivity to ouabain and cAMP: we found ME_2 more susceptible than MF_2 for both drugs. In order to obtain ouabain resistant cell lines ME_2 was cultivated in the presence of 2×10^{-5} *M* ouabain and MF_2 in the presence of 2×10^{-4} *M* ouabain (see Results). The media were changed every three days. ME_2 was trypsinized each week, while MF_2 was trypsinized twice a week. After ME_2 and MF_2 cells were adapted to grow in the presence of these amounts of drug (the growth curves in the presence of, respectively 2×10^{-5} *M* and 2×10^{-4} *M* being identical to the growth curves in absence of drug), the concentration of ouabain was doubled. By this procedure, each of these two cell lines were made progressively more and more resistant until ME_2 was able to grow in the presence of 2×10^{-4} *M* and MF_2 with 2×10^{-3} *M* ouabain (this stage was reached after 50 passages for ME_2 and 40 passages for MF_2). Then a cloning procedure previously described (17) was used as follows: the cells were diluted and seeded in Falcon flasks on small bits of glass. Single colonies were removed from single glass bits with trypsin, replated, grown and again cloned two more times. The same general procedure was employed to generate But_2-cAMP resistant cell lines. Table 1 is a summary of the properties of wild type cell lines (ME_2, MF_2) and clones selected for ouabain resistance (ME_2OR_2, MF_2OR_2) and cAMP resistance (ME_2CAR_1, MF_2CAR_1). Prior studies have shown that these cloned lines were stable for their drug resistant properties (18).

The preparation of crude membranes and the procedure for Na^+/K^+ ATPase assay were described by Jørgensen *et al.* (19). Kinetic studies were performed by transferring aliquots at varying times to test tubes where the reaction was stopped and inorganic phosphate measured (20). Adenylate cyclase activity was measured by the procedure of Kebabian *et al.* (21) modified to use Norit S-G charcoal (100g/ml in 2% bovine serum albumin) to bind the free cyclic nucleotide. Cyclic AMP analysis utilized cells growing in 25 cm^2 plastic flasks at the appropriate density. These cells were exposed to ouabain for 24 h and harvested by addition of 2 ml 5% TCA after the plates were washed two times with 3 ml PBS at 37°C. The procedure for cAMP measurement which followed was patterned after the method described by Brown (23). Protein was measured by the method of Lowry *et al.* (24).

TABLE 1. *Properties of wild type cells and cells selected for ouabain and dibutyryl cAMP resistance*

Cell Line	Morphology	Density Dependent Inhibition of Growth	# of Cells $\times 10^{-4}/cm^2$ at Confluency	Doubling Time (h)
ME_2	Epithelial	+	5.0 ± 1.2	24
ME_2OR_1	Epithelial	+	10.0 ± 2.0	18
ME_2CAR_1	Fibroblastic	-	26.0 ± 2.0	15
MF_2	Fibroblastic	-	32.0 ± 4.0	12
MF_2OR_2	Fibroblastic	-	40.0 ± 6.0	12
MF_2CAR_1	Fibroblastic	-	32.0 ± 4.0	15

RESULTS

Growth and Cross Resistance Studies

For each drug added to the medium we determined the drug concentration which inhibited growth by 50 percent. $G_{1/2}$ is defined as the molarity of the drug which gives 50 percent fewer cells when compared to the control after five days for fibroblastic derived cell lines (MF_2, MF_2OR_2 and MF_2CAR_1) and seven days for epitheliod derived cell lines (ME_2, ME_2OR_2 and ME_2CAR_1) in culture. From Table 2, it can be seen that $G_{1/2}$ is always lower for the ME_2 line

TABLE 2. *The relative susceptibility of the different cell lines to the different drugs expressed by their $G_{1/2}$*

Cell Line	Membrane Active Agents		
	Ouabain (*M*)	cAMP (*M*)	Con A (μg/ml)
ME_2	2×10^{-5}	2×10^{-5}	25
ME_2OR_1	2×10^{-4}	5×10^{-4}	50
ME_2CAR_1	10^{-4}	2×10^{-3}	50
MF_2	2×10^{-4}	5×10^{-4}	100
MF_2OR_2	2×10^{-3}	2×10^{-3}	100
MF_2CAR_1	5×10^{-4}	2×10^{-3}	N.T.*

**N.T. = Not tested.*

than for MF_2 whatever the selective drug is. For each selected cell line we observed a high degree of resistance to the drug used for selection when compared to the wild type cell. As seen in Figure 1A and 1B, the ME_2 and MF_2 wild type cells are much more sensitive to ouabain than their ouabain resistant counterparts, ME_2OR_1 and MF_2OR_2. Moreover, we observed a cross resistance for the other

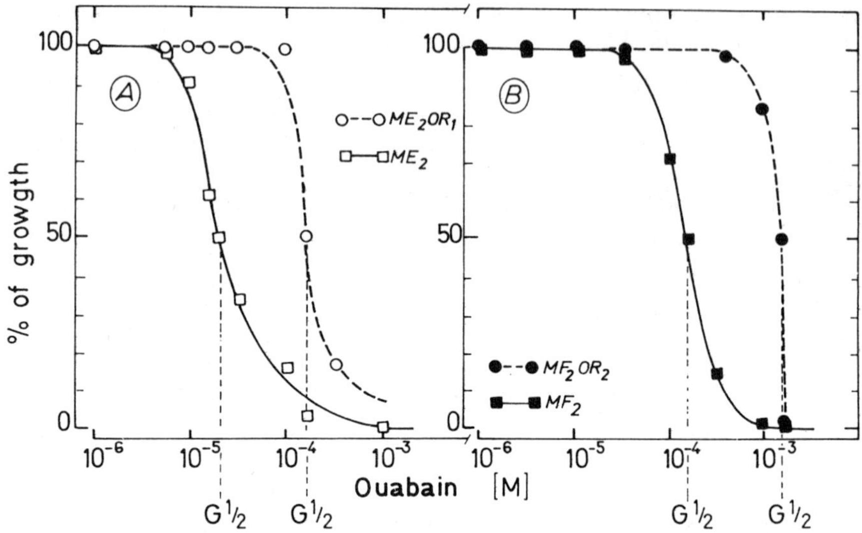

FIGURE 1. *Inhibition of growth by ouabain.*
A - □ - □ , ME_2; o--o, ME_2OR_1
B - ■ - ■ , MF_2; ●--●, MF_2OR_2

tested drugs: e.g., the ouabain resistant line ME_2OR_1, was 25 times more resistant to But_2-cAMP and 2 times more resistant to theophylline and to Concanavalin A (Con A) than the ME_2 line. The resistance to But_2-cAMP was four times higher in MF_2OR_2 than in MF_2. A Con A resistant cell line (ME_2Cr) (25) and a theophylline resistant cell line (ME_2Tr) (26) exhibited the same properties: i.e., both were found also resistant to ouabain and cAMP (results not shown here). Thus, cross resistance to different drugs appears to be a general rule for our mutants as determined by growth studies.

cAMP Levels Before and After Ouabain Treatment

Basal cAMP levels were determined at low cell density and confluency for the different cell lines (low densities were 1/10 to 1/3 of the number of cells found at confluency, see Table 1). The results are summarized in Figure 2. It can be seen that the density dependent ME_2 cells have a higher basal cAMP level than nondensity dependent MF_2 cells. ME_2OR_1 had a lower content of cAMP than ME_2, while ME_2CAR_1 exhibited the same amount as in ME_2. Both MF_2OR_2 and MF_2CAR_1 had the same contents of cAMP as in MF_2.

We observed a striking effect by ouabain after 24 h treatment on the wild type cells: cAMP levels were significantly elevated at 10^{-4} *M* for ME_2 (higher concentrations were toxic) and 10^{-3} *M* for MF_2. In contrast, none of the drug resistant cell lines derived from ME_2 or MF_2 were susceptible to the ouabain effect (10^{-3} or 10^{-4} *M*) on basal cyclic AMP levels, nor were there any effects using

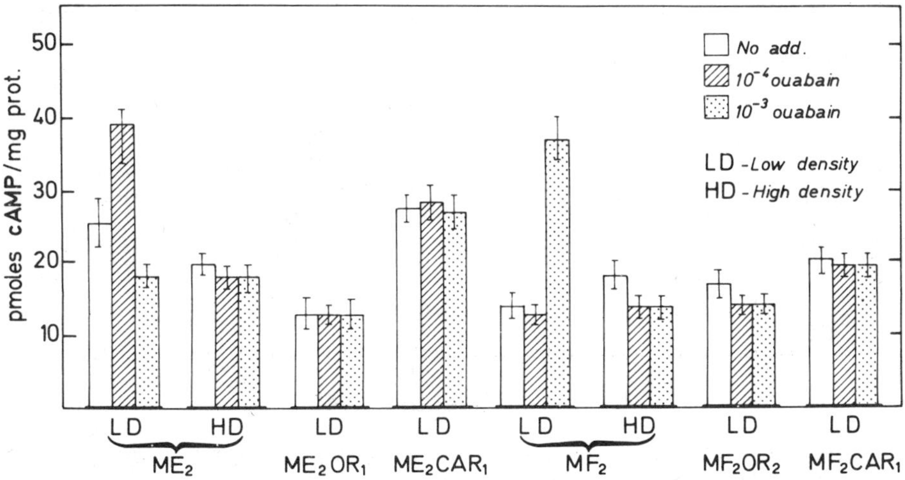

FIGURE 2. *Basal and ouabain-modulated cAMP levels.*

high density wild type cells. Ouabain treatment enhanced cAMP levels only in low density wild type cell lines and had no effect on the drug resistant cell lines.

Na^+/K^+ ATPase Activity Studies

We studied the inhibition of the Na^+/K^+ ATPase activity by ouabain in a crude preparation of plasma membranes isolated from the different cell lines. The enzymes from wild types and from drug resistant cell lines differ in their susceptibility to ouabain inhibition according to (a) their ME_2 or MF_2 origin and (b) the selective agent.

Wild type cells. Ouabain concentrations from 10^{-8} to 10^{-3} *M* were tested on Na^+/K^+ ATPase activity in each cell line. The ouabain concentration for 50 percent inhibition of the original enzyme activity is referred to as $E_{1/2}$. From Figure 3, it can be seen that in fact the MF_2 cell line appears to have two different enzymes, each exhibiting a different dose response to ouabain ($E_{1/2}$ = 8×10^{-7} *M* and 9×10^{-5} *M*). ME_2 has a homogeneously responding enzyme ($E_{1/2}$ = 3.8×10^{-6} *M*).

Ouabain resistant (OR) cell lines. The $E_{1/2}$ is much higher for ME_2OR_1 (1.6×10^{-4} *M*) than for ME_2 (3.8×10^{-6} *M*) (Figure 4), while in both instances the specific Na^+/K^+ ATPase enzyme activities are about the same (20 μmoles/h/mg of protein). MF_2OR_2 ($E_{1/2}$ = 6×10^{-5} *M*) apparently retained the ouabain sensitivity for the enzyme that exhibited the least responsive effect in the MF_2 curve ($E_{1/2}$ = 9×10^{-5} *M*) (Figure 5).

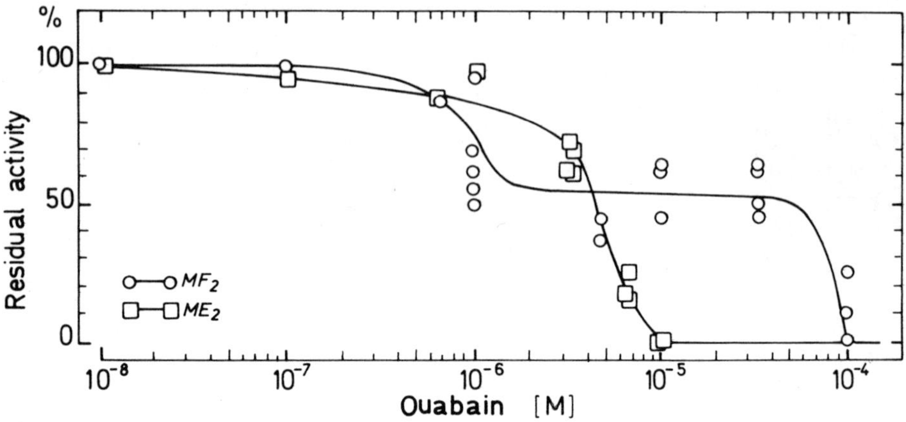

FIGURE 3. *Inhibition of plasma membrane Na^+/K^+ ATPase in ME_2 and MF_2 cell lines.*

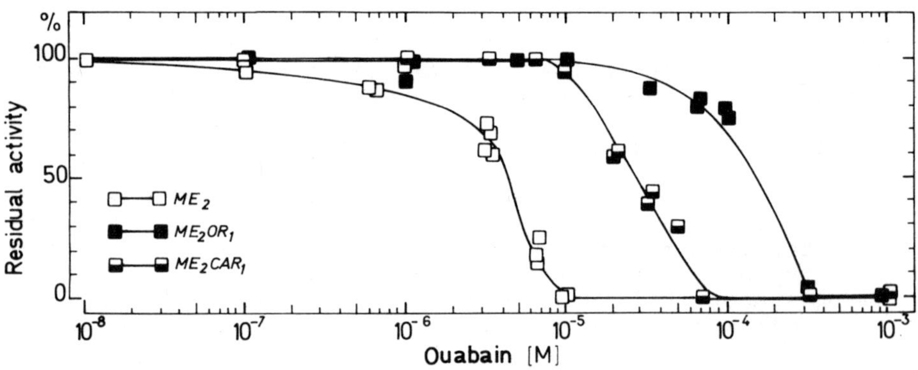

FIGURE 4. *Inhibition of plasma membrane Na^+/K^+ ATPase in ME_2, ME_2OR_1 and ME_2CAR_1.*

But$_2$-cAMP selected (CAR) cell lines. The $E_{1/2}$ of ouabain for the ME_2CAR_1 cell line (2.4×10^{-5} *M*) is intermediate between the $E_{1/2}$ of the wild type ME_2 and the ME_2OR_1 cells (Figure 4). Similarly, the $E_{1/2}$ for MF_2CAR_1 (1.2×10^{-5} *M*) is intermediate between the $E_{1/2}$ of the wild type MF_2 and the MF_2OR_2 cells (Figure 5). Table 3 is a summary of these experiments showing the differential ouabain sensitivity of the Na^+/K^+ ATPase from the parental and drug selected lines. In crude membrane preparations sensitivity of the Na^+/K^+ ATPase to ouabain is decreased in both the But$_2$-cAMP selected cell lines and ouabain selected cell lines as compared to the wild type cells.

Adenylate cyclase studies. Table 4 is a summary of our studies using the wild type and ouabain resistant cell lines as a source of adenylate cyclase. The basal activity

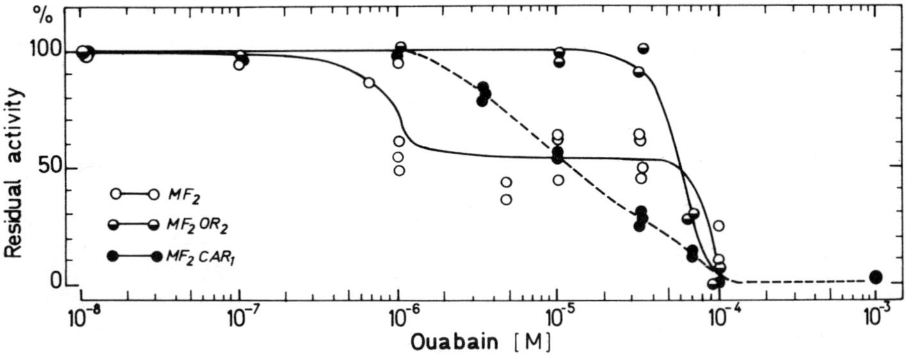

FIGURE 5. *Inhibition of plasma membrane Na^+/K^+ ATPase in MF_2, MF_2OR_2 and MF_2CAR_1.*

TABLE 3. *The relative sensitivity of Na^+/K^+ ATPase from different cell lines to ouabain*

Cell Line	Ouabain (*M*) Needed to Inhibit Enzyme 50% ($E_{1/2}$)
ME_2	$3.8 \pm 0.6 \times 10^{-6}$ *M*
ME_2OR_1	$1.7 \pm 0.2 \times 10^{-4}$ *M*
ME_2CAR_1	$2.4 \pm 0.4 \times 10^{-5}$ *M*
MF_2	$8.0 \pm 1.0 \times 10^{-7}$ *M* and $7.0 \pm 0.7 \times 10^{-5}$ *M*
MF_2OR_2	$6.0 \pm 0.5 \times 10^{-5}$ *M*
MF_2CAR_1	$1.2 \pm 0.1 \times 10^{-5}$ *M*

TABLE 4. *Effect of ouabain on adenylate cyclase of ouabain sensitive and resistant cell lines*

	Adenylate Cyclase Activity (pMoles cAMP Formed/Min/Mg Protein)		
Cell Line	No Additions	10^{-3} *M* Ouabain	10^{-4} *M* Ouabain
ME_2	1.2 ± 0.3	3.7 ± 0.5	2.9 ± 0.8
ME_2OR_1	5.8 ± 1.2	6.0 ± 0.8	5.5 ± 1.1
MF_2	4.3 ± 1.3	9.3 ± 1.8	5.7 ± 1.3
MF_2OR_2	4.8 ± 0.9	5.2 ± 1.2	4.5 ± 1.2

of the ME_2 enzyme is the lowest of the cell lines tested, and both ME_2 and MF_2 are responsive to ouabain as illustrated by an elevated adenylate cyclase specific activity. The cell lines selected for resistance to ouabain (as measured by the growth studies) are also unresponsive to ouabain as measured by adenylate cyclase activity.

DISCUSSION

Cross Resistance

In this report we have described several cell lines selected for resistance to ouabain and But_2-cAMP. These resistant cell lines exhibited cross resistance to agents to which they were not previously exposed. Evidence for cross resistance is biological and biochemical in nature:

1. Tissue culture studies have shown (a) variants obtained with one selective agent are also resistant to the other drugs, and (b) addition of ouabain for 24 h to the culture leads to increased basal cyclic AMP levels in the wild type epithelioid (ME_2) and fibroblastic (MF_2) cells, while remaining low in resistant cell lines (whatever the selective agent was, ouabain or But_2-cAMP).
2. *In vitro* enzyme studies have shown that the Na^+/K^+ ATPase and adenylate cyclase activities of resistant cell lines are less responsive, if at all, to the ouabain effect than the wild type cells. Cell lines specifically resistant to ouabain (14-16,27,28), cyclic AMP (29), Con A (30) and adenosine (31) have been previously described, but were either not tested, or shown not to exhibit cross resistance.

In the case of ouabain resistant cell lines, the discrepancy could be due to the different method of selection used; i.e., we selected by a long-term treatment of the cells with progressively increasing amounts of ouabain without mutagens, all other described ouabain resistant cell lines were obtained by a one-step selection method after mutagenesis (14-16). It is difficult to decide from the selective procedure for resistant variants if the resistance is a single or multiple step process. Because we isolated from a number of clones with different levels of resistance for a given drug we favor the idea of multiple mutations. However, the only evidence for a genetic mutation is the stability of the resistance after the cells were cultivated for 50-80 generations without drugs. An epigenetic effect on the membrane structure characterized by changes in protein-protein or protein-lipid interactions would not be expected to be so stable. We do not know if the observations reported here hold only for tumor cells: the karyotype previously described (17) shows that the modal number of chromosomes was 36 for MF_2 and 63 for ME_2, but such studies have not yet been done with the resistant cell lines.

Cross resistance has been previously interpreted in terms of reduced permeability (32) or modulation of a regulatory gene (33,34). The cross resistance between ouabain and Con A might be explained by membrane enzymes being glycoproteins (35): thus, a single mutation could involve multiple binding sites and/or modulation of several enzyme activities making the cell resistant to many drugs. We hypothesize our variants may involve one mutation affecting several biochemical parameters.

It has been shown that several membrane enzymes require lipids in their vicinity (36,37) and that a change in the lipid moiety modifies the enzyme activity (38). Moreover, it has been suggested that changes in lipid composition of plasma membranes may induce changes in the biological properties of the cell (39). Thus, we hypothesize that in our variants a mutation occurs at a structural or regulatory gene involving membrane lipid composition.

Plasma Membrane Enzyme Coordination

Adenylate cyclase and Na^+/K^+ ATPase are plasma membrane associated enzymes which utilize ATP as substrate. Another characteristic these enzymes have in common is hormone sensitivity but the response is generally in opposite directions. Insulin stimulates ATPase (40) while reportedly inhibiting adenylate cyclase (41). On the other hand, catecholamines (42) and glucocorticoids (43) which stimulate adenylate cyclase inhibit the ATPase (44, 45). Cholera toxin is another agent that stimulates adenylate cyclase (46) while inhibiting the Na^+/K^+ ATPase (47,48). A possible feedback inhibitory role for cyclic AMP on ATPase is supported by reports of the inhibition of Na^+/K^+ ATPase by cyclic AMP (49-51). Likewise, ouabain has been shown to elevate cyclic AMP levels in rat brain slices (52,53), and a similar effect also occurs in cultured hamster fibroblasts (Bannai and Sheppard, unpublished data). All of these independent and isolated observations lead us to speculate that these enzymes may be reciprocally coupled (6) possibly via interaction with the phospholipids of the plasma membrane (25). Our current results further substantiate this possibility by showing that ouabain, the classic inhibitor of Na^+/K^+ ATPase is an adenylate cyclase stimulant.

ACKNOWLEDGEMENTS

We thank David Brus, E. Faivre and V. Zilberfarb for excellent technical assistance and R. J. M. Dubert for stimulating discussions. D. Charlemagne belongs to the scientific groups C.N.R.S. (ER 102) and I.N.S.E.R.M. (group U-116, Dr. Jollès). These studies were supported by U. S. Public Health Service grants HL 06314 and Ca 16228, a contract from la Délégation Générale à la Recherche Scientifique (74.7.0222 and 75.7.0787). J.R.S. was partially supported by an exchange fellowship from the Centre National de la Recherche Scientifique-National foundation.

REFERENCES

1. Dutton, R. W. and Mishell, R. I. 1967. Cell populations and cell proliferation in the *in vitro* response of normal mouse spleen to heterologus erythrocytes. J. Exp. Med. *126*:443-454.
2. Diener, E. and Paetkau, V. H. 1972. Antigen recognition: early surface receptor phenomena induced by binding of a tritium-labeled antigen. Proc. Nat. Acad. Sci. USA *691*:2364-2368.
3. Nowell, P. C. 1960. Phytohaemagglutinin: an initiator of mitosis in cultures of normal human leukocytes. Cancer Res. *20*:462-466.
4. Andersson, J. and Melchers, F. 1973. Induction of immunoglobulin M synthesis and secretion in bone marrow-derived lymphocytes by locally concentrated Concanavalin A. Proc. Nat. Acad. Sci. USA *70*: 416-420.
5. Cuatrecasas, P. 1969. Interaction of insulin with the cell membrane: the primary action of insulin. Proc. Nat. Acad. Sci. USA *63*:450-457.
6. Sheppard, J. R. RC585 Cyclic AMP and the growth of transformed cells M39. In: *Membranes and Viruses in Immunopathology*, Day, S. and Good, R. A. (eds.). Academic Press, New York, 1972, pp. 249-263.
7. Pardee, A. B. 1971. The surface membrane as a regulator of animal cell division. In Vitro *7*:95-104.
8. Wallach, D. F. H. 1968. Cellular membranes and tumor behavior: a new hypothesis. Proc. Nat. Acad. Sci. USA *61*:868-874.
9. Hershko, A., Mamont, P., Shields, R. and Tomkins, G. M. 1971. Pleiotrypic response. Nature New Biol. *232*:206-211.
10. Paraf, A., Guérin-Marchand, C., Aubry, J. and Dubert, J. M. 1975. Mechanism of immune stimulation and tolerance based on a hypothesis on the cell division mechanism directed by the cell membrane. Ann. Immunol. *126*:423-430.
11. Tomkins, G. M. 1974. The metabolic code. Science *189*:760-763.
12. Sheppard, J. R. and Bannai, S. Cyclic AMP and cell proliferation. In: *Control of Proliferation in Animal Cells*, Baserga, R. and Clarkson, B. (eds.). Cold Spring Harbor Laboratory, Cold Spring Harbor, New York, 1974, pp. 571-579.
13. Eckhart, N., Dulbecco, R. and Burger, M. 1971. Temperature dependent surface changes in cells infected or transformed by a thermosensitive mutant of a polyoma virus. Proc. Nat. Acad. Sci. USA *68*:283-286.
14. Thompson, L. H. and Baker, R. M. Isolation of Mutants of cultured mammalian cells. In: *Methods in Cell Biology*, Prescott, D. M. (ed.), Vol. 6. Academic Press, New York, 1973, pp. 210-281.

15. Mayhew, E. 1972. Effect of AMP and related compounds on glycogen content of tetrahymena. J. Cell Physiol. *79*:441-452.
16. Baker, R. M., Brunette, D. M., Mankovitz, R., Thompson, L. H., Whitmore, G. F., Sminovitch, L. and Till, E. 1974. Ouabain-resistant mutants of mouse and hamster cells in culture. Cell *1*:9-21.
17. Legrand, E., Moyne, M. A., Paraf, A. and Duplan, J.F. 1972. Studies on mouse plasmocytoma cells MOPC 173 grown *in vitro*: general properties of fibroblastic and epitheloid cell lines. Ann. de l'institut Pasteur *123*:641-660.
18. Paraf, A., Faivre, E., Zilberfarb, V., Zachoroski, A. and Lelievre, L. 1975. Selection de ligne cellulaires resistantes á l'ouabine. C. R. Acad. Sci. Paris *281*:567-570.
19. Jørgensen, P. L. and Skou, J. C. 1969. Preparation of the highly acrine (Na^+ + K^+) ATPase from the outer medulla of rabbit kidney. Biochem. Biophys. Res. Commun. *37*:39-46.
20. Marsh, B. B. 1959. The estimation of inorganic phosphate in the presence of adenosine triphosphate. Biochem. Biophys. Res. Commun. *32*:39-46.
21. Kebabian, J. W., Petzold, G. L. and Greengard, P. 1972. Dopamine-sensitive adenylate cyclase in caudate nucleus of rat brain and its similarity to the "dopamine receptor." Proc. Nat. Acad. Sci. USA *69*:2145-2149.
22. Sheppard, J. R. 1972. Difference in the cyclic adenosine 3',5' monophosphate levels in normal and transformed cells. Nature New Biol. *236*:14-16.
23. Brown, B. L., Ekins, R. P. and Albano, J. D. M. Saturation assay for cAMP using endogenous binding protein. In: *Advances in Cyclic Nucleotide Research*, Greengard, P. and Robinson, A. G. (eds.), Vol. 2. Raven Press, New York, 1972.
24. Lowry, O. H., Rosebrough, N. J., Farr, A. L. and Randall, R. J. 1951. Protein measurement with the Folin phenol reagent. J. Biol. Chem. *193*:265-275.
25. Guérin, C., Prigent, B., Moyne, M. A. and Paraf, A. 1973. La differenciation du plasmacytome nurin MOPC 173: obtention de variants en culture cellulaire. Bull. du Cancer *59*:367-379.
26. Paraf, A., Faivre, E., Zilberfarb, V., Lelièvre, L. and Zachowski, A. 1975. Selection de lignes cellulaires resistantes a l'AMP cyclique. C. R. Acad. Sci. Paris *281*:683-685.
27. Mankovitz, R., Buchwald, M. and Baker, R. M. 1974. Isolation of ouabain resistant human diploid fibroblasts. Cell *3*:221-226.
28. Rosenberg, H. M. 1975. Variant HeLa cells selected for their resistance to ouabain. J. Cell Physiol. *85*:135-142.
29. Coffino, P., Gray, G. W. and Tomkins, G. M. 1975. Cyclic AMP: a non-essential regulator of the cell

cycle. Proc. Nat. Acad. Sci. USA *72*:878-882.
30. Ozanne, B. and Lurye, E. Isolation and characterization of two classes of Con A resistant SV 3T3 cells. In: *Control of Proliferation in Animal Cells*, Baserga, R. and Clarkson, B. (eds.). Cold Spring Harbor Laboratory, Cold Spring Harbor, New York, 1973, pp. 177-190.
31. McBurney, M. C. and Whitmore, G. F. 1975. Mutants of Chinese hamster cells resistant to adenosine. J. Cell Physiol. *85*:87-100.
32. Ling, V. and Thomson, L. H. 1974. Reduced permeability in CHO cells as a mechanism of resistance to colchicine. J. Cell Physiol. *83*:106-116.
33. Sun, N. C., Chang, C. C. and Chu, E. H. Y. 1975. Mutant hamster cells exhibiting a pleotropic effect on carbohydrate metabolism. Proc. Nat. Acad. Sci. USA *72*:469-473.
34. Wright, J. A. 1973. Resistant mutants of mouse and hamster cells in cell culture. Cell Biol. *56*: 666-675.
35. Kyte, J. 1972. Properties of the two polypeptides of sodium- and potassium-dependent adenosine triphosphate. J. Biol. Chem. *247*:7642-7652.
36. Wheeler, K. P. and Whittam, R. 1970. ATPase activity of sodium pump needs phosphatidylserine. Nature *225*:449-450.
37. Pohl, S. L., Krans, H. M. J., Kozyreff, V., Birnbaumer, L. and Rodbell, M. 1971. The glucagon sensitive adenyl-cyclase system in plasma membranes of rat liver. J. Biol. Chem. *246*:4447-4454.
38. Warren, G. B., Toon, P. A., Birdsall, N. J. M., Lee, A. G. and Metcalfe, J. C. 1974. Complete control of the lipid environment of membrane bound proteins: applications to a Ca^{++} transport system. FEBS Lett. *41*:122-124.
39. Horwitz, A. F., Hatten, M. B. and Burger, M. M. 1974. Membrane fatty acid replacements and their effect on growth and lectin induced agglutinability. Proc. Nat. Acad. Sci. USA *71*:3115-3119.
40. Hadden, J. W., Hadden, E. M., Wilson, E. E., Good, R. A. and Coffey, R. G. 1972. Direct action of insulin on plasma membrane ATPase activity in human lymphocytes. Nature New Biol. *235*:174-176.
41. Illiano, G. and Cuatrecasas, P. 1972. Modulation of adenylate cyclase activity in liver and fat cell membranes by insulin. Science *175*:906-908.
42. Sutherland, E. W. and Rall, T. W. 1957. The properties of an adenylate ribonucleotide produced with cellular particles, ATP, Mg^{++}, and epinephrine glucagon. J. Amer. Chem. Soc. *79*:3608.
43. Coffey, R. G. and Middleton, E., Jr. 1973. Release of histamine from rat mast cells by lysosomal cationic proteins. Int. Arch. Allergy Appl. Immun. *45*:593-611.

44. Tabakoff, B. 1974. Inhibition of sodium, potassium and magnesium activated ATP-ases by acetalaldehyde and biogenic aldehydes. Res. Commun. Chem. Path. Pharmacol. *7*:621-624.
45. Luly, P., Barnabei, D. and Trice, E. 1972. Hormonal control *in vitro* of plasma membrane bound (Na^+-K^+) ATPase of rat liver. Biochim. Biophys. Acta *282*: 441-451.
46. Schafer, D. E., Lust, W. D., Sircan, B. and Goldberg, N. D. 1970. Elevated concentration of cAMP in intestinal mucosa after treatment with cholera toxin. Proc. Nat. Acad. Sci. USA *67*:851.
47. Parkinson, D. K. 1972. Localization of the action of cholera toxin on adenyl cyclase in mucosal epithelial cells of rabbit intestine. J. Clin. Invest. *51*:2292-2298.
48. Hirschdrn, N. 1968. Na^+.K^+ stimulates ATPase of the small intestine of man: studies in cholera and other diarrheal diseases. J. Lab. Clin. Med. *72*:28-39.
49. Mozsik, G. 1969. Some feedback mechanisms by drugs in the interrelationship between the active transport system and the adenylate cyclase system localized in the cell membrane. Eur. J. Pharm. *7*: 319-327.
50. Chaudhary, A. H. and Frenkel, A. W. 1970. Effects of adenosine 3',5'-cyclic monophosphoric acid on certain light-induced reactions and ATPase activity of isolated chromatophores from *Rhodospirillum rubum*. Biochem. Biophys. Res. Commun. *39*:238-246.
51. Tria, E., Luly, P., Tomasi, V., Trevisani, A. and Barnabei, O. 1974. Modulation by cyclic AMP *in vitro* of liver plasma membrane (Na^+-K^+) ATPase and protein kinase. Biochim. Biophys. Acta *343*: 297-306.
52. Huang, M., Schimizu, H. and Daly, J. W. 1972. Accumulation cAMP in incubated slices of brain tissue. J. Med. Chem. *15*:462-466.
53. Shimizu, H., Creveling, C. R. and Daly, J. W. 1970. Cyclic adenosine 3',5' monophosphate formation in brain slices: stimulation by batrachotoxin, ouabain, veratridine and potassium ions. Mol. Pharm. *6*:184-188.

Index

Adenovirus-12, 173
 effect on cAMP level, 178
 induction of DNA synthesis, 174
 restriction fragments, 174
 specific proteins in BHK21, 175
Adenylate cyclase
 in AD12 infected BHK21, 179
 in bacteria, 4,7,27
 and catabolite repression, 4, 14,27,30,37,44
 inhibition by T-antigen, 179
 stimulation by prostaglandins, 183
ATPase and adenylate cyclase regulation, 254,279,284

β-galactosidase, 5,43
Bidirectional control, 143

Calcium
 and DNA synthesis, 110,112
 as a positive intracycle regulator, 114
Calcium-cAMP couplet and neuplastic transformation, 120,121,123
cAMP
 compartmentalization, 157,159
 inhibition of cellular DNA synthesis, 175
 and frequency of T-antigen, 177
 killing of BHK21 cells, 177,178
cAMP levels
 in bacteria, 4
 and ouabain treatment, 280
 resistant mutants, 270,273
Catabolite insensitive cells, 5
Catabolite sensitive cells, 5,44
Caulobacter, 3,36,37,44
Cell-cell contact and information transfer, 136
Cell-cell interaction, 134
Cell-cell signals, 132
Cell-cycle, 36,43,55
 control, 82
CGMP
 assay, 210
 and clinical diagnosis, 204
 and effect of periodate, 211
 excretion in urine, 202-204
 metabolism, 211
 modulation of lymphocyte function, 207,208
 as a proliferative activator, 100
 and transplantable tumors, 202
 and Ying-Yang hypothesis, 146
Chalones, 91
Cholera-enterotoxin, 137,267
 and adenylate cyclase, 137
 and cAMP level, 138
 and *E. coli*, 20,22,29
Choleragenoid, 139
 and toxin binding sites, 139

Deathless mutants, 275
Directional growth control, 163
Disialogangliosides, 136
DNA
 replication, 4,36
 synthesis, 4,36,43
 transcription, 3

Escherichia coli, 3,12,27
  enterotoxin, 137,138
    competition with cholera toxin, 138
    and intestinal adenylate cyclase, 138,152
    and phosphodiesterase, 152

Flagellum, 36

G-1 block, 60
G-2 block, 60
G-1 phase, 61
Gangliosides, 133,134,146
Ganglioside GM1, 20,133,135,137
  and activation of adenylate cyclase, 161
  and cyclic nucleotide metabolism, 161
  and inhibition of cholera toxin binding, 139
  and neuraminidase action, 139
  in normal and malignant cells, 133
  in relation to cell growth, 136
Gene expression, 38
Glucose transport, 7,30
GM-3, 134
Growth regulation
  biochemical basis of, 275
  models of, 273,274
Guanylate cyclase
  activation by NaN_3, 197
  assay of, 209
  and cell division, 146
  effect of Ca^{++}, 193
  effect of Mg^{++}, 197
  inhibition by cholera toxin, 144
  in situ level, 196
  particulate and soluble forms, 192
  protein activator, 197
  stimulation by fibroblast growth factor, 144

HeLa cells, 58
Hematopoietic cells, 82
Hemopoietic cells, 81
Hepatoma cells, 58
Histone kinase, 76
Hydroxy urea
  and amethoptrim double block, 58
  cytocidal action of, 83,85

Immature gangliosides, 135
Indomethacin
  and calcium mobilization, 111
  and DNA synthesis, 108
  and prostaglandin synthesis, 109
Inducible enzymes, 3,4,45
Insulin action
  in bacteria, 16
  and growth stimulation, 149

Lateral mobility of membrane proteins, 133
Lection agglutinability
  of normal cells, 133
  of transformed cells, 133

Membrane fluidity, 133,136,137
Metabolic regulator, 132
Metabolic signal, 132
Methyl-3-isobutyrl xanthine (MIX), 59

Ornithine decarboxylase, 102,106

Periodate oxidation
  and CGMP accumulation, 211-213
  and CGMP metabolism, 211
  and guanylate cyclase activity, 216,218
Phosphodiesterase
  in AD12 infected cells, 184
  bacterial, 4,5
  CGMP, 144
  and regulation of cell-cycle, 88,90
Phosphotransferase, 7,30
Pili, 36
Pleiotropic mutants, 38,45
Polyamines and cAMP action, 104,106
Proliferative activation, 98
Protein kinase, 67,225
  and cAMP, 226
  catalytic subunits, 238
  cyclic nucleotide dependency, 232
  cytoplasmic, 237
  fractionation, 228-231
  intracellular distribution, 226,238
  mutants of, 271
  in porcine ovary nuclei, 228,246
  substrate specificity, 233
  translocation, 225,243

Protein phosphatase, 76

Receptor
behavior, 132
and cell division regulation, 132,133
unidirectional, 163

Sialic acid, 20
S-phase, 59,82
Sporulation, 3
Stalked cell, 36
Swarmer cell, 36
Synchronous cultures, 38,57

T-antigen, 175
Transglycosylation, 136,137
Transmembrane movement, 115,116
and control of hepatocyte proliferation, 117
and proliferation of *Physarum polycephalum*, 118
and proliferation of *Streptococcus lactis*, 117
Transplantable tumors
guanylate cyclase distribution, 191-193
increased cAMP level, 191
Tryptophanase, 5

Unidirectional negative turn off signal, 163

Walsh inhibitor, 68,69,73